Food Digestion and Absorption

Its Role in Food Product Development

Food Chemistry, Function and Analysis

Series editors:
Gary Williamson, *Monash University, Australia*
Alejandro G. Marangoni, *University of Guelph, Canada*
Graham A. Bonwick, *AgriFoodX Limited, UK*
Catherine S. Birch, *AgriFoodX Limited, UK*

Titles in the series:

1: Food Biosensors
2: Sensing Techniques for Food Safety and Quality Control
3: Edible Oil Structuring: Concepts, Methods and Applications
4: Food Irradiation Technologies: Concepts, Applications and Outcomes
5: Non-extractable Polyphenols and Carotenoids: Importance in Human Nutrition and Health
6: Cereal Grain-based Functional Foods: Carbohydrate and Phytochemical Components
7: Steviol Glycosides: Cultivation, Processing, Analysis and Applications in Food
8: Legumes: Nutritional Quality, Processing and Potential Health Benefits
9: Tomato Chemistry, Industrial Processing and Product Development
10: Food Contact Materials Analysis: Mass Spectrometry Techniques
11: Vitamin E: Chemistry and Nutritional Benefits
12: Anthocyanins from Natural Sources: Exploiting Targeted Delivery for Improved Health
13: Carotenoid Esters in Foods: Physical, Chemical and Biological Properties
14: Eggs as Functional Foods and Nutraceuticals for Human Health
15: Rapid Antibody-based Technologies in Food Analysis
16: DNA Techniques to Verify Food Authenticity: Applications in Food Fraud
17: Advanced Gas Chromatography in Food Analysis
18: Handbook of Food Structure Development
19: Mitigating Contamination from Food Processing
20: Biogenic Amines in Food: Analysis, Occurrence and Toxicity
21: Nutrition and Cancer Prevention: From Molecular Mechanisms to Dietary Recommendations
22: Health Claims and Food Labelling
23: Nutraceuticals and Human Health: The Food-to-supplement Paradigm
24: Nutritional Signalling Pathway Activities in Obesity and Diabetes
25: The Chemistry and Bioactive Components of Turmeric
26: Foodomics
27: Food Proteins and Peptides: Emerging Biofunctions, Food and Biomaterial Applications
28: Handbook of Antioxidant Methodology: Approaches to Activity Determination
29: Fats and Associated Compounds: Consumption and Human Health

30: Oral Processing and Consumer Perception: Biophysics, Food Microstructures and Health
31: Development of Trans-free Lipid Systems and their Use in Food Products
32: Advanced Spectroscopic Techniques for Food Quality
33: Berries and Berry Bioactive Compounds in Promoting Health
34: Metabolism of Nutrients by Gut Microbiota
35: Bio- and Nano-sensing Technologies for Food Processing and Packaging
36: Edible Fungi: Chemical Composition, Nutrition and Health Effects
37: Chemistry and Nutritional Effects of Capsicum
38: Pigmented Cereals and Millets: Bioactive Profile and Food Applications
39: Dietary Supplements with Antioxidant Activity: Understanding Mechanisms and Potential Health Benefits
40: Handbook of Cheese Chemistry
41: Flavour and Consumer Perception of Food Proteins
42: Food Digestion and Absorption: Its Role in Food Product Development

How to obtain future titles on publication:
A standing order plan is available for this series. A standing order will bring delivery of each new volume immediately on publication.

For further information please contact:
Book Sales Department, Royal Society of Chemistry, Thomas Graham House, Science Park, Milton Road, Cambridge CB4 0WF, UK
Telephone: +44 (0)1223 420066, Fax: +44 (0)1223 420247,
Email: booksales@rsc.org
Visit our website at www.rsc.org/books

Food Digestion and Absorption
Its Role in Food Product Development

By

C. Anandharamakrishnan
CSIR - National Institute for Interdisciplinary Science and Technology (NIIST), India
Email: anandh@niist.res.in

Jeyan Arthur Moses
National Institute of Food Technology, Entrepreneurship and Management (NIFTEM-T), India
Email: moses.ja@iifpt.edu.in

and

S. Priyanka
Indian Institute of Food Processing Technology, India
Email: Priyankasethupathy@gmail.com

ROYAL SOCIETY OF CHEMISTRY

Food Chemistry, Function and Analysis No. 42

Print ISBN: 978-1-78801-858-6
PDF ISBN: 978-1-83916-242-8
EPUB ISBN: 978-1-83916-243-5
Print ISSN: 2398-0656
Electronic ISSN: 2398-0664

A catalogue record for this book is available from the British Library

The Royal Society of Chemistry is a charity, registered in England and Wales, Number 207890, and a company incorporated in England by Royal Charter (Registered No. RC000524), registered office: Burlington House, Piccadilly, London W1J 0BA, UK, Telephone: +44 (0)20 7437 8656.

Visit our website at www.rsc.org/books

Printed in the United Kingdom by CPI Group (UK) Ltd, Croydon CR0 4YY, UK

Preface

Food product development is often a tedious process, involving many stages and parameters that have to be considered. The need for novel food products always exists and industries are striving to develop products that balance nutrition, consumer acceptability, and costs. During new food product development, technologists and others often focus on the food structuring component. Indeed, this is vital because this decides the market for and life of the product. However, when consumed, foods remain in the oral cavity for only a few seconds to minutes and then spend the rest of their processing period along various sections of the gastrointestinal tract (GIT). Importantly, just like processing conditions and their impact on foods, the environment of the human digestive system has significant implications for the fate of foods and the nutrients present in them. Therefore, what is equally vital is a focus on food destructuring and the understanding of what happens to a food once we have consumed it. A strong understanding of this aspect will certainly be of benefit to food technologists, nutritionists, and health professionals.

Given this, the book *Food Digestion and Absorption: Its Role in Food Product Development* begins with information on the human digestive system, starting with the oral cavity, followed by the stomach and small and large intestines. Each of the relevant chapters can help the reader to visualize the environment and changes that food undergoes along the GIT. Supporting researchers in understanding these aspects better, various *in vitro*, *in vivo*, and *in silico* approaches have been explored, simulating the food digestion and absorption processes; these are presented in the first chapter. Subsequent chapters deal with food structures, the impact of food processing, and destructuring processes. These are supported with content on designing foods with new food ingredients and a specific case of how encapsulation serves as a structuring approach. The last chapter deals with aspects of and

Food Chemistry, Function and Analysis No. 42
Food Digestion and Absorption: Its Role in Food Product Development
By C. Anandharamakrishnan, Jeyan Arthur Moses and S. Priyanka

Published by the Royal Society of Chemistry, www.rsc.org

considerations in developing novel personalized foods. In addition, the book covers interesting content on gut–brain–microbiota interactions, nutrient absorption mechanisms, and state-of-the-art research in this field, on a global scale.

We strongly believe that this book will serve as an excellent resource for multidisciplinary professionals focusing mainly on food product development. With changing trends in consumer behavior and food intake patterns, this resource provides a detailed picture of the need for focusing equally on food destructuring processes and the fate of food in the human digestive system.

C. Anandharamakrishnan, Jeyan Arthur Moses, and Priyanka Sethupathy

Contents

Food Chemistry, Function and Analysis No. 42
Food Digestion and Absorption: Its Role in Food Product Development
By C. Anandharamakrishnan, Jeyan Arthur Moses and S. Priyanka

Published by the Royal Society of Chemistry, www.rsc.org

CHAPTER 1

The Human Oral Cavity and Oral Processing of Foods

1.1 Introduction

The oral processing of foods is a multistage dynamic process occurring in the oral cavity that simultaneously encompasses sensory perception, mechanical destruction, biochemical changes, and colloidal destabilization of consumed food. The brain and human neural system are efficient at coordinating the sensory perception of food.[1,2] Ingested food experiences various processes, such as mechanical deformation and flow, decrease in size, saliva lubrication/wetting, temperature changes leading to phase modification, changes in consistency and rugosity, food particle clustering to form a bolus that can be swallowed readily, and residue formation in and coating of the mouth.[3,4] The oral processing of foods is a concurrent process of sensory perception and food breakdown.[5–7] The nervous system and the brain regulate the entire process.[8] Oral processing of foods can be categorized into three classes: oral physiology, oral physics, and oral psychology.[1]

Oral physiology examines the different responses and alterations that occur in the oral cavity while chewing and swallowing food, including jaw movement, forces generated by the orofacial muscles, salivary secretion, tongue sliding, tongue pressure against the hard palate, and stimulation of receptors such as the mechanoreceptors and thermoreceptors present in the mouth.[9,10] Oral physics examines how applied physical and mechanical forces cause food to deform, fracture, and undergo microstructural changes, and also how those changes affect the oro-sensory perception of the texture of food.[11,12] With the aid of the neurological system's feedback processes, sensory psychology focuses on the perception and measurement of sensory

Food Chemistry, Function and Analysis No. 42
Food Digestion and Absorption: Its Role in Food Product Development
By C. Anandharamakrishnan, Jeyan Arthur Moses and S. Priyanka

Published by the Royal Society of Chemistry, www.rsc.org

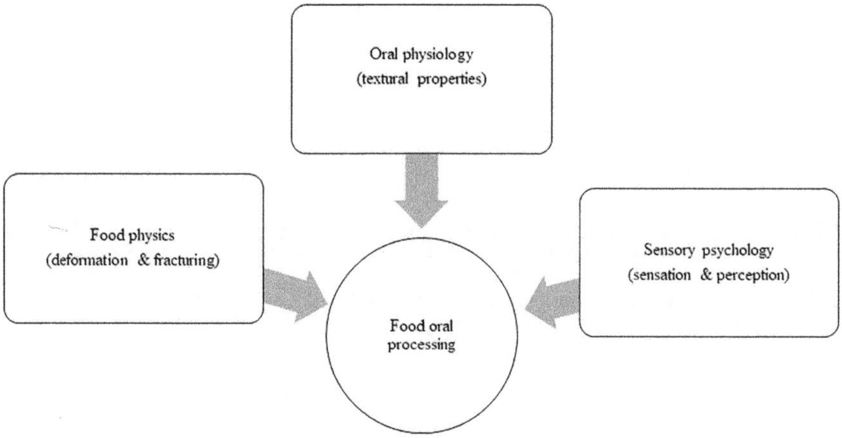

Figure 1.1 Key categories of food oral processing research. Reproduced from ref. 1 with permission from Elsevier, Copyright 2014.

aspects, including the sensation of mouthfeel properties.[13,14] Figure 1.1 depicts various categories of food oral processing.

Recent research in the field of sensory psychology has been focused on neurological facets to comprehend how sensory stimuli and brain signals interact.[15] Some studies provide a deeper understanding of the brain's "reward pathways" which are activated during mastication and texture perception and are captured with the help of functional magnetic resonance imaging (fMRI).[16–18] Nevertheless, some previous investigations were focused additionally on the motor-control signals required to fracture solid foods that are soft and hard in texture.[19,20] This chapter attempts to provide an overall and comprehensive picture of the anatomy and physiology of the oral cavity, various stages and dynamic changes that occur during the oral processing of food, and different sensorial, analytical, and instrumental techniques used to understand the complex and intricate mechanisms of oral processing of foods and sensory perceptions. The chapter gives a brief overview of anatomical and physiological features of the oral cavity and oral processing of different food matrices, and explains the significance of oral processing studies and oro-sensory perception in detail.

1.2 Anatomical and Physiological Features Involved in Oral Processing

1.2.1 Oral Cavity/Mouth

This is the starting point of the gastrointestinal tract (digestive tract). Major functions of the oral cavity include mastication, swallowing, gustation (taste perception), olfaction (flavor perception), and texture perception, apart from other activities such as breathing and social and behavioral functions (speaking, smiling, yawning).[21] The oral cavity consists of the upper and lower lips anteriorly, the cheeks covering both sides laterally, the top covered by a soft and hard palate, and the tongue at the base. Further, teeth are embedded in the dentition supported by bones, the maxilla (upper jaw) and mandible (lower jaw), which are located adjacent to each other in the oral cavity.[22] The oral cavity's end section is a pharyngeal arch structure made up of the hyoid bone, muscles, and larynx cartilage.[23] The hard and soft palate is covered by a thin layer of mucosa, lips that aid in the ingestion of food, and the pharynx that links the oral and nasal cavities to the food pipe (esophagus), which transports the food bolus to the stomach. Figure 1.2 illustrates the various parts of the human oral cavity.

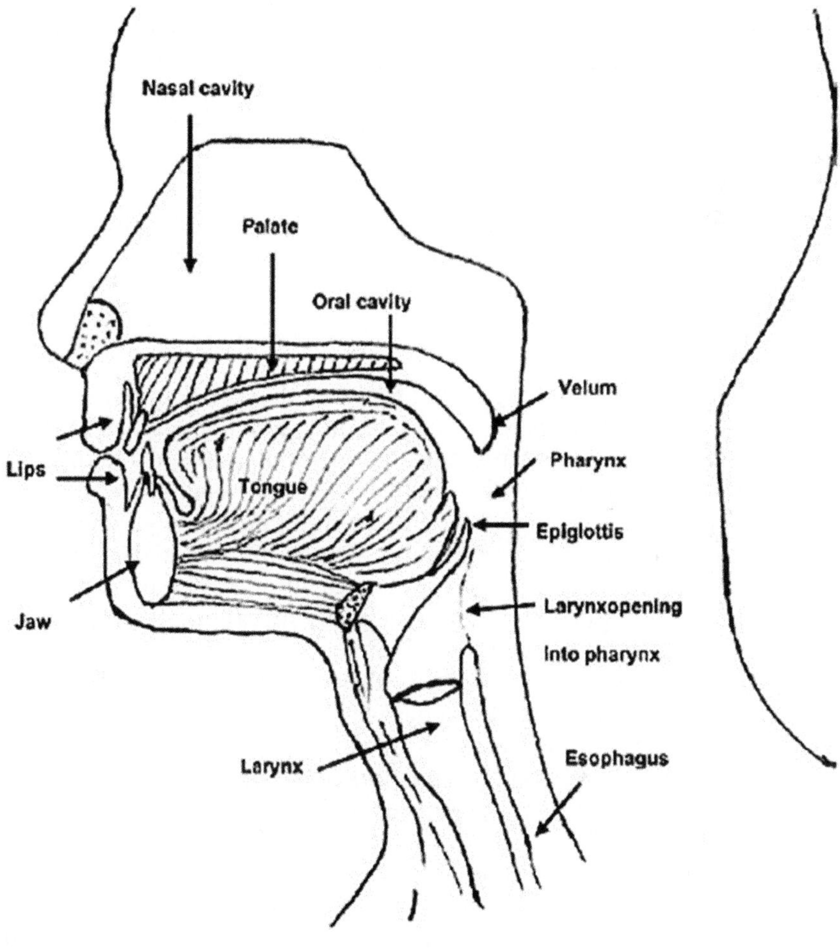

Figure 1.2 Parts of the human oral cavity. Reproduced from ref. 151 with permission from Elsevier, Copyright 2009.

1.2.2 Teeth

Humans have diphyodonty (two generations of teeth) such as the primary deciduous dentition and the secondary permanent dentition. There are around 28–32 teeth in normal adults, which can be classified into four types: incisors, canines, premolars, and molars.[22] Incisors (cutting teeth) are thin with blade-like crown structures, which are mainly used for cutting the ingested food. Canines have a single cone-shaped cusp on their crowns; they are known as piercing/tearing teeth. Premolars are bicuspid teeth that are unique to permanent dentition. Molars consist of more than one cusp, usually four or five, on a broader and flatter biting surface. The molariform teeth, which include premolars and molars, are referred to as grinding teeth.[22] The tooth structure can be divided into the upper crown and lower root; the latter is embedded in the jaw bone and held firmly by the gums. The upper crown is the visible region of teeth that is covered with enamel, which is the toughest tissue in the human body and comes in direct contact with food during mastication.[24] Teeth are the major component used for the size reduction of the ingested food (chewing). The mastication efficiency of an individual is greatly dependent on their teeth. Thus a significant decline in mastication performance is noted in individuals with removable dentures, implant-supported prostheses, and the loss of post-canine teeth.[25,26] Further, studies have found that individuals do not chew their food properly after losing posterior teeth and end up swallowing larger food particles due to poor mastication function.[27]

1.2.3 Tongue

The tongue is the muscular organ that is attached to the floor of the mouth and occupies the major space within the oral cavity. The tongue gets its support from the hyoid bone from the bottom and it is attached to the inner surface of the mandible near the midline.[22] The predominant work of the tongue is to deliver the food/food fragments appropriately between the teeth for mastication, aid in lubrication with saliva, and assist the transportation of food to the back of the oral cavity to facilitate swallowing. Further, the tongue plays a crucial part in the ingestion of food.[21] Movement of the tongue during mastication aids in the migration of food from the front of the mouth to the back of the throat, which supports swallowing and the formation of a bolus. The superficial surface of the tongue consists of many taste buds that can detect five major tastes: salty, sweet, bitter, sour, and umami. Further, chemoreceptors and mechanoreceptors on the surface of the tongue sense the chemical nature and mechanical properties of food, respectively.[28]

The dorsum of the tongue is convex and it can be subdivided into two parts: the palatal part (anterior two-thirds) and the pharyngeal part (posterior one-third). The tongue is divided symmetrically into two halves by a median sulcus and it terminates in a depression called the foramen cecum. The sulcus terminalis is a shallow V-shaped groove that is the junction where the palatal and pharyngeal parts meet. Massive amounts of papillae cover the palatal portion of the tongue, whereas the pharyngeal portion is smoother and has more lymph follicles and muciparous glands (lingual tonsil).[29] Circumvallate papillae, foliate papillae, fungiform papillae, and filiform papillae are the four different kinds of papillae. Taste buds are present in the first three papillae, but not in the filiform papillae. However, it is responsible for transducing touch, temperature, and nociception.[24]

The circumvallate papillae are situated ahead of the sulcus terminalis and foramen cecum, creating an array on either side in the shape of an inverted "V". Foliate papillate are slit-like folds of mucosa containing over 100 taste buds, which are positioned on both the sides of the tongue at the lateral border. The palatal region of the tongue (anterior two-thirds part) is covered with copious amounts of filiform papillae, which look like dwarf-whitish, hair-like conical elevations. The fungiform papillae are isolated deep reddish mushroom-shaped structures that are interspersed between the filiform papillae.[22] The fungiform papillae are most numerous at the tip of the tongue; in addition they are also present on both sides and also scattered unevenly and frugally over the dorsum. Filiform papillae are keratinized whereas fungiform papillae are non-keratinized. Additionally, taste buds, which are essential components of the gustatory sense, are dispersed unevenly throughout the mucous membranes of the mouth and tongue.[21,30]

1.2.4 Saliva and Salivary Glands

Saliva is the lubricant found in the oral cavity and it comprises around 98% water. The remaining 2% consists of other components such as enzymes (α-amylase and lingual lipase), electrolytes (calcium, bicarbonate, magnesium, sodium, chloride, potassium, phosphate), proteins (enzymes, albumin), glycoproteins, immunoglobulins, mucins, and other peptides. It also contains minor amounts of urea, glucose, and ammonia.[31,32] The major role of saliva is to provide lubrication and buffering, maintain tooth integrity, protect against bacteria, assist taste perception, and aid in digestion.[33,34] Figure 1.3. details various significant functions of saliva. The pH of saliva can range between 5.3 and 7.8. In general, humans will produce around 600 mL of saliva per day. The secretion of saliva is higher during mastication and less pronounced during sleep.[35] There are three major salivary glands in the oral cavity, namely parotid, submandibular, and sublingual glands, which are responsible for more than 90% of the saliva secreted, the remainder of the saliva being secreted by minor glands.[22] The details of various salient features of salivary glands are provided in Table 1.1.

1.2.5 Orofacial Muscles

Many muscles present in the head and neck region are responsible for oral processing functions. However, muscles such as the masseter, temporalis,

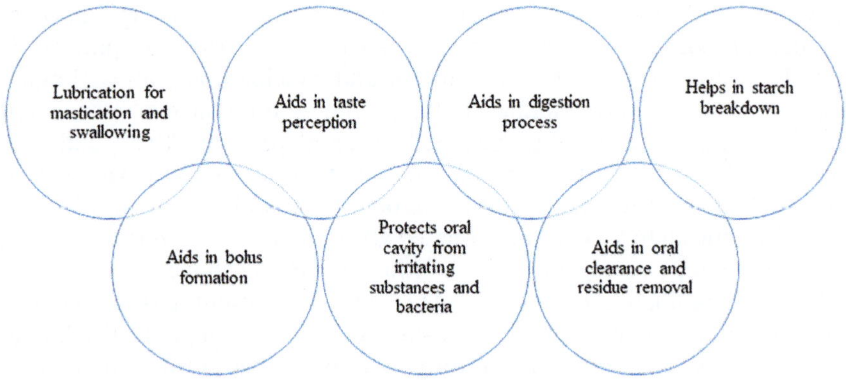

Figure 1.3 Functions of saliva.

and medial and lateral pterygoid are jointly known as "muscles of mastication".[22] The lateral and medial pterygoid muscles are positioned deeper within the infratemporal fossa, while the masseter and temporalis muscles are situated superficially on the face. The muscles of mastication work jointly to move the mandible in various directions at the temporomandibular joint during the mastication process. The mandibular division of the trigeminal nerve supplies all the muscles involved in mastication with their innervation. The muscles present in the suprahyoid region include digastric, mylohyoid, and geniohyoid muscles, which are positioned on the floor of the oral cavity. These muscles run from the hyoid bone to the mandible in the form of sheets of parallel fibrous tissues. Further, the digastric muscle is closely associated with the functions of mastication muscles as it is the principal muscle involved in jaw movement. The muscles present in the suprahyoid region function as jaw intruders and jaw elevators (mouth opening and closing muscles).[29]

The muscles of the tongue are classified into two types, intrinsic and extrinsic muscles. The three fiber groups of the intrinsic muscles, namely transverse, longitudinal, and vertical, are confined to the substance of the tongue and alter its shape.[21] The extrinsic muscles are divided into four groups, the genioglossus, hyoglossus, styloglossus, and palatoglossus. The extrinsic muscles originate from areas outside the tongue such as the skull and hyoid bone, and then spread into the body of the tongue. The extrinsic muscles are for the bodily movement of the tongue.[22] The soft palate is held in place with the help of fibrous palatine aponeurosis. There are pairs of four muscles, the levator veli palatini, the palatoglossus, the tensor veli palatini, and the palatopharyngeus uvulus muscles. All four of these muscles join at the aponeurosis and they are responsible for alterations in the shape and position of the soft palate. Further, all four muscles play a crucial part in opening and shutting of the airway during swallowing, with the levator veli palatine serving as the soft palate's chief elevator.[23]

Table 1.1 Salient features of salivary glands.

Salivary gland	Characteristics	Location	Major components	Ref.
Parotid	The largest of the major salivary glands and its secretion is serous in nature. It is pyramidal in shape and also associated with lymph nodes	Situated anterior to the ear and the sternocleidomastoid muscle in the retromandibular fossa. Parts of the superficial lobe cover the ramus of the mandible and the posterior part of the masseter muscle	α-Amylase is secreted by the acinar cell of the parotid gland. It also produces histidine- and proline-rich proteins	22 and 152
Submandibular	This is a mixed gland but its secretion is primarily serous	Positioned in the posterior part of the submandibular triangle. Additionally, the mandible's body and the anterior and posterior digastric muscle bellies serve as the submandibular triangle's limiting structures. A large part of the gland lies on the mylohyoid muscle	It mainly produces glycosylated proteins (mucins) and histidine- and proline-rich proteins	21 and 153
Sublingual	This is the smallest of the three major salivary glands. It is a mixed gland but the majority of its secretions are mucous elements	Located between various muscles present in the oral cavity such as the geniohyoid muscle, intrinsic muscles of the tongue, mylohyoid muscle, and hypoglossal muscle (medially). It also lies alongside the sublingual fossa of the mandible	It mainly secretes glycosylated proteins (mucins)	22, 152 and 153
Minor glands	There are around 8–30 mixed minor salivary glands. Their duct system empties into the sublingual fold in the floor of the mouth	These minor glands are dispersed throughout the oral mucosa such as labial, palatial, lingual, and buccal mucosa. The minor glands chiefly consist of mucous acinar cells	They mainly secrete mucus, which plays a crucial role in mucosa lubrication. Additionally, they make up a significant proportion of the entire secretion of salivary proteins	153 and 154

1.2.6 Oral Receptors

The oral cavity is packed with many nerve endings which are responsible for the significant sensitivity towards various stimuli such as tactile, taste, and olfaction. Through the use of very specific structures that serve as receptors for appropriate stimuli, the body has several groupings of sensory cells that are extremely sensitive to particular stimuli. The main function of receptors is to monitor and collect vital information regarding various stimuli (temperature, chemical composition) of a given environment and convey it to the central nervous system (CNS). Further, the complex coded data of stimulus in the CNS are then processed in the brain into a perception that is then utilized to produce suitable actions. The receptors must be both selective and sensitive to the corresponding stimulus. The receptors in the human body can be broadly classified into four, based on their sensitivity to physical energy, namely electromagnetic, thermal, chemical, and mechanical. The oral cavity contains all types, except for photoreceptors sensitive to electromagnetic radiation.[36] The receptors of the oral cavity aid in the sensing and perception of food that is consumed. The oral cavity consists of taste, olfactory, and tactile receptors (mechanoreceptors, proprioceptors, thermoreceptors, and nociceptors) to sense the various attributes of food consumed. Table 1.2 summarizes various significant features of oral receptors for the sensory stimuli taste, tactile and olfaction.

The three cranial nerves, namely trigeminal, glossopharyngeal, and facial nerves, innervate the oral region and transmit sensory information to the brain. The chorda tympani branch of the glossopharyngeal nerve and facial nerve innervates the taste buds in the posterior one-third and anterior two-thirds of the tongue, respectively. Thus, both of them are responsible for the sensation of taste.[36] Further, tactile sensations (proprioception, temperature, and nociception) are detected by the trigeminal nerve (three branches), which innervate maximum areas of the orofacial region. The receptors of the oral cavity convey tactile information to the CNS with the aid of the trigeminal somatic sensory system.[37]

In addition, the pons, pontomedullary, and rostral medulla are the levels at which the trigeminal, chorda tympani, and glossopharyngeal nerves, respectively, enter the brainstem. The spinal nucleus and the principal nucleus are the two significant parts of the trigeminal brainstem complex and they are responsible for processing heat and painful stimuli, and mechanosensory stimuli, respectively.[24] The sensory signals (taste, vision, olfaction, and touch) from various sensory regions in the human body are conveyed to the complex multimodal regions of the brain, namely the orbitofrontal cortex.[38,39] Further, their sensory signals are processed and combined to form the comprehensive result which is perception.[40]

1.3 Stages of Oral Processing and Oro-sensory Perception

The oral processing of food comprises six stages, namely (A) first bite, (B) size reduction, (C) granulation, (D) bolus formation, (E) swallowing,

Table 1.2 Significant features of oral receptors for the sensory stimuli taste, tactile and olfaction.

Stimulus type	Nerves	Oral senses	Receptors: Channels	Receptors: Morphology, afferent and characteristics	Receptors: Location and distribution	Significant features	Ref.
Taste	Chorda tympani (VII) Glossopharyngeal (IX)	Sour Salt Sweet Bitter Umami	TRPP3 Epithelial Na1 channel (ENaC) Less-selective cation channel T1R3/T1R dimer T1R1/T1R3 dimer T2Rs	Taste receptors are transmembrane proteins that attach to the chemical molecules that produce the taste sensations of sweet, bitter, and umami and admit the ions that cause the taste sensations of salty and sour	The apical portion of taste cells is where the receptors are found	The taste (gustatory) receptor system is crucial in food consumption and its primary function is to perceive taste, identify food rich in nutrients, and exclude toxic constituents	36, 155 and 156
Tactile	Trigeminal (V)	Mechanoception	DEG/ENac, TRPs, CNG channels, SLP3	Aβ (also some C) afferent fibers Merkel cells (SA I) are sensitive to curvature, edges, and points and aid in the perception of form and texture Ruffini endings (SA II) are sensitive to skin stretch and perceive the tongue's shape and direction of movement of objects. Thus, they assist in the perception of larger particle sizes and also the food bolus	The receptors are present in soft tissues in the mouth, including the mucosa, tongue, and lips	Mechanoreceptors help in the delivery of sensory feedback and also transmit information regarding the rheology and texture of the food in the oral cavity. Further, they play a crucial role in the safe handling and ingestion of food	36 and 157

Table 1.2 (*Continued*)

Stimulus type	Nerves	Oral senses	Receptors: Channels	Receptors: Morphology, afferent and characteristics	Receptors: Location and distribution	Significant features	Ref.
				Meissner corpuscles (FA1) are highly sensitive to skin motion but insensitive to static force, but their exact role in oral processing is not clear			
		Thermoreception	TRPM8 (cold) TRPV3 and TRPV4 (warm) TRPA1 (noxious cold) TRPV1 and TRPV2 (noxious heat)	The receptors are located at the free nerve endings of the afferent fibers	They are present in soft tissues of the mouth (gingiva, tongue, palate, and mucosa) and also in the tooth pulp and dentine	The perception of food temperature plays a significant role in food liking and acceptance. For instance, coffee (hot/cold, not luke warm), and ice-cream (cold). Thermoceptive systems additionally serve a defensive purpose	

		Nociception Pungency	TRP, ASIC, K^+ and ion^- ligand gated TRPV1 (capsaicin) TRPM8 (menthol) TRPA1 (garlic) TRPV3 (thymol)	The receptors are located at the free nerve endings of the afferent fibers Aδ: rapidly adapting type and aids with the detection of pain due to mechanical and thermal stimuli. C: mainly responds to noxious stimulation from various sources	They are present in soft tissues of the mouth (gingiva, tongue, palate, and mucosa) and also in the tooth pulp and dentine	The nociceptive channels aid in the perception of pungent stimuli. Additionally, they prevent mouth burns and biting the lips, tongue, or cheeks	
Olfaction (odor)	Olfactory nerve (I)	Olfactory senses	G-protein-coupled receptors	Olfactory receptor neurons (ORNs)	They are found in the olfactory epithelium that covers the nose's roof	Plays a vital role in the appreciation, identification, and determination of food's safety and freshness. Further, aids food's overall perception	24 and 155

and (F) residue or mouth coating.[41] During the first bite, the bulk of the food is deformed in the mouth and textural characteristics (mechanical or rheological), including fracturability, hardness, and viscosity, are recognized. Size reduction is characterized by tribological and rheological principles and it occurs as a result of prolonged chewing or grinding. The tribological effect is brought on by the meal's relative rubbing on the oral surfaces, which leads to the release of oil or moisture from the food, the secretion of saliva from the salivary glands, and dryness. The size reduction stage allows the perception of mechanical features such as gumminess, chewiness, and adhesiveness, and also geometric tribology dominant traits including roughness, grittiness, and powderiness. In granulation due to capillary bridging, saliva causes particles to cluster together. During the bolus formation stage, the granular particles tend to clump together because of the tongue's shearing and squeezing action against the hard palate, and also due to the inclusion of saliva, which results in the creation of a soft, cohesive bolus that can be swallowed. Different mouthfeel characteristics, including slipperiness, smoothness, and creaminess, can be felt as the bolus forms. The key factors that affect the swallowing process are the bolus's moisture content, particle size, and cohesiveness. The term "residue" refers to mouth coatings that are still present after swallowing the bolus and they are accountable for mouthfeel qualities such as astringency. Significant features of the oral processing of food are presented in Figure 1.4.

1.4 Various Mechanisms Involved in the Oral Destruction of Food

Food is mostly destroyed in the mouth by three complicated mechanisms that occur simultaneously: mechanical breakdown, biochemical and

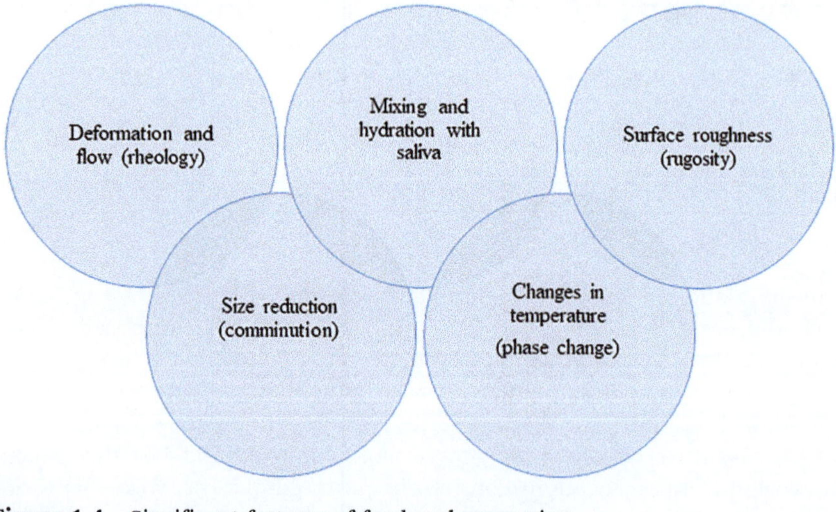

Figure 1.4 Significant features of food oral processing.

enzymatic interaction, and colloidal destabilization.[2] Additionally, they are also the primary reason for the perception of texture in the oral cavity.

1.4.1 Mechanical Destruction

Solid foods undergo mechanical breakdown in the oral cavity, which results in a reduction in size and, along with continual salivation, a swallowable bolus consistency is reached. Semisolid and viscous fluid meals are often modified in numerous ways during oral processing, including temperature modification resulting in melting, chemical breakdown, shearing, and reduction in viscosity. Semisolids and high-viscosity liquids, on the other hand, undergo very little or no mastication since they are already in a state that can be swallowed.[42]

Mechanical deformation makes it easier to perceive a variety of sensory phenomena, including textural characteristics, bursts of flavor and aroma, and the taste of consumed meals.[43,44] Food is mechanically destroyed as a result of several motions and pressures released by various areas of the mouth cavity. These include the relative sliding speeds and pressures produced by the tongue against the hard palate, and the chewing and biting action of teeth that are embedded in the jaws (incisors assist in cutting, premolars and molars help in fine grinding, and canines support ripping). All solid meals, including both hard and soft foods, are subjected to mastication (biting and chewing), which reduces the size of the food's particles and turns it into a bolus that is soft and easy to swallow. A non-flowable solid meal is therefore transformed into a flowable bolus by mechanical breakdown with the aid of saliva, which supports the transition of the ingested food from the mouth to the gastric chamber through the esophagus for digestion. This is crucial because texture perception during oral processing is dynamic. Food undergoes bulk deformation and fracture at the earliest stage of oral processing, and texture perception can be described as a rheological regime. The impression of texture perception during the later phases of oral processing is mostly a tribology-dominated regime that is substantially influenced by the surface characteristics of the food and oral surfaces, wear and friction at the tongue–food, tongue–food–palate, and palate–tongue interfaces, and the lubricating nature of saliva.[45]

1.4.2 Colloidal Destabilization of Food Structure

Salivary glands in the human oral cavity create a complex heterogeneous, colloidal, clear fluid known as saliva. Saliva is normally composed of 98% water and 2% additional inorganic and organic chemicals including mucin, glycoproteins, amylase, lingual lipase, electrolytes, and antibacterial agents.[46] There are four distinct structural phases for saliva: components such as salivary micelles and less water-soluble proteins in the salivary network, a mucin–protein continuous scaffold-like network, bacterial, lipid, and epithelial cells, and an electrolyte–water continuous phase.[47]

The colloidal nature of human saliva has been validated using microscopic images of freeze-dried saliva.[48] In addition to lubricating the mouth and reducing dry mouth and discomfort, saliva also protects oral surfaces such as the teeth and other surfaces from microbial attack. Additionally, saliva also acts as a buffering agent, lubricates solid and semi-solid foods, aids in making a swallowable cohesive bolus, and facilitates the removal of oral residues. Saliva as a lubricant also maintains oral health by protecting the teeth and other oral surfaces from bacterial attack, preventing dry mouth and irritation.[49,50]

Saliva comes immediately in contact with food after ingestion, making it an essential element of oral processing and texture perception. Additionally, the interaction between food and saliva leads to colloidal destabilization, which has a substantial impact on several mouthfeel qualities.[50] The oro-sensory perception of various mouthfeel qualities such as creaminess,[49] astringency,[51] and fatty perceptions such as oiliness and greasiness[49,52] is greatly impacted by the colloidal destabilization of emulsions.

Four fundamental mechanisms have been described in relation to the oral destabilization of food emulsions, namely: (1) bridging flocculation, (2) depletion flocculation, (3) salt-induced aggregation, and (4) coalescence that results from mouth shear and saliva incorporation.[50,53] The concepts of bridging and depletion flocculation are the most widely accepted among these mechanisms. Mucin rich in proline is the vital constituent for bridging and depletion flocculation mechanisms to occur in the mouth. Additionally, mucin has a gel-like consistency and is primarily responsible for saliva's lubricating properties.[54]

A study showed how the charge of the surfactant employed to stabilize food emulsions affects the oral stability of emulsions.[55] In the oral environment, strongly negatively charged emulsions remain completely stable because of strong electrostatic repulsion effects and only experience dilution effects from saliva inclusion. Weakly negatively charged and neutral emulsions undergo reversible depletion flocculation as a result of osmotic pressure generated from their interaction with salivary protein (mucin). In contrast, as a result of bridging flocculation, positively charged emulsions experience rapid phase separation and permanent destabilization in the oral cavity, which is brought on by the attachment of mucins and other large salivary proteins to the emulsion surfaces.[55] Therefore, the mechanism of colloidal destabilization plays a significant role in the perception of food texture, particularly for emulsion-based foods and for liquid foods such as drinks.[2,56]

Generally, a stable emulsion that has been well distributed will have a creamy or smooth mouthfeel, whereas an emulsion that has been destabilized will generate an altogether different texture impression, demonstrating roughness/dryness.[48] Emulsions can occasionally have an oily or greasy mouthfeel as a result of the coalescence of oil droplets brought on by severe flocculation.

The decrease in lubrication by oral colloidal destabilization is caused by the accumulation of salivary protein with tannins and polyphenols. Astringency is felt as a result of reduced surface lubrication and increased friction caused by the disappearance of the mucin (lubricating protein).[57]

1.4.3 Biochemical and Enzymatic Interactions

Lingual lipase and α-amylase are the two key enzymes found in saliva that play a vital role in the digestion of lipids and starches, respectively, and also in oral sensory perception. The amylose and amylopectin's 1-4-linkages of starch molecules are hydrolyzed by the α-amylase to generate maltose sugar. As a result, two important effects are observed when starch-rich foods are broken down in the mouth and come into contact with saliva: oral thinning of the food, which occurs when the food is viscous, and the perception of a sweet taste as a result of sugar synthesis. The oral thinning phenomenon of starchy and sugary foods has been demonstrated in several studies.[58] The viscosity of custard was reduced by a factor of 10 upon exposure to saliva for less than 10 s.[59] Additionally in that study, under two *in vitro* settings, saliva and water, the impact of oral deterioration on two distinct gel types consisting of tapioca starch and whey protein isolate was investigated. A discernible reduction in viscosity was seen in samples with saliva during the course of a normal eating process. Therefore, caution must be exercised when selecting emulsifying agents that are starch-based in the manufacture of food emulsions since they may flocculate upon exposure to the oral environment, which may be induced by the enzymatic breakdown mediated by α-amylase.[2]

Lingual lipase is a further significant enzyme found in human saliva. Because of its hydrophobic nature, it can enter fat globules. Triglyceride molecules are hydrolyzed, yielding glycerol and the appropriate fatty acids. According to some research, the presence of free fatty acids is what causes oro-sensory sensations, including the sense of creaminess and fat.[60] Furthermore, it has been demonstrated that fatty meals are associated with tactile or physical sensations based on data from neural imaging.[61] However, some disagree with this viewpoint, arguing that the level of salivary lipase is low and the quantity of free fatty acids produced to provide a sense of fattiness is very inadequate.[2]

1.5 Effect of Oral Processing on Various Food Matrices

The food matrix/structure plays a significant part in the oral processing fate of ingested food. The structure of food is a systematic organization of various constituents such as molecules to form a food product along with multidimensional spatial scales such as nano-, micro-, and macroscopic levels. The structure or matrix of food is immensely intricate, ranging from self-oriented natural foods of plant and animal origin to those that are formulated, predesigned, and produced by food processors. There are various food matrices, such as solid, liquid, gel, colloidal, emulsion, cellular, dense, viscoelastic, porous, and fibrous. The term matrix is commonly used when one component is embedded in another.[62] Further, foods can be classified into various categories based on how they are perceived in the human oral cavity (texture). Based on the texture, foods are commonly categorized into hard solids, soft solids, semi-solids, and liquids (thin and viscous).[63,64] Table 1.3 categorizes the food products based on their structure in relation to oral processing, sensory and mechanical properties.

Oral processing is significantly influenced by food structure and type of matrix.[65] The kind of matrix, filler, and individuals will determine the particle size distribution, average particle size, and breadth of the size distribution.[66] When food is processed orally, the food matrix breaks down, allowing released chemicals to interact with salivary enzymes.[67]

Table 1.3 Categories of food products based on their structure in relation to oral processing, sensory and mechanical properties.

Food structure	Oral processing	Rheology and tribology (mechanical)	Sensory properties	Examples
Liquids Thin fluids and thick fluids	No chewing before swallowing	Rheology (flow curves, viscosity, viscoelasticity, yield stress), tribology. They can flow	Thick/thin, smooth/rough, creamy, mouth coating	Milk, beverages, yogurt, drinks
Semi-solids	Mainly squeezed between the tongue and palate, molars are not used	Rheology (flow curves, viscosity, viscoelasticity, yield stress), tribology. They do not flow	Hardness, particle characteristics, cohesiveness, adhesiveness, mouth coating	Pudding, custard
Soft solids	Early chewing between the molars	Viscoelasticity, fracture, post-fracture–force deformation, fracture mechanics. Usually they are not crispy		Cheese, processed meat
Hard solids	Chewing/ crushing in between the molars produces typical acoustic sounds	Viscoelasticity, fracture, post-fracture–force deformation, fracture mechanics. Usually they are crispy	Hardness, particle characteristics, cohesiveness, adhesiveness, mouth coating, and crispiness	Crackers, raw vegetables, apples

For example, astringency is produced by the interaction of polyphenols with proline-rich proteins and α-amylase hydrolyzes starch into glucose and dextrin. Additionally, the matrix and microstructure of the food impact the level of interactions. Additionally, during oral processing, the disintegration of the food matrix and hydration assist in releasing taste, flavor, and aroma components that will diffuse into the saliva.[68–70]

Various compounds in the food matrix influence the degree of release and interaction, *e.g.* the presence of polysaccharides and carbohydrates will reduce the aroma release due to the increase in viscosity of the matrix or interaction with flavor compounds directly.[71] The presence of protein will lead to various molecular interactions such as ionic bonding, hydrogen bonding, and hydrophobic bonding. The presence of lipids will separate the oil- and aqueous-based aroma compounds.[72] Further, the increased hardness and chewiness (mechanical properties) of the matrix will result in a prolonged mastication time, increased lubrication, and extended flavor perception with less intensity.[73] Recent research on the oral processing of various food matrices is summarized in Table 1.4.

1.6 Oro-sensory Perception of Texture

1.6.1 Texture

Texture perception is influenced by material characteristics, including the food's physical state, mechanical and surface properties, microstructure, and composition, and also system characteristics, including saliva's lubricating properties, the force that the tongue and teeth exert on the food, and the friction and lubrication that occur between them. Oral processing of foods involves various functions such as jaw movements, muscle activities, and tongue movements, all of which contribute to the conversion of food to a bolus that is safe for swallowing. Concurrently, oral receptors aid in the perception of texture, taste, and flavor while the food structure transforms into a bolus. However, the perception of food texture during oral processing is highly complex and dynamic as the structure of foods changes continually owing to mechanical and biochemical breakdown along with the lubricating effect of saliva.

It is generally known that texture perception in the mouth depends on mechanical characteristics, which are a mix of rheology and tribology, rather than just rheology. Oro-sensory perceptions (texture) and their dominant regimes are summarized in Table 1.5. Furthermore, rheology dominates the early stages of oral processing when there is bulk deformation, but tribology dominates during the latter stages when there are surface contacts, lubrication, and friction between the tongue, food, palate, and bolus. Additionally, there is a period in between during which both rheology and tribology have a vital role in texture perception.[74] Figure 1.5 illustrates the interrelationship between food oral processing, oro-sensory perception and its dominant regimes.

Table 1.4 Summary of recent research on food oral processing.

Food product	Method for determination of oral processing behavior	Significance of the study	Ref.
Liquid, semi-solid, and solid foods	Video analysis for mastication. Mastication features – average bite size (g), number of chews and swallows, eating rate ($g\,s^{-1}$), consumption time (s), and chewing rate (chews s^{-1})	Age, gender, and ethnicity affect oral processing. Elderly persons consume foods for longer (s) and with lower eating rates ($g\,s^{-1}$) than young consumers. Males have a larger bite size (g) and higher eating rate ($g\,s^{-1}$) for solids than females. Bite size (g) and eating rate ($g\,s^{-1}$) were lower for Asian consumers than Caucasian consumers	158
Drinkable, spoonable, and chewable foods	Viscosity measurement for liquids and semi-solids. Uniaxial compression of solids. Video analysis for mastication	Oral processing behavior (mastication features) of consumers is strongly correlated with the rheological and mechanical properties of the foods. Liking, familiarity, and consumption frequency also impact the oral processing behavior of consumers to a lesser extent compared with the rheological and mechanical properties of the foods	159
Bread (carrier) and firm cheese, cheese spread, and mayonnaise as toppings	Video analysis for mastication. Images of expectorated bolus, moisture, and saliva content. Mechanical properties of bolus using a texture analyzer. Dynamic sensory perception using progressive profiling	Oral processing behavior and texture perception were predominately dependent on the carrier. The addition of topping to the carrier facilitated bolus formation and reduced the consumption time and saliva incorporation	160
Meat (shredded and seasoned beef)	*In vitro* mastication using an AM^2 masticator. Granulometry – image analysis (D_{50}) and sieve analysis	This study aids in understanding the mastication behaviors of meat. It also validates the granulometry determination by comparing the *in vitro* and *in vivo* mastication of meat	67

Cooked ham	*In vivo* mastication. Bolus properties – particle area (a_{50}) and number using image analysis. Sensory evaluation – Temporal Check-All-That-Apply (TCATA)	Characterized the dynamics of bolus formation and sensory perception during the consumption of commercial cooked ham products	161
Chocolate-chip biscuit	Motion capture recording of oral activity (Optitrack 3D motion capture system). Mastication behaviors – total oral activity duration, total number of cycles, chewing frequency, and vertical and lateral displacement (mm). Sensory evaluation – texture TDS (temporal dominance of sensations) and flavor TDS	Oral processing of two types of biscuits with minor compositional differences showed substantial variation. Oral activity analysis with and without TDS showed many differences due to the panelists' focus and attention towards the analytical task	143
Emulsion-filled gels	Fracture properties of gels by uniaxial compression tests. Electromyography (EMG) studies – mastication behavior. Bolus properties. Number, size, and roundness of bolus fragments – image analysis. Mixing behavior of gel fractions – image texture analysis. Mechanical properties – two-cycle penetration test. Dynamic texture perception – TDS	The mechanical properties of the gel were affected by the type of gelling agent (agar/gelatin), a combination of layers and concentration (low/high). TDS and oral processing depend on gel composition	162
Bread (baguette, baked bread, and steamed bread)	Mastication – electromyography monitoring. Muscular activity – mean amplitude (μV), maximum amplitude (μV), peak area (μV s), total peak area (μV s). Temporal parameters – burst duration (s), chewing time (s), number of chewing cycles, chewing frequency (s^{-1}).	Mastication properties and swallowing consistency were greatly affected by the texture and structure of the bread. The strongest muscular activity was needed for porous bread (baguette) rather than dense bread (steamed bread), which required a longer chewing time. This phenomenon could not be explained by a power law	163

Table 1.4 (*Continued*)

Food product	Method for determination of oral processing behavior	Significance of the study	Ref.
	Bolus properties – saliva impregnation, particle size distribution – image analysis and laser diffraction, bolus texture (back extrusion rig)		
Emulsion-filled semi-solid gels Agar, gelatine, agar–gelatine, sunflower oil, sugar, and vanilla flavor	Mechanical compression – oil release from gel. Sensory evaluation – qualitative descriptive analysis (QDA), progressive profiling (PP), TDS. Bolus properties – two-cycle penetration test, saliva content (gravimetric), particle size (image analysis), amount of fat release	Dynamic sensory perception of emulsion gels correlated well with fracture properties of the gel and bolus properties	164 and 165
Brittle cereal foams (rye puff and flakes)	Structural characteristics – X-ray microtomography (XMT) system and textural properties. Electromyography measurements. Bolus properties – saliva impregnation, particle size distribution. Starch hydrolysis index	Structural and textural properties greatly influenced the mastication properties but did not have much influence on the starch hydrolysis index. Further, the inclusion of rye fiber increased the hydrolysis rate and increased the number of smaller sized particles. Hence the oral disintegration process and bolus particle size have a significant role in the starch digestion rate	166
Soft cereal foods (sponge cake and brioche)	Oral comfort assessment (questionnaire) – general comfort, ease of bolus formation, pain feeling, texture, and flavor of the product. Chewing parameters such as chewing duration (s), chewing cycle, and chewing frequency (Hz). Bolus characterization – capillary rheometry (apparent viscosity), bolus hydration ratio	Stimulated saliva incorporation is the major factor that affects a majority of bolus properties and oral comfort in the elderly. Further, an increase in fat content leads to lessening of the effect of saliva incorporation and bolus hydration. Thus, modeling oral processing could aid in the design of food for the elderly and people with dysphagia	167 and 168
Soft fermented and steamed Indian food (idli)	*In vitro* and *in vivo* oral mastication parameters such as bolus granulometry, microstructure, texture, and rheology. Mathematical modeling	Both *in vitro* and *in vivo* bolus parameters such as particle size, rheology, and texture showed similar trends and good correlation	169

Table 1.5 Oro-sensory perceptions and their dominant regimes. Adapted from ref. 45 with permission from Elsevier, Copyright 2012.

Major sensation regime	Textural properties
a. Mechanics/rheology	Hardness
Early part of oral processing	Brittleness
Bulk properties	Springiness
b. Rheology/tribology	Smoothness
Bulk rheology and surface tribology, sensory intensity	Slipperiness
	Creaminess
c. Tribology	Thickness
Later stage of eating surface and lubrication behavior	Consistency
	Astringency

1.6.2 Oral Rheology and Tribology

1.6.2.1 Oral Rheology

Rheology is the study of the deformation and flow of matter, which involve forces, deformation, and time. The fundamental rheological tests aid in understanding various intrinsic food properties such as strength, deformability, modulus, toughness, and elasticity, which can be sensed during the processing of food in the mouth. This indicates that classical rheological and fracture mechanics properties will provide us with information on texture perception whereas oral processing and also molecular mechanisms explain structure–function relationships. Based on the state and structure of materials/foods, their rheological properties are governed by various factors and laws. The rheological properties of solids (ideal elastic) and fluids (ideal viscous) are governed by the relationship between stress and strain (time–rate independent) and stress and strain rate (time dependent), respectively.[75] Solids obey Hooke's law, which states that the relationship between stress (applied force per unit area) and strain (deformation per unit length) is linear, whereas fluids obey Newton's law, which states that the relationship between stress and strain rate is linear. However, most foods behave as both solid-like and fluid-like, which is known as viscoelastic behavior. The Deborah number (De) is often used in rheology to characterize the nature of food (solid/fluid) with respect to time. For solid foods, De is expected to be large ($De \gg 1$).[76] Figure 1.6 depicts the classification of foods based on their rheological behavior.

Oral rheology focuses on how food deforms and flows under the influence of various factors such as bite force, degree of chewing, addition of saliva, and bolus consistency (optimum to swallow) within the oral cavity.[24] Further, the oral rheology of foods has a significant impact on the rate of transfer of aroma and taste components to the retronasal cavity and chemoreceptors on the tongue, respectively,[77] and the tactile sensation (mechanical) originating from the resistance to shear enforced by the teeth, tongue, and hard palate.[33] Further, the consistency of the food is continuously altered during oral

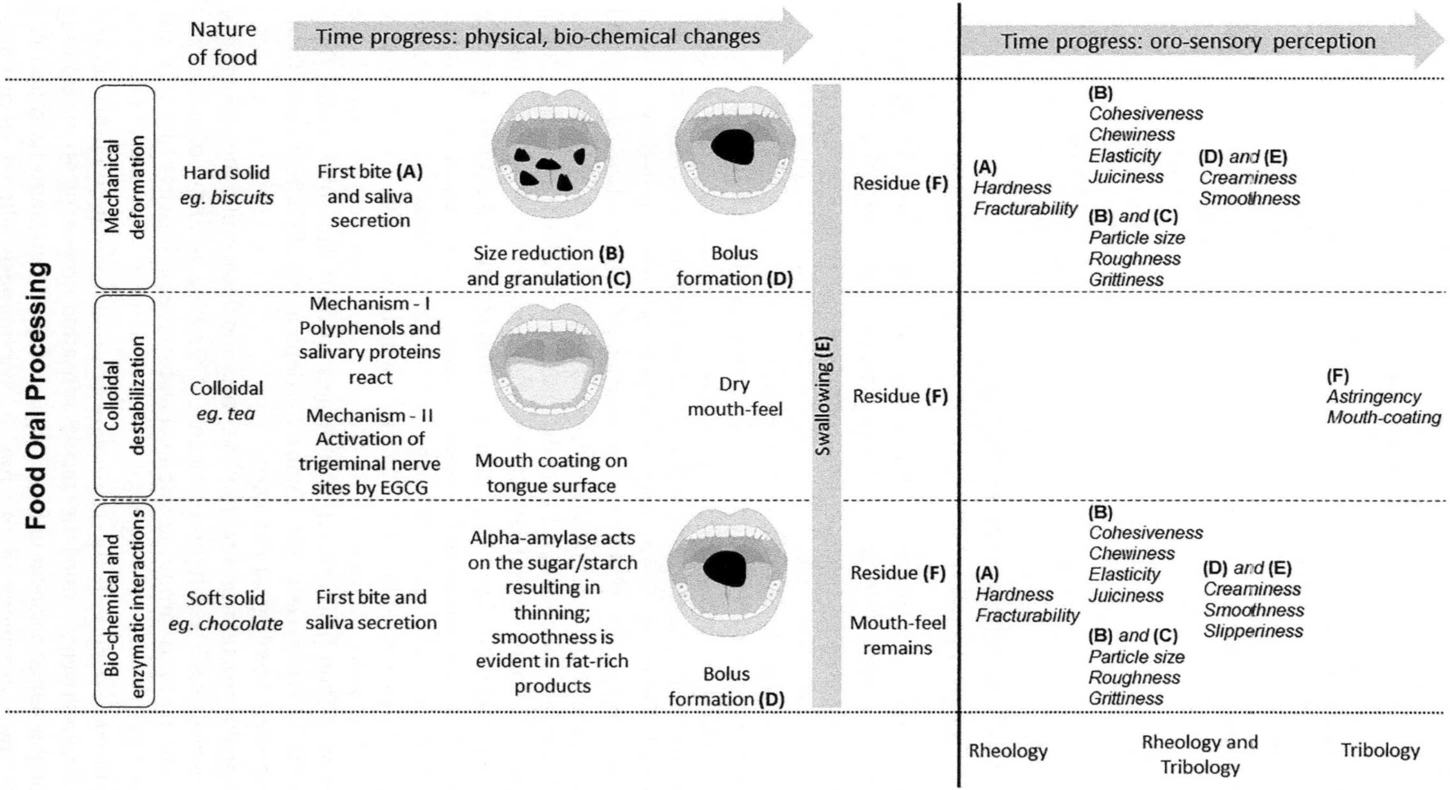

Figure 1.5 Inter-relationship between food oral processing, oro-sensory perception, and its dominant regimes. Reproduced from ref. 33 with permission from Taylor & Francis, Copyright 2021.

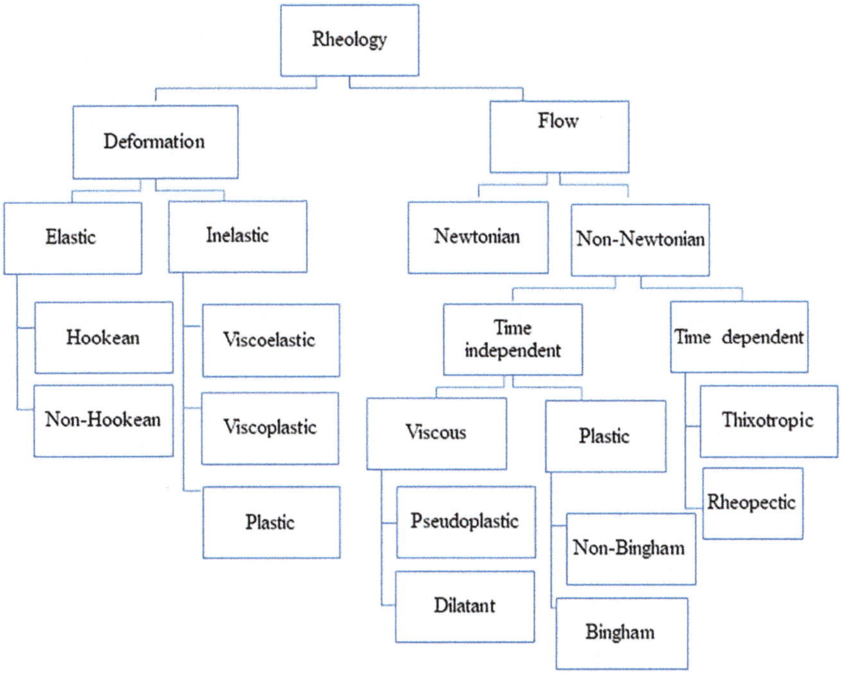

Figure 1.6 Classification of foods based on their rheological behavior.

processing from first bite to swallow owing to mastication and mixing of saliva. Hence the knowledge of the rheological properties of a bolus (food–saliva mixture) throughout oral processing is essential to understand its influence on the sensory perception of foods. However, the understanding of the oral rheological behavior of foods is very tedious and also it is difficult to predict how and to what extent it will impact the sensory perception of food. Table 1.6 provides a brief overview of various techniques and instruments used to investigate the rheological properties (food and bolus) and associated sensory perception.

Rheological investigations at a shear rate of 50 s^{-1} on liquid and semi-solid meals when assessed showed a link between the viscosity and perceived thickness of different fluid foods.[78] Additionally, extremely thin fluids and highly viscous fluids demonstrated greater association at shear rates of 10 and 1000 s^{-1}, respectively.[78] Rheological tests such as storage modulus and apparent yield stress measurements showed a favorable correlation with texture perceptions such as firmness for soft foods, *e.g.* mayonnaise and yogurt.[79] Using instrumental methods and other physiological tests as objective techniques, it was found that viscosity measurement at a shear rate of 50 s^{-1} did not have any correlation with complex sensory perceptions including fattiness, stickiness, smoothness, and creaminess for soft foods (such as custard).[59,80] In reality, the contact of the bolus with the oral surfaces, its surface characteristics, and its rheological characteristics all play a role in complicated sensory experiences.[81]

Table 1.6 Rheological techniques and instruments used in investigating the mechanical properties (food and bolus) and their associated sensory perceptions.

Sample	Rheological technique	Mechanical parameters	Instrumental features	Textural and oral parameters	Inference	Ref.
Caramel formulations (agar/gelatine/corn syrup)	Small-amplitude oscillatory rheology. Creep recovery. Pressure-sensitive tack testing	Storage moduli (G'), loss moduli (G''), and phase angle (δ). Initial compliance (J_0), maximum compliance (J_{max}), retardation time (λ_{ret}), and creep recovery percentage (crp). Task force	Controlled-stress rheometer with 25 mm diameter parallel plates. Universal testing machine to test mechanical adhesion	Hardness. Cohesiveness. Stickiness. Tooth packing. Tooth adhesiveness	Hardness was correlated positively whereas cohesiveness was associated inversely with creep recovery percentage. Further, the rheological properties determined in creep recovery were more highly correlated with sensory texture than oscillatory testing. Stickiness and tooth adhesiveness correlated positively with pressure-sensitive adhesive force. However, tooth packing was not linked with any mechanical property	170
Bead-layered hydrogels. Calcium/sodium alginate beads sandwiched between layers of κ-carrageenan (κC) gel matrix	Uniaxial single compression tests and puncture tests (food)	Fracture force Fracture stress Puncture force	A TA-TX2 Texture Analyser with a Volodkevitch Bite Jaw probe and cylindrical 59 mm platen probe was used for the puncture and compression tests,	Hard Chewy Smooth Slippery Pasty Gritty	This study showed that sensory perception of texture by humans (layered hydrogels) was the opposite compared with large deformation measurements (compression and puncture tests).	171

			respectively, with a 50 kg load cell		The reason for the contrasting results could be due to the role of saliva during *in vivo* oral processing of the samples, the heterogeneous structure of the model foods, and the lack of saliva in analysis.	
Polysaccharide hydrogels (cylindrical shape). Gelatin gels. κ-Carrageenan/ι-carrageenan/konjac glucomannan/xanthan gels	Large-deformation compression and extension tests Extrusion test Extensional and shear rheology	Young's modulus Fracture stress and fracture strain (compression and extension) Extrusion pressure	An XT.T2 Texture Analyzer with a 100 mm diameter aluminum plunger (compression), gel rings (extension), and syringe (extrusion) was used. A Rheostress 600 cone (2°) and plate geometry (Thermo Scientific) rheometer was used for the shear testing	Hardness (cutting effort) Firmness Extensibility	The actual sensory data did not have an excellent correlation with the instrumental parameters in the case of gelatine gel as it was observed to be less firm and less hard than assessed from instrumental examinations (conducted at 22 °C below its melting temperature), which is due to its lower melting temperature. Nevertheless, all of the polysaccharide gels had melting points that were higher than the body temperature.	172

Table 1.6 (*Continued*)

Sample	Rheological technique	Mechanical parameters	Instrumental features	Textural and oral parameters	Inference	Ref.
					According to this study, there is a strong association between extrusion pressure and fracture stress in extension, compression, and shear, and also between extrusion pressure and Young's modulus ratio and fracture strain	
Group 1: semi-skimmed milk, organic fruit purée, tomato ketchup, delice chocolate dessert, crème fraîche, honey. Group 2: orange juice, ready-to-eat custard, mayonnaise, chocolate mousse, Nutella hazelnut spread, strawberry jam. Group 3: yogurt drink, ready-to-eat jelly, Greek yogurt, fromage frais, Italian ricotta cheese, and smooth peanut butter	Flow behavior Apparent viscosity Fluid stretching tests	Apparent viscosity Maximum tensile force (F_{max}) and the work until the maximum force (W_{max})	The temperature was 37 °C and a Bohlin C-VOR rheometer with cone-and-plate geometry (cone angle 4° and plate diameter 40 mm) was employed. A TA.XTplus Texture Analyser equipped with a 30 kg load cell and 40 mm diameter probe made of stainless steel was used	Ease/difficulty of swallowing. Oral exposure time	According to this study, there is a strong association between apparent shear viscosity and swallowing convenience, highlighting the significance of shear deformation in bolus preparation and swallowing. Furthermore, the study found a considerably larger association between fluid stretching behavior and the sensory difficulty of swallowing. Hence bolus stretchability is a crucial factor in inducing the swallowing response	173

Emulsion-filled agar/gelatine gels and their bolus	Two-cycle penetration test	Gel properties (fracture, strain, stress). Bolus texture (first peak force, resilience, adhesiveness, and flowability)	A TA.XTplus Texture Analyser (Stable Micro Systems – SMS) with a 20 mm diameter cylindrical steel probe was used	Firm, sticky, elastic, moist, refreshing, grainy, creamy, melting	At the beginning of oral processing, the mechanical characteristics of the bolus, such as first penetration peak force and flowability, were connected to sensory firmness. Additionally, there were relationships between bolus characteristics and texture perception during the middle stage of oral processing, *e.g.* the perception of moist and refreshing was connected to flowability, and the elastic and sticky sensation was related to resilience and adhesiveness. At the end stage of oral processing, the perceived texture sensation creamy or grainy was related to high bolus flowability and broken down particles, respectively, in the case of emulsion-filled gels boluses	165

Table 1.6 (*Continued*)

Sample	Rheological technique	Mechanical parameters	Instrumental features	Textural and oral parameters	Inference	Ref.
Thickened fluid sample thickeners were used (xanthan gum, tapioca starch, carboxymethylcellulose gum)	Flow behavior Apparent viscosity	Yield stresses Shear rate Elastic modulus G', viscous modulus G'' Apparent viscosity Density	A universal stress rheometer with a 40 mm cone-and-plate geometry was used. Measurements were made gravimetrically by weighing 10 mL of fluid in a sterile disposable 10 mL slip-tip syringe	Oral cohesiveness Propulsion effort Stickiness Oral residue	The study found strong correlations between fluid viscosities and oral cohesiveness ($r = 0.97$) at a shear rate of 10 s^{-1}. In addition, there was a positive correlation between fluid viscosities and propulsion effort ($r = 0.97$) at a shear rate of 50 s^{-1}. Further, strong positive correlations were found between viscosities at 100 s^{-1} and perceived stickiness ($r = 0.78$) and oral residue ($r = 0.80$). However, the yield stress was not a direct indicator of the sensory attribute studied	174

Biscuits and their boluses	Back extrusion. Consecutive double compression	Flow behavior index Consistency coefficient Peak force adhesiveness Cohesiveness	An Instron 5543 tensile tester with a 1000 N load cell and a piston rig with a stainless-steel cylindrical plunger (diameter 17.6 mm)	Hard, crunchy/crispy, crumbly, dry, and sticky	This study indicated that the dominant texture perception in the initial stages of mastication was well correlated with the fracture properties of the biscuits. Midway through the mastication process, there was a sensory change from dry to sticky, which raised the "cohesiveness" of the boluses and lowered their consistency coefficient, peak force, and "adhesiveness". During the swallowing stage, the boluses displayed shear thinning/pasty behavior, and stickiness was the dominant texture experience	175

1.6.2.2 Oral Tribology

"Tribology", which means rubbing or sliding, is described as "the science of friction, wear, and lubrication between interacting surfaces that are in relative motion, with or without the presence of a medium separating them". It is a very ancient discipline that has been widely utilized in the automotive industry,[82] engineering,[83] polymer technology,[84] nanotechnology,[85] medicine,[86] and cosmetics.[87] Only lately has its importance in food applications been recognized. Several fundamental, empirical, and imitative tests may be used to predict the texture that will be experienced during the early stages of oral processing. This explains why tribological studies are necessary; by combining the strengths of existing methods, an ideal instrumental method of texture measurement must be developed that can simulate oral processing and quantify textural properties as perceived in the oral cavity. Substantial efforts have been made in soft tribology in recent years to comprehend and quantify the physics of oral tribological phenomena. However, more widespread studies and research must be conducted.

During the oral processing stage, food directly engages with the body and is deformed as the texture is perceived. Over time, surface characteristics become more important than bulk features. Further, in texture perception, both oral tribology and lubrication have a vital role. The major factors involved in oral tribology are illustrated in Figure 1.7. During the oral processing of food, the oral cavity's interacting surfaces, such as the teeth, hard palate–tongue, teeth–food, and food–tongue–palate interfaces, tongue–teeth,

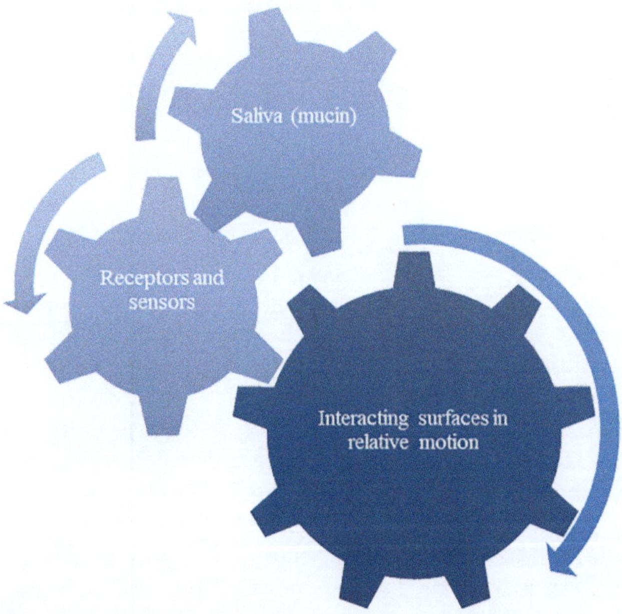

Figure 1.7 Oral tribology.

food–tongue, and lips are all in motion.[45] Additionally, the tongue lacks the hard palate's smoothness, and its roughness is caused by the existence of papillae which are 100 μm in height and diameter.[88] A thin layer of saliva coats every component of the oral cavity, ranging in thickness from 5 μm on the hard palate to roughly 25 μm on the tongue.[34,89] This offers a defense against microbial assaults and also friction and wear, such as surface irritation or dryness.[32]

The lubricating role of saliva in the mouth is frequently overlooked during tribological examination, which results in significant variations in the recorded friction values. In the mouth cavity, saliva works very well as a lubricant, interacting with different elements of the food system and preventing wear by producing a coating of salivary proteins that is resistant to wear.[90] Additionally, saliva-lubricated smooth surfaces have a friction coefficient of about 0.02.[91] The tribological characteristics of saliva are significantly influenced by several factors, including surface roughness, the presence of surfactants, ionic strength, and applied force.[91] It is recommended to coat or absorb a small layer of saliva on the device's surface before the addition of a food sample to mimic the lubricating properties of saliva in the mouth.[92] Additionally, during the oral processing stage, the tongue glides across the palate and squeezes it with a force ranging from 0.01 to 90 N and a sliding speed of up to 200 $mm\,s^{-1}$.[28] However, the tongue's exerted pressures and sliding rates change depending on the food type, taking into account the impacts of structure and hardness, and also the change in time and location on the tongue. Therefore, it is strongly advised to evaluate the tribological characteristics of food under a variety of loads and sliding speeds to gain further insights into oral lubrication and processing.[93] These are significant factors that are not taken into account in traditional textural measuring techniques. To understand better how textures are perceived, it is important to understand how foods behave under various conditions of friction and lubrication since these properties change over time, based on the type of food and also depending on how the tongue moves.[93]

According to recent research on the oral processing of food, oro-sensory perceptions of texture and frictional measures are connected; nonetheless, it is extremely difficult to measure and identify their empirical links. Furthermore, as tribology is a "system" property, great care should be taken to optimize additional elements such as properties of the food sample, inclusion of saliva, sliding speed, and selection of the tribo-pair as these components are crucial in estimating the correlations amongst friction coefficients and texture. Thus, establishing a link between friction coefficients and oro-texture perception is perhaps the most evident problem. The main cause of this problem is that the friction coefficient depends on several factors in addition to the food matrix, including the instrumentation, lubrication, testing procedure, sliding speed, applied load, tribo-pair, and surface characteristics. Numerous elements of food tribology have advanced significantly over the past 10 years, including the creation of diverse tribological systems with various tribo-pairs and a wide range of loads and sliding speeds. However, in measuring the complexity of oral processing, the optimization of "system" parameters remains a significant difficulty. In addition to the disadvantages already described, it is also difficult to comprehend the proper entrainment rates at which the correlation between texture-related sensory qualities and friction coefficient could be elucidated. To acquire a comprehensive knowledge of the link between oral processing and sensory perception, an integrative approach incorporating several additional characterization techniques such as rheology, microstructure, and texture analysis along with tribology is essential. Table 1.7 summarizes recent tribological studies of foods with a special focus on food texture.

1.7 Perception of Taste

1.7.1 Taste Sensation

Taste (gustation) is also a significant sensation that is perceived during the oral processing of foods. The sensation of taste is perceived when the active chemical compounds in the food are released due to mastication and hydration with saliva and interact with the taste receptors present in the taste buds.[24] Taste buds are specialized onion-shaped structures that consist of 50–150 taste receptors.[94] The taste buds are mainly located on the upper surface of the tongue (dorsum) and other parts of the oropharynx, such as the larynx, pharynx, and epiglottis.[95] There are five basic well-recognized and approved taste modalities, namely sweetness, sourness, saltiness, bitterness, and umami (savory).[44] Various regions of the tongue are responsible for the sensation of different gustatory preferences such as back (bitter), posterior lateral edges (sour and bitter), anterior lateral edges and tip (salt), and front (sweet) due to the presence of specific taste receptors in the taste buds that are distributed among different papillae in the tongue epithelium.[95]

1.7.2 Sixth Taste Modality: Oleogustus or Starchy?

Further, recent research suggests that there are other taste modalities, namely oleogustus (perception of fat) and starchy (perception of oligosaccharides), which have significant potential to become the sixth taste sense. Further, there are numerous theories related to the taste perception of fat, such as free fatty acids (high-fat foods) as a chemosensory signal for dietary fat,[96] lipid-binding protein CD36 (located in the tongue epithelial cells) that could function as both a scavenger and receptor of lipid,[97] and the role of lingual lipase in the enzymatic hydrolysis of triglycerides.[60] A study observed that the lingual lipase is active during the oral processing of high-fat foods, which requires a complex oral processing effort.[96] However, it has also been reported that lingual lipase does not contribute to the detection of oral fats. Further, a recent study showed that oligosaccharides can be detected without using sweet receptors.[98]

Table 1.7 Recent tribological studies of foods. Reproduced from ref. 33 with permission from Taylor & Francis, Copyright 2021.

Textural parameter	Sample	Tribometer setup	Sliding surface	Contact pressure/N	Sliding speed/ $mm\,s^{-1}$	Temperature/°C	Inference	Ref.
Creaminess	Homogenized milk with a fat content of 0.06 and 8.68%	Mini traction machine	Neoprene O-ring, silicone, neoprene, and Teflon disk	5	5–500	20	At low speed ($<100\ mm\,s^{-1}$), the friction coefficient and creaminess were found to be significantly correlated	176
Mouth coating and clearance	Chocolate	Ball-on-three plates triborheometer cell	Steel ball and polyurethane plates	0.5	0.001–420	37	The friction behavior of melted chocolate demonstrated that the degree of mouth coating is greatly influenced by how different textures were perceived	177
Slipperiness Powderiness Stickiness Sliminess Film forming	Polysaccharide-gelled protein particle dispersions with various sizes, shapes and hardnesses	Mini traction machine	Neoprene O-ring, neoprene rubber disk	2 and 5	5–500	30 (±1)	Slipperiness and friction coefficient were strongly associated, although there was no obvious relationship between the other attributes (filmy, sticky, and slimy). Powdery perception was correlated with	178

Table 1.7 (*Continued*)

Textural parameter	Sample	Tribometer setup	Sliding surface	Contact pressure/N	Sliding speed/ $mm\,s^{-1}$	Temperature/°C	Inference	Ref.
							particle hardness and shape	
Graininess Creaminess Thickness	Skim-milk yogurt enriched with whey protein–pectin mixtures	Ball-on-three-plates triborheometer cell	Stainless-steel ball and elastic pad (styrene–butadiene rubber)	3	0.001–1000	10	Complexes of whey protein and pectin can be utilized to replace fat	179
Chalkiness Thickness Smoothness	Acid milk gels and saliva	Double ball-on-disc	Polypropylene balls; whey protein isolate gel plates	2.1	0.016–100	25	Increased spoon and mouth viscosity, lumpy appearance, and chalkiness, in addition to diminished smoothness and gritty mouth coating, were all associated with an increased friction coefficient	180
Fat-related perceptions	Sunflower oil-in-water emulsions, flavored and unflavored milk, stirred yogurt, and emulsified white sauce	Pin-on-plate and plate-on-ring triborheometer cell	Acrylonitrile–butadiene	—	10	—	According to tribology data, the friction factor may be employed as an instrumental measure for fat perception	181
Creaminess and fat-related perceptions	Microbubbles in liquids with a thickener (xanthan), liquids without	OTC	Polydimethylsilane (PDMS)–glass	0.5	10–80	Room temperature	Oil-in-water emulsions have superior lubrication qualities to protein solutions and	182

	thickener, and agar/gelatin gel						microbubble dispersions. Microbubbles might be used in food products as a substitute for fat and to change the texture	
Astringency	Saliva: astringent solution (1 : 1) Tannins Caffeine Four types of wines	A simple device attached to a TA.XTplus Texture Analyser	PDMS	0.57	0.025–10 0.075	28	Positive correlation between humans and instruments High astringency perception was directly correlated with high friction coefficient	183
Smoothness	Saliva: emulsion (1 : 1) Sunflower oil Oil mass fraction (1, 5, 10, 20, and 30 wt%) Whey protein isolate (WPI) (−ve) WPI (+ve) Lysozyme (+ve) Modified starch (neutral)	Ball-on disk tribology configuration	PDMS Stainless-steel base	0.57	0.1–30	28	Smoothness perception was related to friction between the tongue and the palate. Oil/water emulsions stabilized by various emulsifiers can be distinguished for smoothness perception based on oil mass fraction and kind of emulsifier	184

1.7.3 Factors Influencing Taste Sensation

The taste sensation is very unique and complex because its perception and intensities vary between individuals based on the age, sex, genetic variants of receptors, and taste preferences. A study was conducted to analyze the impact of age, sex, and genetics on the taste acuity of relatively healthy, European populations between the ages of 18 and 80 years. A decrease in the perception of all taste stimuli, especially bitter and sour tastes, was observed with increasing age, females sensed all the tastes more than men (except for umami), and TAS2R38-rs713598 polymorphism was strongly linked with bitter taste perception and also altered the sex differences for bitterness based on the genotype.[99]

Further, the composition, texture, and structure of foods, the temperature of foods, the degree of mastication, and hydration by saliva also play a crucial role in the gustatory perception and intensity of various taste and flavor (odor–taste–texture) modalities.[39,100–102] In addition, individuals also vary in their sensitivity to certain compounds; for instance, individuals specifically sensitive to phenylthiocarbamide (a bitter-tasting compound) are acknowledged as supertasters. Further, it has been found that approximately 25% of the population are "supertasters", a further 25% are "non-tasters", and the remaining 50% are medium tasters.[103] Also, the supertasters are more sensitive to other taste stimuli.[104] An oral processing study was conducted using bread as a model food, with three different varieties of bread in terms of structure and/or composition but with the same salt content. The denser bread was perceived as less salty and exhibited a simple texture perception sequence compared with the others. The study acknowledged the influence of structure and texture on the perception of salty taste.[105]

1.8 Multi-sensory Flavor Perception

Flavor perception is a complex combination of multiple sensations such as gustation (taste), olfaction (retronasal and orthonasal), and oral–somatosensation (tactile, trigeminal, temperature, and pain).[106] In addition, visual (food color) and auditory (food-eating sound) signals also play a significant role in the flavor perception of foods.[107] The olfactory sensory system, especially both the retronasal system (which involves posterior nares to detect aromas emanating from the foods while we consume them) and the orthonasal system (a conventional system that is associated with the inhalation of external odors) play a crucial role in the detection of food flavors.[108] Further, various oral processing parameters (mastication, salivation, and swallowing), breathing, and, most importantly, individual differences (the way in which they orally process the foods and retronasal release of aroma compounds) have a significant influence on the perception of flavor and its intensity.[103]

1.8.1 Olfactory–Gustatory and Oral–Somatosensory Interactions

The sense of taste and smell plays a vital part in multisensory flavor perception. Indeed, several studies reported that olfactory cues are more critical than taste signals, as they contribute around 80–90% towards human flavor perception.[109] A few studies have provided evidence that there is a significant enhancement of the perception of olfactory stimuli (both retronasal and orthonasal) when the gustatory signal is maintained at either a sub- or supra-threshold level.[110–113] However, this kind of multisensory integration was observed to be dependent on certain combinations of surfactants and tastants used (*i.e.* stimulus congruency – the degree to which pairing of two stimuli in a food product is acceptable),[114,115] and on the cultural background of the participants studied.[103,116]

Furthermore, the oral-somatosensory system provides us with tactile cues (temperature and texture of food) that appear to play an important role in multisensory flavor perception.[107,117] A study highlighted the key role played by texture (mouthfeel) in driving multisensory flavor perception.[118] The study found that regardless of whether the odor was provided orthonasally or retronasally, as the viscosity of the solution (milk) increased the perceived flavor intensity (creamy) decreased.[118] Further, a few studies have elegantly demonstrated the cross-modal interactions between temperature and taste sensations.[119] Around 33–50% of the population experiences "thermal taste" illusions such as variation of the temperature of the tongue (increase or decrease) can provoke basic taste senses (sweetness, sourness, saltiness, and bitterness).[102] Further, thermal tasters (people who perceive the illusion of thermal flavor) tend to perceive all oral stimuli (sweet, sour, salty, bitter, and astringent) as more intense compared with non-thermal tasters.[120]

1.8.2 Auditory Contributions

The auditory cues experienced during the consumption of food play a vital role in the multisensory perception of flavors and also texture.[121] Most of the published research has made an effort to establish a correlation between subjective scores and objective/mechanical measurements for various foods (potato chips), especially for sensorial attributes such as crispiness/crunchiness and/or freshness. Further, it was found that the auditory cues heard during the oral processing of potato chips contribute around 15% to the perception of their crispness and/or freshness.[122] In a study conducted with potato chips, recorded sounds of the auditory feedback of the oral processing of chips (biting) were played in headphones. When the overall sound level was amplified and/or when just the high-frequency sounds were intensified, the panelists reported that chips were both significantly crisper and fresher. Similarly, when the overall intensity of the sound was reduced and/or when the high-frequency sounds were attenuated, the panelists scored the crisps as significantly staler and softer.[122] Subsequently, another study was conducted to demonstrate that modifying the loudness of popping sounds of carbonated fizzy beverages could significantly impact participants' ratings.[123] Further, it is very important to note that external environmental noise and background music also have a significant impact on our eating and drinking behaviors.[124,125]

1.8.3 Visual Interactions Influencing Flavor Perception

Over the years, several studies have reported that altering the color of foods to either appropriate, inappropriate, or absent has a major role in influencing people's decisions regarding the taste, aroma, and flavor of the foods being evaluated. However, other factors such as opacity and varied intensities of color (colored beverages) have been less researched and also gave mixed results. A study demonstrated the effect of color on the taste perception of consumers.[126] The impact of red-, yellow-, and green-colored solutions on the detection thresholds and taste sensitivity for sweet, sour, bitter, and salty tastes (the four basic tastes) was investigated. The study revealed that the participants could only identify the taste of colored solutions at higher concentrations of tastants compared with uncolored solutions. Colorants did not affect the taste detection thresholds of the salt solutions, probably because salty food comes in a range of colors. However, in the case of a sweet solution, the taste sensitivity of participants varied based on the colors such as yellow (decreased), green (increased), and red coloring (no effect). In the case of bitter taste, yellow and green coloring of the solution had no effect whereas a substantial decrease in participants' sensitivity was observed for a red-colored solution. For sour taste, both yellow- and green-colored solutions decreased the sensitivity of participants and red color had no significant effect.[126]

Further, various studies investigated the impact of visual cues on the discrimination of olfactory stimuli (if presented in their appropriate color, inappropriate color, and uncolored),[127] identification of flavored beverages (both fruit and non-fruit flavors),[128] and also on the odor intensity.[129] Moreover, a study was conducted to explore the impact of color on how people perceive the aroma of wine (red and white wine) and validated the domination of visual signals over orthonasal olfaction.[130] In addition, several studies were conducted to identify the impact of visual contribution directly on flavor perception by subjecting participants to various combinations (pairs) of flavor and color signals,[131] for instance, appropriate pairing (a cherry-flavored drink colored red), inappropriate pairing (a lime-flavored drink colored red), and incomplete pairing (either colorless with flavor or flavorless with color).[132]

1.8.4 Role of Cognitive Neuroscience

The olfactory and gustatory stimuli are captured by the sensory receptors and transmitted to the piriform cortex and primary taste cortex, respectively;

and both of the sensory signals are projected to the orbitofrontal cortex.[133,134] Currently, available evidence indicates that the orbitofrontal cortex plays a crucial role in modulating the multisensory interactions in flavor perception.[135] On the other hand, the oral–somatosensory signals such as proprioception, chemical irritation, texture, nociception, and temperature are sensed by the oral receptors during the oral processing of food. The tactical signals are transmitted to the brain through the trigeminal nerve, which projects directly to the primary somatosensory cortex. Hence the oral texture sensation is also represented in the orbitofrontal cortex.[133] Some researchers used functional magnetic resonance imaging (fMRI) to observe the neural interactions in the orbitofrontal cortex for congruent and incongruent pairings of various olfactory and gustatory stimuli.[136,137] They observed enhanced neural activity in the orbitofrontal cortex and increased pleasantness on evaluating congruent pairs of the olfactory–gustatory stimulus compared with incongruent combinations. Further, several studies applied various cognitive neuroscience techniques such as neurophysiology, neuroimaging, psychophysics, and computational modeling, in addition to conventional sensory techniques, to understand the reasons behind the complex sensory perception of foods.[121,138]

1.9 Sensory Characterization of Foods

Sensory evaluation is the scientific discipline that is used to "evoke, measure, analyze, and interpret reactions to those characteristics of food as they are perceived by the senses of sight, smell, taste, touch, and hearing". The sensory evaluation/profiling of foods is mandatory to determine consumer acceptability, product quality control, storage and shelf-life, processing changes, product development, and product reformulation.[139] There are two different approaches to ascertain the sensory characteristics of foods, namely instrumental techniques (objective methods) and sensory evaluation by human panelists (subjective methods).[140] Further, when both sensory evaluation techniques (objective and subjective) are combined in the characterization of food products, we obtain a great deal of information and a holistic picture regarding the product. However, the proper selection of instrumental and sensory methods is crucial to obtain a valid result about the sensory features of a product. Table 1.8 summarizes sensory profiling methods and instrumental and analytical techniques used in investigating different sensory perceptions.

1.9.1 Sensory Profiling: Static and Dynamic Techniques

Sensory profiling techniques can be broadly classified into two categories based on the period of evaluation, namely static and dynamic sensory evaluation techniques. Various conventional sensory evaluation techniques, such as descriptive and difference tests, consider sensory attributes as a static phenomenon (sensation).[141] However, we know that the oro-sensory

Table 1.8 Summary of sensory profiling methods and instrumental and analytical techniques used in investigating different sensory perceptions.

Sensory parameter	Food type	Scope of the study	Sensory attributes	Sensory profiling method[a]		Instrumental and analytical methods[a]	Ref.
				Static	Dynamic		
Texture	Emulsion-filled gels	To understand the oral processing and dynamic texture perception of composite foods by using emulsion-filled gels	Firm, sticky, elastic, moist, refreshing, grainy, creamy, melting	QDA	PP and TDS	Texture analyzer (uniaxial single compression tests). Mechanical bolus properties using two-cycle penetration tests	162 and 164
	Low-calorie chocolate	To compare and validate the results of the sensory profile obtained by QDA with TDS results of low-calorie chocolates	Hardness Fracturability	QDA	TDS	Texture analyzer (penetration test). Tribometer (steel-on-PDMS tribo-pairing)	185 and 186
	Thin porridge	To study the impact of the type of composite flour and pH on the sensory quality of thin porridges	Firmness Consistency Cohesiveness Index of viscosity	Descriptive sensory analysis	—	Texture analyzer (back-extrusion method)	187
	Beef burgers	To investigate the impact of salt (NaCl) reduction and its size changes on the instrumental texture and temporal sensory profile, and also to determine the temporal drivers of liking beef burgers	Hardness Springiness Cohesiveness Chewiness	—	TCATA	Texture profile analysis (double compression test)	147

	Ice creams	To understand how oral processing behavior influences the dynamic sensory perception of ice-creams	Fruitiness Sweetness Chewiness Iciness Smoothness Coldness Firmness	—	TDS	Texture analysis (penetration test)	188
Aroma	Cumin (*Cuminum cyminum* L.)	To characterize the aroma-active compounds of cumin	Odor profiling (woody, pleasant, floral, lemony, spicy, green, herbal, cooling, cumin-like, and earthy)	QDS (sniff testing method)	TI	GC–olfactometry GC–MS Electronic nose analysis	189
	White bread	To characterize the aroma release and perception of white bread during oral processing	Sour Flour-like Fermentation-like Creamy Roasty Sweet	—	TDS	GC–IMS	190
Taste	Wheat sourdough bread	To characterize the sour taste of sourdough bread and the evolution of its perception during bread tasting. To identify instrumental variables that could aid in predicting the sour taste of bread	Sour taste	QDA	PP	pH, TTA, organic acid content, and density measurements	191

Table 1.8 (*Continued*)

Sensory parameter	Food type	Scope of the study	Sensory attributes	Sensory profiling method[a]		Instrumental and analytical methods[a]	Ref.
				Static	Dynamic		
	Tea	To evaluate pu-erh tea (various grade levels and ages) based on taste properties or chemical composition	Sweet and mellow, astringent, heavy and thick, sour, fresh and brisk bright and tasty taste Soft and smooth scorched flavor and old flavor	Attribute different test QDA	—	Chemical analysis (total free amino acids, polyphenols, caffeine, soluble sugar, pH) and ET	192
Flavor	Orange juices	To evaluate the correlation between the sensory and instrumental volatile flavor analyses of orange juices prepared by different processing methods	Orange flesh, orange peel, grapefruit, orange powder, cooked orange, burnt, citrus, yogurt, medicinal, cough syrup sour and bittersweet characteristic aromatics	Descriptive sensory analysis		DHS and GC–MS	193

Texture, taste, and flavor	Dry-cured loins	To investigate the influence of NaCl replacement by KCl (15, 20 and 25%) on the sensory characteristics of dry-cured loins	Overall flavor Saltiness Cured flavor Rancid flavor Bitterness Hardness Juiciness Fibrousness Umami taste Pastiness	Discriminative/ difference analysis FP analysis	TI and TDS	—	194
Texture and flavor	Chocolate-chip biscuits	To evaluate the effect of TDS evaluations of texture and flavor on oral activity during the consumption of two different chocolate-chip biscuits	Texture (hard, dry, crisp, sandy, compact, melting chocolate, non-melting chocolate, and sticks to teeth). Flavor (sweet, bitter, chocolate, buttery, floury, and nutty)	—	TDS	Motion capture recording of oral activity	143
Flavor and aroma	Espresso coffee	To understand the flavor perception of espresso coffee and the release kinetics of aroma compounds	Sweet, sour, bitter, astringent, roasted, burnt, caramel, vegetal and nutty	—	TDS	NS analysis using PTR-TOF-MS	195

[a] QDA, qualitative descriptive analysis; PP, progressive profiling; TI, time–intensity; TDS, temporal dominance of sensations; TCATA, temporal check-all-that-apply; GC, gas chromatography; MS, mass spectrometry; IMS, ion mobility spectrometry; TTA, total titratable acidity; ET, electronic tongue; DHS, dynamic headspace sampling; FP, flash profile; NS, nose space; PTR-TOF-MS, proton transfer reaction time-of-flight mass spectrometry.

perception of texture, flavor, and taste is complex and dynamic in nature. In the case of static sensory profiling, the sensory perception over the entire period of oral processing (first bite–mastication–bolus formation–swallow–aftertaste) is expressed as a single sensation even though the sensation was dynamic over time.[139] In dynamic sensory evaluation techniques such as temporal dominance of sensation (TDS),[142] time–intensity (TI) analysis, progressive profiling (PP), and temporal check-all-that-apply (TCATA), the sensory attributes perceived throughout the various stages of the oral processing of food are measured over time. Also, the sensory information acquired from food oral processing helps us comprehend better the connection between sensory perception (texture, taste, and flavor) and the structure and composition of food. The dynamic evaluation techniques allow sensory panelists to record a series of sensory changes that occur during oral processing due to the dynamic structural changes (mastication and mixing of saliva) and release of flavor, taste, and aroma compounds.[143]

1.9.2 Dynamic Sensory Techniques

The progressive profiling method is a relatively low-resolution dynamic method compared with TDS, TI, and TCATA.[139,144] TI analysis aids in measuring the intensity perception of specific traits over the food consumption period. However, at one time only the perceived intensity of mostly one or a maximum of two sensory attributes [dual-attribute time–intensity (DA-TI)] can be determined effectively.[145] The TI technique is quantitative in nature so it requires extensively trained panel members.[4] The TCATA technique is a new dynamic method for investigating the multidimensional sensory properties of products as they evolve. TCATA is the dynamic extension of the check-all-that-apply (CATA) method.[146] In this method, the selection and deselection of sensory attributes are tracked continuously over time, allowing the assessors to demonstrate the evolution and changes in sensory attributes of products over time. Multiple attributes can be selected simultaneously in TCATA, which describes the sensations that arise either sequentially or concurrently.[147]

The TDS method is a relatively new dynamic methodology that is frequently used to assess how sensory perception changes dynamically during the whole oral processing event, from the initial bite to the swallowing point.[148] TDS helps to illustrate the order of change in about 10 sensory attributes during food consumption at one time. Owing to its novelty, ease of use, and affordability, this approach has attracted a lot of research interest. TDS is the most preferred method compared with the TI method because of its ability to reveal the product perception pattern over time.[149] Further, the TDS technique, apart from providing the sequence of change in qualitative sensory attributes, also provides approximate quantitative data regarding the intensities (maximum or minimum) of dominant sensory attributes.[150] In contrast to TI, the sensory panelist needs less intensive training, and inexperienced panelists can also be used.[4]

1.10 Conclusion and Future Perspectives

In the recent past, there have been many studies on food oral processing to determine the link between food structure and the oro-sensory perception of texture, flavor, and taste. Consumers are extremely aware of health and well-being owing to the prevalence of numerous non-communicable lifestyle disorders such as diabetes, hypertension, cardiovascular disease, and obesity. Both government and food processors are forced to implement changes in food formulations to reduce sugar, salt, and fat content to produce healthy foods without altering their sensory properties. A lot of research is being carried out to develop new products (food structuring) with improved health benefits by choosing healthy ingredients and processing methods to preserve and retain their fullest nutritional and sensory properties. This chapter has summarized how food is processed in the mouth and its governing mechanisms and factors, and also how sensory attributes are perceived and factors that influence the sensory perception of food in the mouth. Additionally, a brief compilation of recent research work related to food oral processing, sensory perception, and various techniques employed in their investigation has been presented and discussed. Nevertheless, there is still a huge need at present and in the future to expand our existing knowledge on the oral processing of food to develop new food products with superior nutritional quality, health benefits, and modified sensory attributes so as to satisfy the needs of various sections of the population.

References

1. J. Chen, Food oral processing: Some important underpinning principles of eating and sensory perception, *Food Struct.*, 2014, **1**(2), 91–105.
2. J. Chen, Food oral processing: Mechanisms and implications of food oral destruction, *Trends Food Sci. Technol.*, 2015, **45**(2), 222–228.
3. M. C. Bourne, Is rheology enough for food texture measurement?, *J. Texture Stud.*, 1975, **6**(2), 259–262.
4. K. D. Foster, J. M. V. Grigor, J. N. Cheong, M. J. Y. Yoo, J. E. Bronlund and M. P. Morgenstern, The role of oral processing in dynamic sensory perception, *J. Food Sci.*, 2011, **76**(2), R49–R61.
5. G. Feron and C. Salles, Food oral processing in humans: links between physiological parameters, release of flavour stimuli and flavour perception of food, *J. Food Stud.*, 2018, **7**(1), 1–12.
6. E. M. Krop, M. M. Hetherington, C. Nekitsing, S. Miquel, L. Postelnicu and A. Sarkar, Influence of oral processing on appetite and food intake–A systematic review and meta-analysis, *Appetite*, 2018, **125**, 253–269.
7. S. Fiszman and A. Tarrega, The dynamics of texture perception of hard solid food: A review of the contribution of the temporal dominance of sensations technique, *J. Texture Stud.*, 2018, **49**(2), 202–212.
8. X. Wang and J. Chen, Food oral processing: Recent developments and challenges, *Curr. Opin. Colloid Interface Sci.*, 2017, **28**, 22–30.

9. G. Carpenter and A. Blissett, Mastication, Salivation and their Interactions during Eating, *Time-Dependent Measures of Perception in Sensory Evaluation*, 2017, p. 27.
10. E. A. Foegeding, C. R. Daubert, M. A. Drake, G. Essick, M. Trulsson and C. J. Vinyard, *et al.*, A comprehensive approach to understanding textural properties of semi-and soft-solid foods, *J. Texture Stud.*, 2011, **42**(2), 103–129.
11. M. D. de Lavergne, F. van de Velde and M. Stieger, Bolus matters: the influence of food oral breakdown on dynamic texture perception, *Food Funct.*, 2017, **8**(2), 464–480.
12. P. J. Lillford, The importance of food microstructure in fracture physics and texture perception, *J. Texture Stud.*, 2011, **42**(2), 130–136.
13. E. B. Goldstein, *Encyclopedia of perception*, Sage, 2009, vol. 1.
14. C. Spence, Mouth-watering: the influence of environmental and cognitive factors on salivation and gustatory/flavor perception, *J. Texture Stud.*, 2011, **42**(2), 157–171.
15. K. Oatley, *Perceptions and representations: The theoretical bases of brain research and psychology*, Routledge, 2017.
16. Y. Liu, J.-H. Gao, H.-L. Liu and P. T. Fox, The temporal response of the brain after eating revealed by functional MRI, *Nature*, 2000, **405**(6790), 1058.
17. S. T. Francis and S. Eldeghaidy, Imaging methodologies and applications for nutrition research: what can functional MRI offer?, *Proc. Nutr. Soc.*, 2015, **74**(2), 89–98.
18. A. Quintero, E. Ichesco, C. Myers, R. Schutt and G. E. Gerstner, Brain activity and human unilateral chewing: an fMRI study, *J. Dent. Res.*, 2013, **92**(2), 136–142.
19. M. Onozuka, M. Fujita, K. Watanabe, Y. Hirano, M. Niwa and K. Nishiyama, *et al.*, Mapping brain region activity during chewing: a functional magnetic resonance imaging study, *J. Dent. Res.*, 2002, **81**(11), 743–746.
20. T. Takahashi, T. Miyamoto, A. Terao and A. Yokoyama, Cerebral activation related to the control of mastication during changes in food hardness, *Neuroscience*, 2007, **145**(3), 791–794.
21. L. J. Pereira, Oral cavity, *Food Oral Processing: Fundamentals of Eating and Sensory Perception*, 2012, pp. 1–14.
22. B. K. B. Berkovitz, B. J. Moxham, R. W. A. Linden and A. J. Sloan, *Master Dentistry Volume 3 Oral Biology E-Book: Oral Anatomy, Histology, Physiology and Biochemistry*, Elsevier Health Sciences, 2010, vol. 3.
23. R. Z. German and J. B. Palmer, Anatomy and development of oral cavity and pharynx, *GI Motility online*, 2006, DOI: 10.1038/gimo5.
24. J. Chen and L. Engelen, *Food Oral Processing: Fundamentals of Eating and Sensory Perception*, John Wiley & Sons, 2012.
25. A. der Bilt, Assessment of mastication with implications for oral rehabilitation: a review, *J. Oral Rehabil.*, 2011, **38**(10), 754–780.

26. A. van der Bilt, Human oral function: a review, *Braz. J. Oral Sci.*, 2002, **1**(1), 7–18.
27. K. Ikebe, K. Matsuda, R. Kagawa, K. Enoki, T. Okada and M. Yoshida, *et al.*, Masticatory performance in older subjects with varying degrees of tooth loss, *J. Dent.*, 2012, **40**(1), 71–76.
28. K. M. Hiiemae and J. B. Palmer, Tongue movements in feeding and speech, *Crit. Rev. Oral Biol. Med.*, 2003, **14**(6), 413–429.
29. H. Gray, *Anatomy of the human body*, Lea & Febiger, Philadelphia, 1918, Bartleby.com, 2000, 2001.
30. J. Chandrashekar, M. A. Hoon, N. J. P. Ryba and C. S. Zuker, The receptors and cells for mammalian taste, *Nature*, 2006, **444**(7117), 288–294.
31. W. M. Edgar, Saliva: its secretion, composition and functions, *Br. Dent. J.*, 1992, **172**(8), 305–312.
32. S. P. Humphrey and R. T. Williamson, A review of saliva: normal composition, flow, and function, *J. Prosthet. Dent.*, 2001, **85**(2), 162–169.
33. P. Sethupathy, J. A. Moses and C. Anandharamakrishnan, Food Oral Processing and Tribology: Instrumental Approaches and Emerging Applications, *Food Rev. Int.*, 2020, 1–34.
34. G. Carpenter, Role of saliva in the oral processing of food, in *Food Oral Processing*, Wiley-Blackwell, 2012, pp. 45–60.
35. M. Edgar and S. M. Higham, Saliva and the control of plaque pH, *Saliva and oral health*, 2004, p. 97.
36. L. Engelen, Oral receptors, *Food Oral Processing: Fundamentals of Eating and Sensory Perception*, 2012, pp. 15–38.
37. E. T. Rolls, Taste, Olfactory and Food-texture Processing in the Brain and the Control of Appetite, in *Obesity prevention*, Elsevier, 2010, pp. 41–56.
38. E. T. Rolls, Taste, olfactory, and food texture processing in the brain, and the control of food intake, *Physiol. Behav.*, 2005, **85**(1), 45–56.
39. E. T. Rolls, Taste, olfactory, and food reward value processing in the brain, *Prog. Neurobiol.*, 2015, **127**, 64–90.
40. J. V. Verhagen and L. Engelen, The neurocognitive bases of human multimodal food perception: sensory integration, *Neurosci. Biobehav. Rev.*, 2006, **30**(5), 613–650.
41. J. R. Stokes, M. W. Boehm and S. K. Baier, Oral processing, texture and mouthfeel: From rheology to tribology and beyond, *Curr. Opin. Colloid Interface Sci.*, 2013, **18**(4), 349–359.
42. J. F. Prinz, A. M. Janssen and R. A. De Wijk, In vitro simulation of the oral processing of semi-solid foods, *Food Hydrocolloids*, 2007, **21**(3), 397–401.
43. T. Funami, In vivo and rheological approaches for characterizing food oral processing and usefulness of polysaccharides as texture modifiers-A review, *Food Hydrocolloids*, 2017, **68**, 2–14.
44. D. Liu, Y. Deng, L. Sha, M. A. Hashem and S. Gai, Impact of oral processing on texture attributes and taste perception, *J. Food Sci. Technol.*, 2017, **54**(8), 2585–2593.

45. J. Chen and J. R. Stokes, Rheology and tribology: Two distinctive regimes of food texture sensation, *Trends Food Sci. Technol.*, 2012, **25**(1), 4–12.
46. M. J. Levine, M. S. Reddy, L. A. Tabak, R. E. Loomis, E. J. Bergey and P. C. Jones, *et al.*, Structural aspects of salivary glycoproteins, *J. Dent. Res.*, 1987, **66**(2), 436–441.
47. P.-O. Glantz, Interfacial phenomena in the oral cavity, *Colloids Surf., A*, 1997, **123**, 657–670.
48. R. G. Schipper, E. Silletti and M. H. Vingerhoeds, Saliva as research material: biochemical, physicochemical and practical aspects, *Arch. Oral Biol.*, 2007, **52**(12), 1114–1135.
49. E. Dickinson, On the road to understanding and control of creaminess perception in food colloids, *Food Hydrocolloids*, 2018, **77**, 372–385.
50. A. C. Mosca and J. Chen, Food-saliva interactions: Mechanisms and implications, *Trends Food Sci. Technol.*, 2017, **66**, 125–134.
51. I. García-Estévez, A. M. Ramos-Pineda and M. T. Escribano-Bailón, Interactions between wine phenolic compounds and human saliva in astringency perception, *Food Funct.*, 2018, **9**(3), 1294–1309.
52. E. Guichard, V. Galindo-Cuspinera and G. Feron, Physiological mechanisms explaining human differences in fat perception and liking in food spreads-a review, *Trends Food Sci. Technol.*, 2018, **74**, 46–55.
53. A. Sarkar and H. Singh, Oral behaviour of food emulsions, *Food Oral Processing*, Wiley-Blackwell, UK, 2012, pp. 111–137.
54. I. M. Dawood and S. K. El-Samarrai, Saliva and Oral Health, *Int. J. Adv. Res. Biol. Sci.*, 2018, **5**(7), 1–45.
55. E. Silletti, M. H. Vingerhoeds, W. Norde and G. A. Van Aken, The role of electrostatics in saliva-induced emulsion flocculation, *Food Hydrocolloids*, 2007, **21**(4), 596–606.
56. G. Davidov-Pardo, I. J. Joye and D. J. McClements, Food-grade protein-based nanoparticles and microparticles for bioactive delivery: fabrication, characterization, and utilization, in *Advances in Protein Chemistry and Structural Biology*, Elsevier, 2015, pp. 293–325.
57. H. L. Gibbins and G. H. Carpenter, Alternative mechanisms of astringency–what is the role of saliva?, *J. Texture Stud.*, 2013, **44**(5), 364–375.
58. J. Bridges, J. Smythe and R. Reddrick, Impact of salivary enzyme activity on the oral perception of starch containing foods, *J. Texture Stud.*, 2017, **48**(4), 288–293.
59. R. A. De Wijk, J. F. Prinz and A. M. Janssen, Explaining perceived oral texture of starch-based custard desserts from standard and novel instrumental tests, *Food Hydrocolloids*, 2006, **20**(1), 24–34.
60. R. D. Mattes, Oral fatty acid signaling and intestinal lipid processing: support and supposition, *Physiol. Behav.*, 2011, **105**(1), 27–35.
61. E. T. Rolls, The neural representation of oral texture including fat texture, *J. Texture Stud.*, 2011, **42**(2), 137–156.
62. J. M. Aguilera, The food matrix: implications in processing, nutrition and health, *Crit. Rev. Food Sci. Nutr.*, 2019, **59**(22), 3612–3629.

63. T. van Vliet, G. A. van Aken, H. H. J. de Jongh and R. J. Hamer, Colloidal aspects of texture perception, *Adv. Colloid Interface Sci.*, 2009, **150**(1), 27–40.
64. M. Stieger and F. van de Velde, Microstructure, texture and oral processing: new ways to reduce sugar and salt in foods, *Curr. Opin. Colloid Interface Sci.*, 2013, **18**(4), 334–348.
65. M. Bourne, *Food Texture and Viscosity: Concept and Measurement*, Elsevier, 2002.
66. S. C. Hutchings, K. D. Foster, J. E. Bronlund, R. G. Lentle, J. R. Jones and M. P. Morgenstern, Mastication of heterogeneous foods: Peanuts inside two different food matrices, *Food Qual. Prefer.*, 2011, **22**(4), 332–339.
67. A. Duconseille, O. François, P. Bruno, L. Céline, P. Marie-Agnès and H. Martine, Measuring the effects of *in vitro* mastication on bolus granulometry of shredded meat: A proposal for a new methodological procedure, *Food Res. Int.*, 2019, **116**, 1266–1273.
68. C. Salles, A. Tarrega, P. Mielle, J. Maratray, P. Gorria and J. Liaboeuf, *et al.*, Development of a chewing simulator for food breakdown and the analysis of *in vitro* flavor compound release in a mouth environment, *J. Food Eng.*, 2007, **82**(2), 189–198.
69. A. Tarrega, C. Yven, E. Semon and C. Salles, In-mouth aroma compound release during cheese consumption: Relationship with food bolus formation, *Int. Dairy J.*, 2011, **21**(5), 358–364.
70. M. Repoux, H. Labouré, P. Courcoux, I. Andriot, É Sémon and C. Yven, *et al.*, Combined effect of cheese characteristics and food oral processing on *in vivo* aroma release, *Flavour Fragrance J.*, 2012, **27**(6), 414–423.
71. A. Voilley and I. Souchon, Flavour retention and release from the food matrix: an overview, *Flavour in Foods*, 2006, pp. 117–132.
72. L. Paravisini and E. Guichard, Interactions between aroma compounds and food matrix, in *Flavour: From Food to Perception*, ed. E. Guichard, C. Salles, M. Morzel and A.-M. Le Bon, 2017, pp. 208–234.
73. C. E. Wilson and W. E. Brown, Influence of food matrix structure and oral breakdown during mastication on temporal perception of flavor, *J. Sens. Stud.*, 1997, **12**(1), 69–86.
74. P. T. M. Nguyen, O. Kravchuk, B. Bhandari and S. Prakash, Effect of different hydrocolloids on texture, rheology, tribology and sensory perception of texture and mouthfeel of low-fat pot-set yoghurt, *Food Hydrocolloids*, 2017, **72**, 90–104.
75. J. R. Stokes, 11 'Oral' Rheology, *Food Oral Processing: Fundamentals of Eating and Sensory Perception*, 2012, p. 227.
76. M. A. Rao, *Rheology of fluid, semisolid, and solid foods: principles and applications*, Springer Science & Business Media, 2013.
77. A. J. Taylor, Release and transport of flavors in vivo: physicochemical, physiological, and perceptual considerations, *Compr. Rev. Food Sci. Food Saf.*, 2002, **1**(2), 45–57.

78. F. Shama and P. Sherman, Identification of stimuli controlling the sensory evaluation of viscosity II. Oral methods, *J. Texture Stud.*, 1973, **4**(1), 111–118.
79. M. E. J. Terpstra, A. M. Janssen, J. F. Prinz, R. A. De Wijk, H. Weenen and E. Van Der Linden, Modeling of thickness for semisolid foods, *J. Texture Stud.*, 2005, **36**(2), 213–233.
80. R. A. de Wijk and J. F. Prinz, The role of friction in perceived oral texture, *Food Qual. Prefer.*, 2005, **16**(2), 121–129.
81. E. M. Krop, M. M. Hetherington, M. Holmes, S. Miquel and A. Sarkar, On relating rheology and oral tribology to sensory properties in hydrogels, *Food Hydrocolloids*, 2019, **88**, 101–113.
82. W. K. Shafi, A. Raina, M. I. U. Haq and A. Khajuria, Applications of Industrial Tribology, *International Research Journal of Engineering and Technology*, 2018, **5**(1), 1285–1289.
83. G. Stachowiak and A. W. Batchelor, *Engineering Tribology*, Butterworth-Heinemann, 2013.
84. J. Lind, Tribology of Polymer Composites for Elevated Temperature Applications, PhD thesis, Uppsala University, 2017.
85. S. F. Ermakov and N. K. Myshkin, *Liquid-Crystal Nanomaterials: Tribology and Applications*, Springer, 2018, vol. 267.
86. J. Villanueva, L. Trino, J. Thomas, D. Bijukumar, D. Royhman and M. M. Stack, *et al.*, Corrosion, tribology, and tribocorrosion research in biomedical implants: progressive trend in the published literature, *J. Bio Tribo Corros.*, 2017, **3**(1), 1.
87. S. Duprat-de-Paule, J. Guilbot, A. Roso, S. Cambos and A. Pierre, Augmented bio-based lipids for cosmetics, *OCL*, 2018, **25**(5), D503.
88. H. Ranc, A. Elkhyat, C. Servais, S. Mac-Mary, B. Launay and P. Humbert, Friction coefficient and wettability of oral mucosal tissue: changes induced by a salivary layer, *Colloids Surf., A*, 2006, **276**(1–3), 155–161.
89. R. Pramanik, S. M. Osailan, S. J. Challacombe, D. Urquhart and G. B. Proctor, Protein and mucin retention on oral mucosal surfaces in dry mouth patients, *Eur. J. Oral Sci.*, 2010, **118**(3), 245–253.
90. S. Ployon, C. Belloir, A. Bonnotte, J. Lherminier, F. Canon and M. Morzel, The membrane-associated MUC1 improves adhesion of salivary MUC5B on buccal cells. Application to development of an *in vitro* cellular model of oral epithelium, *Arch. Oral Biol.*, 2016, **61**, 149–155.
91. J. H. H. Bongaerts, D. Rossetti and J. R. Stokes, The lubricating properties of human whole saliva, *Tribol. Lett.*, 2007, **27**(3), 277–287.
92. N. Selway and J. R. Stokes, Insights into the dynamics of oral lubrication and mouthfeel using soft tribology: Differentiating semi-fluid foods with similar rheology, *Food Res. Int.*, 2013, **54**(1), 423–431.
93. C. Pradal and J. R. Stokes, Oral tribology: bridging the gap between physical measurements and sensory experience, *Curr. Opin. Food Sci.*, 2016, **9**, 34–41.
94. N. Chaudhari and S. D. Roper, Review series: The cell biology of taste, *J. Cell Biol.*, 2010, **190**(3), 285.

95. M. A. Hoon, E. Adler, J. Lindemeier, J. F. Battey, N. J. P. Ryba and C. S. Zuker, Putative mammalian taste receptors: a class of taste-specific GPCRs with distinct topographic selectivity, *Cell*, 1999, **96**(4), 541–551.
96. B. V. Kulkarni and R. D. Mattes, Lingual lipase activity in the orosensory detection of fat by humans, *Am. J. Physiol.: Regul., Integr. Comp. Physiol.*, 2014, **306**(12), R879–R885.
97. N. A. Khan and P. Besnard, Oro-sensory perception of dietary lipids: new insights into the fat taste transduction, *Biochim. Biophys. Acta, Mol. Cell Biol. Lipids*, 2009, **1791**(3), 149–155.
98. T. J. Lapis, M. H. Penner and J. Lim, Humans can taste glucose oligomers independent of the hT1R2/hT1R3 sweet taste receptor, *Chem. Senses*, 2016, **41**(9), 755–762.
99. R. Barragán, O. Coltell, O. Portolés, E. Asensio, J. Sorlí and C. Ortega-Azorín, *et al.*, Bitter, sweet, salty, sour and umami taste perception decreases with age: Sex-specific analysis, modulation by genetic variants and taste-preference associations in 18 to 80 year-old subjects, *Nutrients*, 2018, **10**(10), 1539.
100. M. E. Malone, I. A. M. Appelqvist and I. T. Norton, Oral behaviour of food hydrocolloids and emulsions. Part 2. Taste and aroma release, *Food Hydrocolloids*, 2003, **17**(6), 775–784.
101. A. C. Mosca, F. van de Velde, J. H. F. Bult, M. A. J. S. van Boekel and M. Stieger, Taste enhancement in food gels: Effect of fracture properties on oral breakdown, bolus formation and sweetness intensity, *Food Hydrocolloids*, 2015, **43**, 794–802.
102. B. G. Green and P. George, 'Thermal taste' predicts higher responsiveness to chemical taste and flavor, *Chem. Senses*, 2004, **29**(7), 617–628.
103. C. Spence, 10 Multi-sensory Integration and the Psychophysics of Flavour Perception, *Food Oral Processing: Fundamentals of Eating and Sensory Perception*, 2012, p. 203.
104. M. R. Bajec, G. J. Pickering and N. DeCourville, Influence of stimulus temperature on orosensory perception and variation with taste phenotype, *Chemosens. Percept.*, 2012, **5**(3–4), 243–265.
105. M. Panouille, A. Saint-Eve, I. Deleris, F. Le Bleis and I. Souchon, Oral processing and bolus properties drive the dynamics of salty and texture perceptions of bread, *Food Res. Int.*, 2014, **62**, 238–246.
106. ISO, ISO 5492:2008, Sensory Analysis–Vocabulary, International Organization for Standardization Geneva, Switzerland, 2008.
107. C. Spence, Multisensory flavour perception, *Flavour: from Food To Perception*, 2017, pp. 373–394.
108. J. Delwiche, The impact of perceptual interactions on perceived flavor, *Food Qual. Prefer.*, 2004, **15**(2), 137–146.
109. F. Chartier, *Taste Buds and Molecules: The Art and Science of Food, Wine, and Flavor*, Houghton Mifflin Harcourt, 2012.
110. R. A. Frank and J. Byram, Taste–smell interactions are tastant and odorant dependent, *Chem. Senses*, 1988, **13**(3), 445–455.

111. P. Dalton, N. Doolittle, H. Nagata and P. A. S. Breslin, The merging of the senses: integration of subthreshold taste and smell, *Nat. Neurosci.*, 2000, **3**(5), 431–432.
112. J. F. Delwiche and A. L. Heffelfinger, Cross-modal additivity of taste and smell, *J. Sens. Stud.*, 2005, **20**(6), 512–525.
113. J. C. Pfeiffer, T. A. Hollowood, J. Hort and A. J. Taylor, Temporal synchrony and integration of sub-threshold taste and smell signals, *Chem. Senses*, 2005, **30**(7), 539–545.
114. D. Labbe, L. Damevin, C. Vaccher, C. Morgenegg and N. Martin, Modulation of perceived taste by olfaction in familiar and unfamiliar beverages, *Food Qual. Prefer.*, 2006, **17**(7–8), 582–589.
115. H. N. J. Schifferstein and P. W. J. Verlegh, The role of congruency and pleasantness in odor-induced taste enhancement, *Acta Psychol.*, 1996, **94**(1), 87–105.
116. P. A. Breslin, N. Doolittle and P. Dalton, Subthreshold integration of taste and smell: the role of experience in flavor integration, *Chem. Senses*, 2001, **26**(8), 1035.
117. A. S. Szczesniak, Texture is a sensory property, *Food Qual. Prefer.*, 2002, **13**(4), 215–225.
118. J. H. F. Bult, R. A. de Wijk and T. Hummel, Investigations on multimodal sensory integration: texture, taste, and ortho-and retronasal olfactory stimuli in concert, *Neurosci. Lett.*, 2007, **411**(1), 6–10.
119. A. Cruz and B. G. Green, Thermal stimulation of taste, *Nature*, 2000, **403**(6772), 889–892.
120. M. R. Bajec and G. J. Pickering, Thermal taste, PROP responsiveness, and perception of oral sensations, *Physiol. Behav.*, 2008, **95**(4), 581–590.
121. C. Spence, Auditory contributions to flavour perception and feeding behaviour, *Physiol. Behav.*, 2012, **107**(4), 505–515.
122. M. Zampini and C. Spence, The role of auditory cues in modulating the perceived crispness and staleness of potato chips, *J. Sens. Stud.*, 2004, **19**(5), 347–363.
123. M. Zampini and C. Spence, Modifying the multisensory perception of a carbonated beverage using auditory cues, *Food Qual. Prefer.*, 2005, **16**(7), 632–641.
124. C. Spence and M. Zampini, Auditory contributions to multisensory product perception, *Acta Acust. Acust.*, 2006, **92**(6), 1009–1025.
125. C. Spence and M. U. Shankar, The influence of auditory cues on the perception of, and responses to, food and drink, *J. Sens. Stud.*, 2010, **25**(3), 406–430.
126. J. A. Maga, Influence of color on taste thresholds, *Chem. Senses*, 1974, **1**(1), 115–119.
127. R. J. Stevenson and M. Oaten, The effect of appropriate and inappropriate stimulus color on odor discrimination, *Percept. Psychophys.*, 2008, **70**(4), 640–646.
128. M. Shankar, C. Simons, B. Shiv, S. McClure, C. Levitan and C. Spence, An expectations-based approach to explaining the cross-modal

influence of color on orthonasal olfactory identification: The influence of the degree of discrepancy, *Atten. Percept. Psychophys.*, 2010, **72**(7), 1981–1993.

129. D. A. Zwllner and L. A. Whitten, The effect of color intensity and appropriateness on color-induced odor enhancement, *Am. J. Psychol.*, 1999, **112**(4), 585.
130. G. Morrot, F. Brochet and D. Dubourdieu, The color of odors, *Brain Lang.*, 2001, **79**(2), 309–320.
131. M. U. Shankar, C. A. Levitan and C. Spence, Grape expectations: The role of cognitive influences in color–flavor interactions, *Conscious. Cognit.*, 2010, **19**(1), 380–390.
132. C. N. DuBose, A. V. Cardello and O. Maller, Effects of colorants and flavorants on identification, perceived flavor intensity, and hedonic quality of fruit-flavored beverages and cake, *J. Food Sci.*, 1980, **45**(5), 1393–1399.
133. S. A. Simon, I. E. de Araujo, R. Gutierrez and M. A. L. Nicolelis, The neural mechanisms of gustation: a distributed processing code, *Nat. Rev. Neurosci.*, 2006, **7**(11), 890–901.
134. M. G. Veldhuizen, D. J. Nachtigal and D. M. Small, Taste cortex contributes to odor quality coding, *Chem. Senses*, 2009, A8.
135. D. M. Small and J. Prescott, Odor/taste integration and the perception of flavor, *Exp. Brain Res.*, 2005, **166**(3–4), 345–357.
136. D. M. Small, Crossmodal integration–insights from the chemical senses, *Trends Neurosci.*, 2004, **27**(3), 120–123.
137. S. Eldeghaidy, L. Marciani, F. McGlone, T. Hollowood, J. Hort and K. Head, *et al.*, The cortical response to the oral perception of fat emulsions and the effect of taster status, *J. Neurophysiol.*, 2011, **105**(5), 2572–2581.
138. A. J. Taylor and D. D. Roberts, *Flavor Perception*, John Wiley & Sons, 2008.
139. C. F. Ross, Sensory science at the human–machine interface, *Trends Food Sci. Technol.*, 2009, **20**(2), 63–72.
140. H. T. Lawless and H. Heymann, *Sensory Evaluation of Food: Principles and Practices*, Springer Science & Business Media, 2010.
141. S. Shamil, L. J. Wyeth and D. Kilcast, Flavour release and perception in reduced-fat foods, *Food Qual. Prefer.*, 1991, **3**(1), 51–60.
142. P. Sethupathy, S. R. Priyadarshini, J. A. Moses and C. Anandharamakrishnan, Matrix-dependent oral processing, orosensory perception, and glycemic index of chocolate bars, *J. Food Process. Preserv.*, 2021, **45**(12), e16067.
143. A. Rizo, I. Jimenez-Pérez, A. Camacho-García, S. Fiszman, P. Pérez-Soriano and A. Tárrega, Impact of texture TDS and flavour TDS tasks and of chocolate-chip biscuit characteristics on oral processing features, *Food Qual. Prefer.*, 2019, **76**, 109–117.
144. F. R. Jack, J. R. Piggott and A. Paterson, Analysis of textural changes in hard cheese during mastication by progressive profiling, *J. Food Sci.*, 1994, **59**(3), 539–543.

145. G. B. Dijksterhuis and J. R. Piggott, Dynamic methods of sensory analysis, *Trends Food Sci. Technol.*, 2000, **11**(8), 284–290.
146. J. C. Castura, L. Antúnez, A. Giménez and G. Ares, Temporal Check-All-That-Apply (TCATA): A novel dynamic method for characterizing products, *Food Qual. Prefer.*, 2016, **47**, 79–90.
147. J. D. Rios-Mera, E. Saldaña, M. L. M. Cruzado-Bravo, M. M. Martins, I. Patinho and M. M. Selani, *et al.*, Impact of the content and size of NaCl on dynamic sensory profile and instrumental texture of beef burgers, *Meat Sci.*, 2020, **161**, 107992.
148. N. Pineau, A. G. de Bouillé, M. Lepage, F. Lenfant, P. Schlich and N. Martin, *et al.*, Temporal dominance of sensations: What is a good attribute list?, *Food Qual. Prefer.*, 2012, **26**(2), 159–165.
149. R. Di Monaco, C. Su, P. Masi and S. Cavella, Temporal dominance of sensations: A review, *Trends Food Sci. Technol.*, 2014, **38**(2), 104–112.
150. P. Sethupathy, S. K. Sivakamasundari, J. A. Moses and C. Anandharamakrishnan, Effect of varietal differences on the oral processing behavior and bolus properties of cooked rice, *Int. J. Food Eng.*, 2020, **17**(3), 177–188.
151. J. Chen, Food oral processing—A review, *Food Hydrocolloids*, 2009, **23**(1), 1–25.
152. E. J. Bialek, W. Jakubowski, P. Zajkowski, K. T. Szopinski and A. Osmolski, US of the major salivary glands: anatomy and spatial relationships, pathologic conditions, and pitfalls, *Radiographics*, 2006, **26**(3), 745–763.
153. C. Llena-Puy, The Role of Saliva in Maintaining Oral Health and as an Aid to Diagnosis, *Med. Oral Patol. Oral Cir. Bucal*, 2006, **11**(5), E449–E455.
154. A. M. Pedersen, A. Bardow, S. B. Jensen and B. Nauntofte, Saliva and gastrointestinal functions of taste, mastication, swallowing and digestion, *Oral Dis.*, 2002, **8**(3), 117–129.
155. J. J. Feher, *Quantitative Human Physiology: An Introduction*, Academic Press, 2017.
156. K. Kohyama, Oral sensing of food properties, *J. Texture Stud.*, 2015, **46**(3), 138–151.
157. P. Haggard and L. de Boer, Oral somatosensory awareness, *Neurosci. Biobehav. Rev.*, 2014, **47**, 469–484.
158. E. C. Ketel, M. G. Aguayo-Mendoza, R. A. de Wijk, C. de Graaf, B. Piqueras-Fiszman and M. Stieger, Age, gender, ethnicity and eating capability influence oral processing behaviour of liquid, semi-solid and solid foods differently, *Food Res. Int.*, 2019, **119**, 143–151.
159. M. G. Aguayo-Mendoza, E. C. Ketel, E. van der Linden, C. G. Forde, B. Piqueras-Fiszman and M. Stieger, Oral processing behavior of drinkable, spoonable and chewable foods is primarily determined by rheological and mechanical food properties, *Food Qual. Prefer.*, 2019, **71**, 87–95.
160. A. van Eck, N. Hardeman, N. Karatza, V. Fogliano, E. Scholten and M. Stieger, Oral processing behavior and dynamic sensory perception

of composite foods: Toppings assist saliva in bolus formation, *Food Qual. Prefer.*, 2019, **71**, 497–509.

161. A. Rizo, E. Peña, A. D. Alarcon-Rojo, S. Fiszman and A. Tárrega, Relating texture perception of cooked ham to the bolus evolution in the mouth, *Food Res. Int.*, 2019, **118**, 4–12.
162. M. D. de Lavergne, C. Tournier, D. Bertrand, C. Salles, F. de Velde and M. Stieger, Dynamic texture perception, oral processing behaviour and bolus properties of emulsion-filled gels with and without contrasting mechanical properties, *Food Hydrocolloids*, 2016, **52**, 648–660.
163. J. Gao, J. X. Wong, J. C.-S. Lim, J. Henry and W. Zhou, Influence of bread structure on human oral processing, *J. Food Eng.*, 2015, **167**, 147–155.
164. M. D. De Lavergne, M. van Delft, F. van de Velde, M. A. J. S. van Boekel and M. Stieger, Dynamic texture perception and oral processing of semi-solid food gels: Part 1: Comparison between QDA, progressive profiling and TDS, *Food Hydrocolloids*, 2015, **43**, 207–217.
165. M. D. de Lavergne, F. Van De Velde, M. Van Boekel and M. Stieger, Dynamic texture perception and oral processing of semi-solid food gels: Part 2: Impact of breakdown behaviour on bolus properties and dynamic texture perception, *Food Hydrocolloids*, 2015, **49**, 61–72.
166. S. A. Alam, S. Pentikäinen, J. Närväinen, U. Holopainen-Mantila, K. Poutanen and N. Sozer, Effects of structural and textural properties of brittle cereal foams on mechanisms of oral breakdown and *in vitro* starch digestibility, *Food Res. Int.*, 2017, **96**, 1–11.
167. M. Assad-Bustillos, C. Tournier, C. Septier, G. Della Valle and G. Feron, Relationships of oral comfort perception and bolus properties in the elderly with salivary flow rate and oral health status for two soft cereal foods, *Food Res. Int.*, 2019, **118**, 13–21.
168. M. Assad-Bustillos, C. Tournier, G. Feron, S. Guessasma, A. L. Reguerre and V. G. Della, Fragmentation of two soft cereal products during oral processing in the elderly: Impact of product properties and oral health status, *Food Hydrocolloids*, 2019, **91**, 153–165.
169. V. Raja, S. R. Priyadarshini, J. A. Moses and C. Anandharamakrishnan, A dynamic *in vitro* oral mastication system to study the oral processing behavior of soft foods, *Food Funct.*, 2022, **13**(20), 10426–10438.
170. T. B. Wagoner, P. J. Luck and E. A. Foegeding, Caramel as a model system for evaluating the roles of mechanical properties and oral processing on sensory perception of texture, *J. Food Sci.*, 2016, **81**(3), S736–S744.
171. E. Stribiţcaia, E. M. Krop, R. Lewin, M. Holmes and A. Sarkar, Tribology and rheology of bead-layered hydrogels: Influence of bead size on sensory perception, *Food Hydrocolloids*, 2020, **104**, 105692.
172. T. Brenner, M. Tomczyńska-Mleko, S. Mleko and K. Nishinari, The extrusion test and sensory perception revisited: Some comments on generality and the effect of measurement temperature, *J. Texture Stud.*, 2017, **48**(6), 487–493.

173. J. Chen and L. Lolivret, The determining role of bolus rheology in triggering a swallowing, *Food Hydrocolloids*, 2011, **25**(3), 325–332.
174. A. I. V. Ross, P. Tyler, M. G. Borgognone and B. M. Eriksen, Relationships between shear rheology and sensory attributes of hydrocolloid-thickened fluids designed to compensate for impairments in oral manipulation and swallowing, *J. Food Eng.*, 2019, **263**, 123–131.
175. A. K. Young, J. N. Cheong, D. I. Hedderley, M. P. Morgenstern and B. J. James, Understanding the link between bolus properties and perceived texture, *J. Texture Stud.*, 2013, **44**(5), 376–386.
176. A. Chojnicka-Paszun, H. H. J. De Jongh and C. G. De Kruif, Sensory perception and lubrication properties of milk: Influence of fat content, *Int. Dairy J.*, 2012, **26**(1), 15–22.
177. A. M. Carvalho-da-Silva, I. Van Damme, W. Taylor, J. Hort and B. Wolf, Oral processing of two milk chocolate samples, *Food Funct.*, 2013, **4**(3), 461–469.
178. A. Chojnicka-Paszun, S. Doussinault and H. H. J. De Jongh, Sensorial analysis of polysaccharide–gelled protein particle dispersions in relation to lubrication and viscosity properties, *Food Res. Int.*, 2014, **56**, 199–210.
179. A. Krzeminski, K. A. Prell, M. Busch-Stockfisch, J. Weiss and J. Hinrichs, Whey protein–pectin complexes as new texturising elements in fat-reduced yoghurt systems, *Int. Dairy J.*, 2014, **36**(2), 118–127.
180. H. S. Joyner, C. W. Pernell and C. R. Daubert, Impact of formulation and saliva on acid milk gel friction behavior, *J. Food Sci.*, 2014, **79**(5), E867–E880.
181. B. Le Calvé, C. Saint-Léger, R. Babas, J. Gelin, A. Parker and P. Erni, *et al.*, Fat perception: How sensitive are we?, *J. Texture Stud.*, 2015, **46**(3), 200–211.
182. T. A. M. Rovers, G. Sala, E. der Linden and M. B. J. Meinders, Potential of Microbubbles as Fat Replacer: Effect on Rheological, Tribological and Sensorial Properties of Model Food Systems, *J. Texture Stud.*, 2016, **47**(3), 220–230.
183. N. Brossard, H. Cai, F. Osorio, E. Bordeu and J. Chen, "Oral" tribological study on the astringency sensation of red wines, *J. Texture Stud.*, 2016, **47**(5), 392–402.
184. R. Upadhyay and J. Chen, Smoothness as a tactile percept: Correlating 'oral' tribology with sensory measurements, *Food Hydrocolloids*, 2019, **87**, 38–47.
185. M. Kiumarsi, D. Majchrzak, S. Yeganehzad, H. Jäger and M. Shahbazi, Comparative study of instrumental properties and sensory profiling of low-calorie chocolate containing hydrophobically modified inulin. Part 1: Rheological, thermal, structural and external preference mapping, *Food Hydrocolloids*, 2020, **104**, 105698.
186. M. Kiumarsi, D. Majchrzak, H. Jäger, J. Song, O. Lieleg and M. Shahbazi, Comparative study of instrumental properties and sensory profiling of low-calorie chocolate containing hydrophobically

modified inulin. Part II: Proton mobility, topological, tribological and dynamic sensory properties, *Food Hydrocolloids*, 2021, **110**, 106144.
187. C. Onyango and G. W. Wanjala, Quality of porridge from sub-Saharan Africa evaluated using instrumental techniques and descriptive sensory lexicon. Part 2: Thin porridge, *African J. Food Sci.*, 2018, **12**(5), 104–114.
188. M. Doyennette, M. G. Aguayo-Mendoza, A.-M. Williamson, S. I. F. S. Martins and M. Stieger, Capturing the impact of oral processing behaviour on consumption time and dynamic sensory perception of ice creams differing in hardness, *Food Qual. Prefer.*, 2019, **78**, 103721.
189. R. Ravi, M. Prakash and K. K. Bhat, Characterization of aroma active compounds of cumin (Cuminum cyminum L.) by GC–MS, E-Nose, and Sensory Techniques, *Int. J. Food Prop.*, 2013, **16**(5), 1048–1058.
190. D. Pu, H. Zhang, Y. Zhang, B. Sun, F. Ren and H. Chen, *et al.*, Characterization of the aroma release and perception of white bread during oral processing by gas chromatography-ion mobility spectrometry and temporal dominance of sensations analysis, *Food Res. Int.*, 2019, **123**, 612–622.
191. H. Clement, C. Prost, C. Rannou, H. Chiron, M. Bonnand-Ducasse and P. Courcoux, *et al.*, Can instrumental characterization help predicting sour taste perception of wheat sourdough bread?, *Food Res. Int.*, 2020, 109159.
192. L. Gao, M. Bian, R. Mi, X. Hu and J. Wu, Quality identification and evaluation of Pu-erh teas of different grade levels and various ages through sensory evaluation and instrumental analysis, *Int. J. Food Sci. Technol.*, 2016, **51**(6), 1338–1348.
193. M. K. Kim, H. W. Jang and K.-G. Lee, Sensory and instrumental volatile flavor analysis of commercial orange juices prepared by different processing methods, *Food Chem.*, 2018, **267**, 217–222.
194. L. Lorido, M. Estévez and S. Ventanas, Fast and dynamic descriptive techniques (Flash Profile, Time-intensity and Temporal Dominance of Sensations) for sensory characterization of dry-cured loins, *Meat Sci.*, 2018, **145**, 154–162.
195. M. Charles, A. Romano, S. Yener, M. Barnabà, L. Navarini and T. D. Märk, *et al.*, Understanding flavour perception of espresso coffee by the combination of a dynamic sensory method and *in vivo* nosespace analysis, *Food Res. Int.*, 2015, **69**, 9–20.

CHAPTER 2

The Human Stomach and Gastric Digestion

2.1 Introduction

The gastric digestion of foods is a very critical and essential process in which the food is broken down through mechanical forces, enzymes, and chemical changes to extract the individual nutrients from the food matrix for absorption and assimilation, which aids in maintaining the health and well-being of humans.[1] The digestive system is made up of a hollow, lengthy, muscular tube beginning at the mouth, followed by the esophagus, stomach, small intestine, and large intestine, and terminating at the anus.[2] The stomach plays four significant roles in food digestion: temporary storage of ingested food, mixing with gastric secretions, reduction of particle size, and emptying the chyme into the duodenum.[3,4] Food reaches the stomach predominantly unaltered in the case of liquids, or as a bolus after significant disintegration in the case of semi-solid and solid foods after the oral processing step (mouth) and esophageal transit.[5,6]

The stomach is a crucial muscular organ in the gastrointestinal (GI) tract that acts as a food storage vessel where the food particles are digested mechanically (size reduction–gastric motility) and chemically (mixed with gastric secretions: acid and enzymes) to form a mixture called chyme.[7,8] The food particles in the stomach are disintegrated mechanically by the repeated grinding action of gastric movements.[9] Further, the ingested food is chemically digested by the gastric juice, which is a mixture of acid (hydrochloric acid, HCl) and enzymes (pepsin and gastric lipase) that facilitates the breakdown of proteins and fats.[10] Peristaltic movement of the stomach walls (*i.e.* antral contraction waves) generates adequate forces and fluid motions that enable the breakdown of food and mixing and propelling of the food

Food Chemistry, Function and Analysis No. 42
Food Digestion and Absorption: Its Role in Food Product Development
By C. Anandharamakrishnan, Jeyan Arthur Moses and S. Priyanka

Published by the Royal Society of Chemistry, www.rsc.org

and the chyme towards the pylorus and into the small intestine.[2,11] The particle size of the gastric contents is reduced to less than 1–2 mm before emptying into the small intestine, through propulsion/retropulsion along with the action of acids and enzymes.[12–14]

The discharge of food from the gastric chamber to the small intestine is controlled by the pyloric sphincter.[15,16] Typically, foods such as liquids and digestible solids remain in the stomach for about 2–3 h before being emptied into the small intestine as chyme for nutrient absorption.[17,18] However, the duration of gastric digestion and gastric emptying varies depending on physicochemical food properties such as structure, viscosity, energy density, and composition.[14,19] Apart from gastric digestion, the stomach also plays a prominent role in the nutrient release and absorption mechanisms in the small intestine through chyme formation and regulating gastric emptying rates. In addition, the stomach performs various engineering unit operations during the gastric digestion process such as storage, grinding, diffusion, mixing, pumping, and sieving.[10]

In this chapter, we discuss in detail the functional anatomy of the stomach, the gastric environment and secretions, gastric physiological processes such as gastric motility, gastric mixing and gastric emptying, and various factors that affect gastric digestion. In addition, we consider in particular detail the mechanical and chemical digestion that food undergoes and how various properties of food, such as composition, structural features (macro- and microstructure changes), and processing methods affect the gastric digestion behavior of foods. Further, various instrumental techniques used to measure and characterize quantitatively and qualitatively the gastric processes and food behavior during gastric digestion are described.

2.2 Functional Anatomy of the Human Stomach

The stomach is the most expanded cylindrical J-shaped organ in the alimentary canal with a bag-like structure and a capacity of 1–1.5 L in adults.[8,20] The stomach connects the terminal region of the esophagus and the start of the small intestine (duodenum) and it is located in the epigastric and upper left side of the abdominal region at the level of the first lumbar vertebra.[8,16] The entire stomach region can be subdivided into two regions, namely (i) the proximal region (cardia, fundus, and one-third of the stomach body), which aids in the storage of the incoming food material (bolus) from the esophagus and also controls gastric emptying of liquids, and (ii) the distal region (two-thirds of the stomach body, antrum, and pylorus), which acts as a grinder, mixer, and siever of solid food, and also acts a pump that aids in propelling the digested food into the small intestine during the gastric emptying process.[21,22] Further, the stomach contains two curvatures (greater and lesser curvatures) and two sphincters (lower esophageal sphincter and pyloric sphincter).[8,23] The gastric wall of the stomach is lined with mucosa, submucosa, muscularis propria, and serosa.[20,24] The anatomical features and key functions of the stomach are illustrated in Figures 2.1 and 2.2.

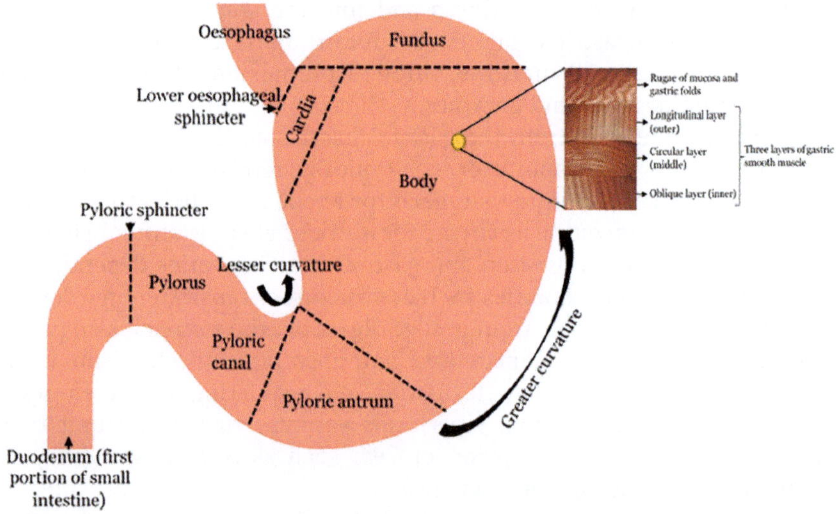

Figure 2.1 Anatomy of the human stomach. Reproduced from ref. 48 with permission from Taylor & Francis, Copyright 2020.

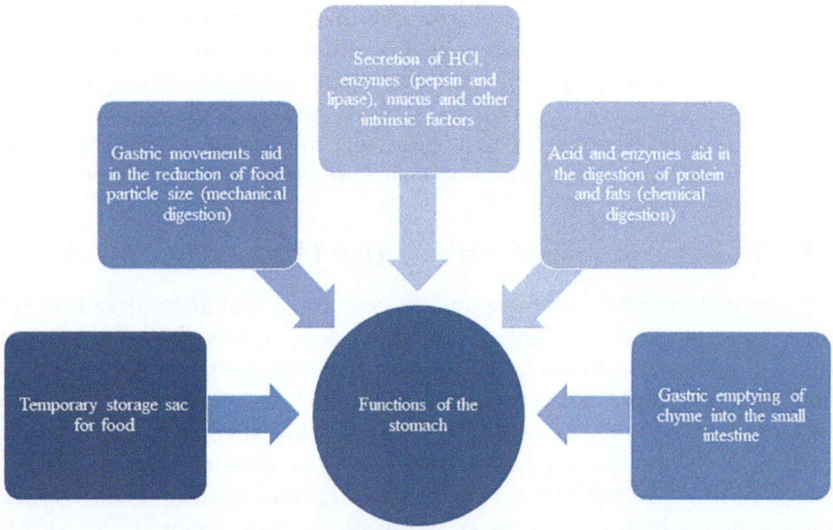

Figure 2.2 Key functions of the stomach.

2.2.1 Cardia, Fundus, and Body of the Stomach

The cardia is the region where the esophagus connects the stomach (gastroesophageal junction) *via* the cardiac orifice, through which the food passes into the stomach.[8,15] Moreover, the cardia encompasses around 5% of the total gastric surface area, which is filled with mucin-secreting cells.[2,25] The lower esophageal sphincter situated at the cardia aids in the prevention of gastroesophageal reflux and helps in emptying the food bolus into the stomach.[26] The fundus is the upper domed portion of the stomach's curvature.[27] Further, the fundus extends above the gastroesophageal junction on the left side of the stomach, making a sharp angle with the distal esophagus forming the cardiac notch of the stomach.[16] The largest portion of the stomach is known as the gastric body/corpus region.[28] The gastric body's interior surface is lined with thick, velvety mucosal folds that enhance the gastric surface area called rugae, which extend distally to the incisura on the lesser curvature.[4,29] Around 75% of the gastric/parietal glands responsible for the secretion of pepsinogen, acids, intrinsic factor, mucus, histamine, and somatostatin are present in the body and fundus regions of the stomach.[25]

2.2.2 Antrum and Pylorus of the Stomach

The antrum of the stomach is a tapered, tube-shaped structure situated exactly below the stomach's body portion.[9] However, the gastric antrum is estimated to be a region from the angular notch along the distal lesser curvature to a point along an inferior line to the distal greater curvature, as it is not distinguishable anatomically.[16] The antrum has a smooth surface and flat mucosa, unlike the gastric body (rugae).[2,30] The antrum harbors mucus-secreting cells, gastrin-secreting G cells, and alkali-secreting surface epithelium and endocrine.[4,15] The food experiences grinding forces and physiological shear stress in the antrum region of the stomach.[22] The antrum empties the chyme into the pylorus region (pylorus canal), which is situated at the terminal region of the stomach, and the pylorus canal links the stomach to the duodenum.

The pylorus canal leads to the pyloric sphincter, which is positioned at the junction of the stomach, and the duodenum is a thickened ring of smooth muscles.[8,15] The pyloric sphincter acts as both a sieve and a pump, which aids in emptying gastric digesta into the small intestine for additional digestion and nutrient absorption.[10] Additionally, it also acts as a control valve that regulates and facilitates the controlled emptying of digested food (chyme) from the gastric chamber into the duodenum for further digestion and absorption.[31] The peristaltic waves generated in the stomach during the digestion process control the opening and closing of the pyloric sphincter.[32] Moreover, the sphincter remains partially open to allow the free flow of water into the duodenum during rest (fasting stage), but it instantaneously closes upon consumption of food to permit gastric digestion to take place.[17]

2.3 Gastric Environment and Secretions

Gastric secretion can be classified into three phases, the cephalic phase, gastric phase, and intestinal phase, based on their stimulation factors, namely signals from the vagus nerve (during ingestion and oral processing of food), vagal reflex, and gastrin–histamine stimulation (upon arrival of food in the stomach), and nervous and hormonal mechanisms (as the food enters the intestine), respectively.[33,34] The specialized cells in the gastric glands of the stomach secrete about 2–3 L of gastric juice per day.[2] Also, gastric secretion exhibits a higher degree of inter- and intra-individual variation.[35] Further, the composition and volume of gastric secretions also vary depending on several factors, such as meal consistency,[36] time after meal consumption (fasting/ingesting conditions),[37] intragastric location,[38] smoking habits,[39] age,[40] stress,[41] and coffee and caffeine intake.[42,43] For instance, a larger quantity of gastric juice is secreted upon ingestion of solid foods compared with liquids.[44] Table 2.1 illustrates significant gastric secretions and their digestive functions.

2.3.1 Gastric Juice: Composition and Function

Gastric juice is a combination of a variety of substances such as digestive enzymes, acids, mucus, intrinsic factor, and hormones.[25] Additionally, gastric juice contains a number of electrolytes, including sodium, bicarbonate, calcium, magnesium, potassium, and phosphate.[2] Further, the key functions of gastric juice are to provide the necessary pH for enzyme activation, to aid in protein digestion, to destroy incoming microorganisms, to stimulate the flow of biliary and pancreatic juices, to protect the gastric lining against acid autodigestion, and to aid in the absorption of various macro- and micronutrients (iron and vitamin B_{12}).[45,46] Gastric juice is acidic in nature, mainly due to HCl secretion, and its pH ranges from 2 to 3.[47] Further, in healthy humans, the stomach in a resting or fasting condition is maintained at an average pH of 2, but during ingestion of food the intragastric pH of the stomach increases owing to the buffering actions of the food consumed.[2] However, owing to the release of gastric acid by the parietal cells of the gastric wall, the pH of the stomach reverts to its normal level.[48] Additionally, gastric juice typically has a temperature of 37 °C, which is ideal for the activity of the gastric enzyme.[49]

2.3.2 Gastric Mucosal Zones and Specialized Cells

The stomach is lined with the gastric mucosa, which comprises several glands containing the specialized cells that are responsible for specific gastric secretions. The glands of the gastric mucosa located at different anatomical parts of the stomach can be classified into three categories, namely cardiac, oxyntic, and pyloric glands, based on the predominant cell types present within the glands.[8] The glands in the cardiac zone are mainly populated with mucus-secreting cells, which secrete the alkaline mucus that is responsible for protecting the gastric epithelium against shear stress and acid.[50] The fundus and body of the stomach is known as the oxyntic zone, which contains a major portion (*i.e.* around 75–80%) of the gastric glands.[4]

Table 2.1 Significant gastric secretions and their digestive functions.

Mucosal zone	Main gastric location	Specialized cell type	Key stimulant	Substance secreted	Significant functions of the secretion
Cardiac zone	Cardia and fundus	Mucosal cells (surface and neck)	Tonic secretion	Mucus and bicarbonate	Serves as a physical barrier between the lumen and the epithelium. It coats and lubricates the gastric surface to provide gastroprotection by buffering against gastric acid to avoid gastric epithelial damage
Oxyntic zone	Fundus and corpus	Parietal cell	Acetylcholine Gastrin Histamine	Hydrochloric acid (HCl, gastric acid)	Aids in the activation of the pepsinogen to form pepsin (proteolytic enzyme). Maintains a favorable acid environment for peptic proteolysis of dietary protein. The acidic environment is bacteriostatic. The acidic gastric condition reduces ferric iron to the more soluble ferrous form, which enhances the absorption of iron in the duodenum
	Fundus and corpus	Parietal cells	Acetylcholine Gastrin Histamine	Intrinsic factor (protein)	Permits intestinal absorption of vitamin B_{12}
	Corpus	Enterochromaffin-like cells	Acetylcholine Gastrin	Histamine	Stimulates gastric acid secretion
	Corpus	Chief cells	Acetylcholine Secretin	Pepsinogen and gastric lipase	Enables the digestion of proteins and fats
	Fundus and corpus	D cells	The acid in the stomach Gastrin Cholecystokinin	Somatostatin	Inhibits gastric acid secretion
Pyloric zone	Antrum	G cells	Acetylcholine Peptides Amino acids	Gastrin	Stimulates gastric acid secretion
	Antrum	Mucous cells (gastric epithelium)	Tonic secretion	Mucus and pepsinogen	Provides a physical barrier between the lumen and the epithelium. Aids in protein digestion

The oxyntic zone is mainly made up of two types of specialized cells, the chief cells (zymogenic or peptic cells) and parietal cells (oxyntic cells), which are responsible for the secretion of pepsinogen (precursor of proteolytic enzyme pepsin) and of hydrochloric acid and intrinsic factor, respectively.[16] The pyloric glands present in the antrum and pyloric region consist mainly of mucus, endocrine, and G cells (secreting hormone gastrin).[51] Further, mucus-secreting surface epithelial cells are located in most of the areas of the gastric mucosa, including the gastric pits. In addition, various enteroendocrine cells, including enterochromaffin cells, enterochromaffin-like cells, ghrelin cells, G cells, and D cells, are distributed along the gastric mucosa and release hormones such as serotonin, histamine, ghrelin, gastrin, and somatostatin, respectively, in an endocrine fashion.[8,16]

2.4 Gastric Motility

The distinct patterns of involuntary contraction and relaxation (motor activity) of the smooth muscle in the muscularis of the gastric wall during the fasting and postprandial periods are referred to as gastric motility.[52] Gastric movements occur as an outcome of complex coordination amongst various intrinsic and extrinsic stimuli. However, gastric motility is fundamentally an intrinsic stimulus (gastric slow waves – myogenic contraction) that is modulated by nerves (enteric and autonomic nervous system) and a variety of gastrointestinal hormones.[53] The interstitial cells of Cajal (pacemaker cells) (ICCs) positioned in the gastric wall along the greater curvature propagating distally and circumferentially from the corpus towards the pylorus are responsible for the gastric slow wave/pacesetter potential, which is the principle source of gastric myoelectrical activity.[4]

Gastric slow waves are an organized electrical rhythmic event that aids in the spontaneous depolarization of smooth muscle cells and provides the onset force for a cyclic pattern of muscle contraction. The slow waves occur regularly at a rate of *ca.* 3 cycles min^{-1}.[54,55] Several gastrointestinal hormones are responsible for controlling gastric motility. Motilin, ghrelin, and gastrin aid in stimulating gastric motility and increase the gastric emptying rate.[45] However, there is another group of hormones, namely glucagon-like peptide 1 (GLP-1), peptide YY, somatostatin, glucagon, and cholecystokinin, all of which originate from the small intestine and hinder gastric movement and delay gastric emptying.[56] Figure 2.3 shows a schematic representation of gastric motility patterns including accommodation and antral contraction waves resulting in mixing and grinding of luminal content.

2.4.1 Different Patterns of Gastric Motility

The gastric motility behavior can be broadly classified into two different patterns: fasting motility and postprandial motility (fed state motility).

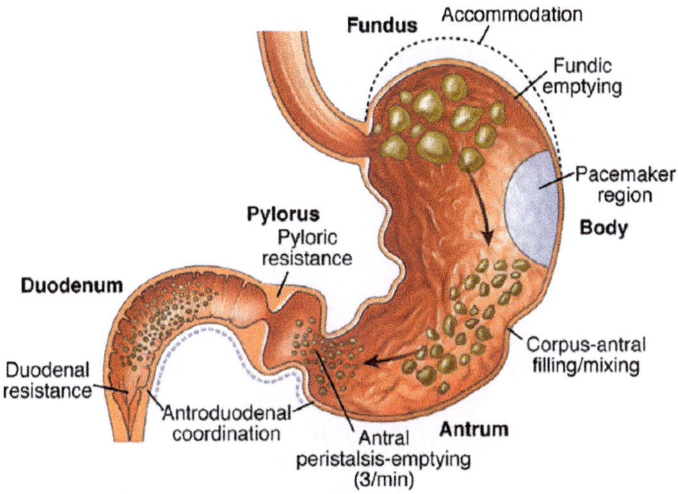

Figure 2.3 Schematic representation of gastric motility patterns including accommodation and antral contraction waves resulting in mixing and grinding of luminal content. Reproduced from ref. 52 with permission from Elsevier, Copyright 2009.

2.4.1.1 *Interdigestive/Fasting Motility*

The characteristic cyclical motor activity (gastric movements) that occurs during the fasting/inter-digestive period is known as the migrating motility complex or migrating motor complex (MMC). The slow waves generated by ICCs cause a cyclic pattern of motility that propagates distally towards the pyloric region.[57] The inter-digestive motility pattern is relatively different from the digestive/postprandial motility. The significant role of inter-digestive motility is to empty the gastric chamber (stomach) by flushing out the remnants such as undigested food, indigestible fibrous food particles, gastric secretions, dead cells, and bacteria.[58] The MMC is popularly known as a "gastrointestinal housekeeper" as it performs the "housekeeping", *i.e.* cleaning by removing all the leftovers after digestion. Further, the inter-digestive motility pattern belongs to the whole GI tract.

2.4.1.2 *Different Phases of Migrating Motility Complex (MMC)*

The inter-digestive motility comprises four different phases of MCC recurring every 90–120 min during fasting.[59] Phase I persists for about 45–60 min; there is little to no contractile activity during this stage and the stomach is mostly quiescent.[60] In phase II, the contractions are characterized as random, frequent, irregular, medium–high amplitude, and progressively strong, and this phase persists for around 30–40 min.[61] Phase III lasts for a very short period of 5–15 min, but it consists of intense contractions (high amplitude and maximum frequency) that travel from the antrum to the distal ileum.[62] The contractions that occur in phase III are known as the "activity front" because it sweeps all the food residues out of the stomach. Phase IV is characterized by rapid inhibition of contractile activity, which usually takes around 1–5 min of preparatory time to merge with the upcoming phase, either the reappearance of phase 1 of MMC or digestive motility.[59]

2.4.1.3 Factors That Affect Migrating Motility Complex (MMC)

There are a few factors that significantly influence MMC such as motilin (a polypeptide hormone produced by M cells of the small intestine), neurons of the enteric nervous system (ENS), and bile acids (gallbladder emptying). During the fasting period, motilin is released in pulses every 90–120 min, and peaks during phase III of the gastric MMC. Motilin plays a significant role in the regulation of MMC by activating the myenteric plexus of the stomach, which then induces the MMC through cholinergic pathways.[4] In addition, during the fasting state, a positive correlation was observed between the levels of bile acids and fluctuations in plasma motilin levels.[59] The factors that regulate MMC are not clearly understood; nevertheless, it is expected to be multifactorial. Moreover, neurons of the ENS are being excited especially during phases II and III of MMC based on an intrinsic timing mechanism, which is blocked by the presence of food in the upper GI tract.[53] Hence it looks like they have an important role in the generation of contractions. The ingestion of food triggers vagal stimulation, which immediately halts the MMC pattern and alters the gastric motor and hormonal activity.

2.4.2 Digestive/Postprandial Motility

Digestive motility is very crucial for the gastric digestion of food. The ingestion of a meal inhibits the cyclical motor pattern of fasting motility and activates significant changes to the gastric motility pattern. The digestive motility pattern includes relaxation of the proximal stomach, and an extended period of contractile activity comprising irregular phasic contraction in the proximal antrum and enhanced tonic and phasic contractile activity in the pyloric region (terminal antrum, pylorus, and pyloric sphincter).[62] Digestive motility occurs as a result of complex synchronization among gastric slow waves, neurohormonal contributions, intrinsic and extrinsic innervation, and properties of ingested substances (structure and calorie content).[53] Further, gastric motility plays a significant role in the production of chyme.

2.4.2.1 Different Phases of Digestive Motility

Digestive motility can be classified into three phases, the filling phase, pumping phase, and grinding phase.[17] The filling phase encompasses the relaxation of the proximal stomach to accommodate the incoming food and shows no contraction. The pumping phase involves both slow tonic contraction of the fundus (pressure pump) and enhanced phasic contractions of the corpus and antrum (peristaltic pump), which aid in the mixing of gastric contents (gastric juice and ingested food) and convey the gastric mixture towards the pyloric region.[63] The grinding phase is characterized by powerful antral contraction waves (ACWs) and pyloric contraction results in the generation of strong retropulsive jets that facilitate further gastric mixing, breakdown of solid particles, and selective gastric emptying.[52] Gastric motility is the key driver responsible for various digestive functions such as gastric accommodation, gastric mixing, grinding of solid foods, and gastric emptying.

2.4.2.2 Gastric Accommodation

Gastric accommodation is gastric muscle tone reduction (relaxation) of the proximal region of the stomach to "accommodate" the ingested food. It supports the stomach in performing its temporary food storage function by expanding the stomach's volume without significantly increasing the intragastric pressure.[60] Gastric accommodation can be categorized into two types based on the initiating stimulus: receptive and adaptive relaxation.[52] Receptive relaxation is stimulated by the oropharynx and esophagus due to eating or dry swallowing or any other mechanical triggers and does not require movement of the food bolus into the stomach. Adaptive relaxation is a comparatively slower process activated by the stimulation of stretch receptors upon stomach and duodenum distension and increased gastric intraluminal pressure.[4] Further, certain properties of the ingested food and nutrient sensing in the duodenum play a significant role in adaptive relaxation.

2.4.2.3 Antral Contraction Waves (ACWs)

The ACWs are generated from the electric signals of the ICC (slow waves) that stimulate contractions of the smooth muscle cells and travel faster circumferentially.[4] The antral peristalsis physically breaks down food particles with mechanical forces of up to 0.65 N.[64] In the case of both solid and liquid meals, the ACWs have an average speed and frequency of about 1.5–3 mm s^{-1} and 2.6–3 waves min^{-1}, respectively.[2] Nevertheless, there is disagreement in the literature regarding the impact of meal composition on gastric motility. Further, the available records suggest that various factors such as food type, volume, and nutrient content affect gastric motility. Since various non-invasive techniques are readily available to monitor motility parameters, comprehensive research must be conducted with a broader diversity of test meals with standardized food structures and composition.

The antral peristalsis or antral contraction waves (ACWs) are the significant motility pattern that is responsible for gastric emptying and gastric mixing, and aid in the disintegration of solid food particles.[52] X-ray images (cinefluoroscopic studies) of gastric motility revealed that the ACWs begin at the corpus of the stomach as a small ring-shaped muscular contraction and travel towards the antrum of the stomach.[4] The initial minor ring of contraction contracts thc pylorus to retain gastric contents, which in turn increases the pressure within the antrum. Further, as the ACWs become stronger they forcefully propel the gastric mixture towards the contracted pylorus, and gastric contents are squirted back into the body of the stomach (retropulsive jet).[61] This complex interplay between pyloric contraction and relaxation (opening and closing) and arriving ACWs supports the regulation of gastric emptying and is responsible for mixing and grinding of gastric digesta.[17]

2.5 Gastric Mixing

Gastric mixing is an intricate phenomenon that encompasses the mixing of ingested food with gastric secretions. Hence it is crucial for the proper digestion of foods and absorption of nutrients. Different food-related characteristics (meal composition, food structure, buffering capacity, physical characteristics, and breakdown rate), and also physiological aspects of the stomach (secretion rate, emptying rate, and motility parameters), influence the gastric mixing process.[25] The process of gastric mixing is controlled by various gastric motor functions, such as tonic contractions of the proximal stomach, antral peristalsis/ACWs, and opening and closure of the pylorus.[61] In addition, gastric mixing significantly influences the residence time of a meal in the stomach and local intragastric biochemical environments such as pH and enzymes after ingestion of foods and drugs.

Gastric mixing processes are studied and understood with the help of *in silico* (computational modeling), *in vitro* (stomach models), and *in vivo* (human and animal trials) approaches.[65–67] Further, gastric mixing can also be measured through the mixing index. Moreover, gastric mixing plays a significant role in various vital applications in the food, medical, and nutrition arenas such as food breakdown, probiotic microorganism viability, inactivation of pathogenic microbes, drug dissolution, and gastroesophageal acid reflux.[25] Hence a deep knowledge of gastric mixing is required for the optimization of medicine and food products for persons with specific diseases such as gastroparesis, Parkinson's disease, gastric ulcers, and infections, and those with impaired gastric secretions and functions.

2.5.1 Different Types of Gastric Mixing

Gastric mixing can be classified into three different types, namely solid–solid mixing (advective mixing of different portions of the solid meal), solid–liquid mixing (advective and diffusive mixing of a solid meal with gastric secretions), and liquid–liquid mixing (advective and diffusive mixing of a liquid meal with gastric secretions).[25] Gastric motility and diffusion both play a significant role in the gastric mixing of solid–liquid and liquid–liquid systems, whereas antral peristalsis is the primary driving factor for the solid–solid mixing process.[66,68]

Diffusion plays a significant role in the mixing of gastric contents (gastric secretions and food) in the proximal stomach region, as the ingested food is stored there for extended periods (2–3 h), being exposed only to tonic contractions. Further, during the initial stages of digestion, an unmixed layer (1.0–3.0 cm) of concentrated gastric acid (pH 1.6–1.8) is present on top of the ingested food in the proximal stomach, which raises the risk of gastro-esophageal reflux disease (GERD) and acid reflux.[69] Nevertheless, in the distal region of the stomach, where powerful phasic contractions drive the gastric mixing process, diffusion might play a limited role. Further, it was also found that most of the gastric mixing occurred in the antrum due to the retropulsive jet-type flow and recirculating eddies created by the antral contraction waves.[67]

2.5.2 Factors That Affect the Gastric Mixing Process

2.5.2.1 Food Properties

The gastric mixing process is commonly studied with the aid of animal models (pigs) and human and computational modeling of the stomach. However, the process of gastric mixing and factors that influence gastric mixing and its impact on digestion have not been characterized thoroughly in the literature. Nevertheless, some studies have been conducted specifically to analyze gastric mixing. Several studies reported that the gastric mixing process is significantly affected by the food texture and structure (soft or rigid foods)[70] and the particle size distribution (fine or coarse particles),[71,72] viscosity (high and low viscosity),[73] and rheological properties (Newtonian and shear-thinning fluids)[74] of the ingested food.

Numerous animal studies using pigs as models have been conducted to study gastric digestion, suggesting that food properties affect the intragastric mixing of solid meals by creating a pH gradient among the proximal and distal regions of the stomach. In one study, pigs fed with finely ground corn and coarsely ground corn (cracked corn) exhibited no pH gradient and alterations in their intragastric pH in different stomach regions (lower pH in the antrum), respectively. The possible reason for the observation of no intragastric pH difference in the case of finely ground meals was assumed to be the outcome of decreased meal viscosity, greater gastric secretions, and thorough mixing.[75] Further, in a study conducted in pigs, Ange *et al.* observed that the pH of the proximal stomach was >3 even 12 h after the meal, signifying that the pH gradient in the stomach persists for an extended period after ingesting solid foods.[76]

In another study, pigs were fed cooked rice (soft food) and diced almonds (rigid food), and the gastric mixing process was studied over a period of 8–12 h of digestion. The gastric mixing was characterized using particle mixing and the intragastric pH distribution and measured using a mixing index. For all of the test meals the intragastric pH gradient was observed for up to 180 min after ingestion of food and the particle mixing pattern was different among soft (cooked white and brown rice) and rigid foods (diced raw and roasted almonds). It was hypothesized that both enhanced gastric secretions and a higher gastric emptying rate are responsible for the rapid gastric mixing of soft foods compared with rigid foods.[70] Further, it is clear from the above-mentioned studies that solid–liquid mixing (intragastric pH) was considerably influenced by the type of meal, a specific section of the stomach, and digestion time. Gastric acid diffusion into boluses of different rice varieties, namely brown and white rice, was explored using a one-dimensional diffusion system.[65] The acid diffused faster in the case of the brown rice bolus compared with the white rice bolus. The effective diffusivity of acid into the bolus was impacted by the rice variety, signifying that food properties have a significant role in both advective and diffusive mixing processes.[65]

In recent times, several non-invasive imaging, numerical, and computational techniques have been widely used to predict and understand the peristaltic flow, pressure fields, diffusion rates, and mixing patterns that occur during gastric digestion. A three-dimensional stomach model was used to study the changes in gastric flow and mixing in the case of liquid foods with different viscosities. It was found that the viscosity of the fluids significantly affected the gastric flow and mixing; in the case of fluids with viscosity higher than that of water, the recirculating eddies and retropulsive jet diminished considerably, but antral mixing was observed for low-viscosity fluids. Additionally, there was less recirculation between the proximal and distal stomachs and localized mixing at the antrum in fluids with higher viscosity. However, the flow and mixing patterns became independent of viscosity when the fluid viscosity rose above a certain threshold value.[9,66]

2.5.2.2 Gastric Physiological Properties

Various gastric physiological processes, such as gastric secretion rate, motility, and emptying, are significantly impacted by several disease conditions and certain medications, and in turn they considerably affect the food digestion and gastric mixing processes.[59,61] Gastroparesis is a condition where gastric emptying of food is delayed due to the weakened myoelectrical activity of the stomach muscles, which alters gastric motility and may reduce gastric mixing.[53] Patients with mild and severe Parkinson's disease show a decreased gastric emptying rate and diminished antral contractions with smaller amplitudes after ingestion of food.[77] The gastric secretion rate is affected by ingested food and individual-related factors and, in addition, clinical conditions such as gastric ulcers (*Helicobacter pylori* infection), intake of specific drugs such as proton pump inhibitors (omeprazole), and H2 receptor antagonists (ranitidine, cimetidine) reduce the acid secretion to significant levels which modify the gastric mixing patterns.[78,79]

The intake of antacids to neutralize acids can also impact gastric mixing. In addition, gastric hypersecretion is possible owing to the withdrawal of drugs such as proton pump inhibitors, and some disease conditions such as Zollinger–Ellison syndrome (the existence of a stomach tumor that secretes gastrin) also profoundly affect gastric mixing and digestion.[80] Hence it is clear from various studies that gastric mixing is a slow and complex process that is affected by food properties and also has clinical relevance. As a result, when creating functional foods for populations with gastrointestinal illnesses, it will be important to consider differences in stomach motility, emptying, and secretion caused by diseases or medicines because these are crucial inputs to the gastric mixing process.

2.6 Gastric Emptying

This is a very complex process in which gastric contents are discharged into the small intestine's duodenum region for further digestion and nutrient absorption. For comparison, the gastric emptying rates are usually measured as the time taken to empty 50% of the ingested contents ($t_{\frac{1}{2}}$).[19] Gastric emptying is an important step in the overall food digestion process due to its control over the digestion and nutrient absorption in the small intestine (*i.e.* faster or slower gastric emptying results in impaired nutrient absorption).[17] Various factors such as food-related factors, biological factors, gastrointestinal features, and several medical conditions affect the gastric emptying rate.

The gastric emptying of foods can be assessed by various techniques, such as scintigraphy, magnetic resonance imaging, ultrasound, stable isotope breathing tests, Wireless Sports Capsule Technology (WMC), and acetaminophen or paracetamol absorption.[62] Table 2.2 shows recent research advances in investigating gastric emptying and various factors that affect gastric emptying during digestion.

2.6.1 Gastric Emptying Process and Its Distinct Phases

The gastric emptying process is classified into four distinct phases, (a) tonic contractions, (b) peristaltic contractions, (c) retropulsion, and (d) emptying.[81] Figure 2.4 illustrates the process of gastric emptying. The transfer of the ingested food towards the distal stomach is facilitated by tonic contractions of the proximal stomach. Further, the peristaltic wave assists in the obstruction of the antral lumen that triggers the contraction of the pylorus, which narrows the opening of the pylorus. The jetting back of chyme into the stomach upon arrival of the peristaltic wave is known as retropulsion. During retropulsion, the gastric contents experience vigorous mixing and antral grinding. Repeated grinding and mixing through peristalsis and

Table 2.2 Recent research advances in investigating gastric emptying and various factors that affect gastric emptying during digestion.

Meal/food	Study methodology	Characterization technique	Key factors	Significance of the study	Ref.
Two isocaloric meals with different food structures: liquid (emulsion) and semi-solid (mixed Gouda cheese in low-fat yogurt) meals	The crossover study comprised 10 male volunteers (age 20–50 years and BMI 19–30) to evaluate the variations in gastric emptying, satiety indicators, and levels of the GI hormone CCK	Imaging of gastric contents (volume, macrostructure, and layering) using magnetic resonance imaging (MRI). Satiety: self-reported using the visual analog scale (VAS) technique. CCK content using a radioimmunoassay	Food structure. CCK secretion	The semi-solid meal had a 30 min longer gastric retention half-time (delayed gastric emptying) compared with the liquid meal. The significant difference in the gastric emptying rates was observed until 50 min after ingestion, and the apparent emptying rate of the liquid meal was more than twice that of the semi-solid meal. The structured semi-solid meal, owing to its higher viscous nature of food boluses, prevented significant early emptying and thus suppressed the rate of CCK secretion and exhibited a heightened sense of satiety and fullness	87
Two semi-solid meals with different viscosities: standard-viscosity meal (chocolate pudding, heavy cream, and marker) and high-viscosity meal (standard-viscosity meal with the addition of 3.3 g of guar gum)	The randomized crossover study consisted of 15 healthy males (age 18–40 years and BMI 20.0–29.9) to evaluate meal duration, eating rate, appetite, gastric emptying, and plasma assay	Appetite questionnaire (visual analog scale). Gastric emptying rate (acetaminophen absorption test). Blood test: insulin, glucose, acetaminophen, and glucose-dependent insulinotropic peptide (GIP)	Viscosity of foods	Viscosity of foods (semi-solid) modulated food consumption behavior (altered eating rate and satiety) and gastric digestion kinetics (gastric emptying rate and glycemic response). Highly viscous meal greatly reduced the gastric	89

				emptying rate and postprandial plasma GIP concentration. The ingestion of highly viscous food increased satiety and fullness and also significantly lowered the eating rate and hunger	
Cooked rice (white and brown) with identical amylose contents	Dynamic *in vitro* rat stomach (DIVRS) model was used to simulate digestion and the static soaking method	Gastric digesta from DIVRS was tested for maltose, pH, gastric retention ratio, particle size distribution, and rheology. Moisture absorption, microstructure, and texture were determined for rice samples during gastric digestion	The rigid bran layer in brown rice acts as a physical barrier to acids and enzymes, and higher gastric digesta viscosity	The bran layer is the crucial factor that significantly affects various physicochemical properties. Brown rice exhibited a lower gastric emptying rate and higher resistance to disintegration under gastric conditions due to the presence of a tough bran layer that hindered hydration by gastric secretions and hydrolysis by gastric enzymes. White rice particles (32%) were retained less in the stomach than brown rice particles (53%). The higher viscosity of brown rice gastric digesta also resulted in lower starch digestion and gastric emptying rates	110
Three isocaloric drinkable yogurts with different compositions and structures	Dynamic *in vitro* gastrointestinal digestion system (DIDGI)	Gastric emptying assessment (gastric emptying half-time and gastric residual volume) by	Composition (protein and fiber) and viscosity	Despite having the same calorie content, the addition of proteins and fibers to yoghurt delayed gastric emptying, which	111

Table 2.2 (*Continued*)

Meal/food	Study methodology	Characterization technique	Key factors	Significance of the study	Ref.
		gamma-scintigraphy on pigs. Sodium dodecyl sulfate polyacrylamide gel electrophoresis (SDS-PAGE) analysis: milk protein hydrolysis		had a substantial impact on gastric digestion. Low-viscosity yogurt (with extra protein and fiber) showed a 20% lower gastric emptying half-time compared with the control yogurt. However, the gastric emptying half-time of high-viscosity yogurt was not statistically significant compared with the control yogurt and also the low-viscosity yogurt, indicating that the viscosity of undigested yogurt did not have a critical role in determining gastric emptying in this study. However, the yogurt viscosity affected the time of residence further in the small intestine but not in the stomach	
Shakes with different viscosities and energy densities (four varieties: thin shake 100 and 500 kcal and thick shake 100 and 500 kcal)	The randomized crossover study consisted of 15 healthy males (age 18–35 years and BMI 18–25) to investigate the impact of viscosity and energy density on gastric emptying	Gastric emptying parameters (gastric volume) were assessed using an MRI scanner. Subjective appetite ratings (hunger, fullness, prospective consumption, desire to eat, and thirst) using VAS	Viscosity and energy density	Gastric emptying was delayed by increasing both the viscosity and energy load. However, the energy load had a profound effect on gastric emptying compared with viscosity. Further, viscosity had a positive	90

				correlation with perceived fullness/satiety due to the mouthfeel and oral exposure and not through the delay of gastric emptying. Thus, a thick shake with 500 kcal had the highest gastric emptying half-time whereas a thick shake with 100 kcal led to higher fullness than the thin 500 kcal shake	
Meals with varied energy and macronutrient compositions (high-fat meal, isoenergetic, and equal mass low-fat meal compared with the high-fat meal)	The randomized, balanced, single-blind study was conducted among nine healthy male volunteers to evaluate satiety and gastric emptying	Sodium [^{13}C]acetate breath test was conducted to measure gastric emptying half-time. Satiety parameters (hunger, thirst, desire to eat, tiredness, fullness, and cold) were measured using 150 mm VAS every 30 min	Energy and macronutrient composition	Gastric emptying at lunch was effectively influenced by the macronutrient and energy content of the food eaten during breakfast due to the extended CCK response to fat. A high-fat meal had a higher gastric emptying half-time than both low-fat meals. Further, the consumption of a high-fat meal at breakfast considerably increased food consumption (*i.e.* macronutrients such as carbohydrates, proteins, and especially fat) at lunch in terms of both energy and mass. However, isocalorific low-fat meals exhibited increased satiety compared with other	103

Table 2.2 (*Continued*)

Meal/food	Study methodology	Characterization technique	Key factors	Significance of the study	Ref.
				meals and resulted in less food consumption during lunch (in terms of both energy and grams of macronutrients)	
Wheat and oat meals, together with a mango beverage. The meals had the same weight but differed in appearance, temperature, caloric density, physical form, and viscosity	Test subjects selected for the study included 10 healthy males (age 22–32 years and BMI 22 ± 1.4 kg m^{-2}) to investigate the effect of meal physicochemical characteristics on gastric emptying and motor functions and its subsequent impact on glycemic response	Gastric motility, gastric emptying, volume of the meal and stomach were measured using MRI scanning	Caloric density, viscosity, and the physical form (structure)	Wheat meal had delayed gastric emptying regardless of having a lower viscosity than oatmeal, which is due to the higher energy density and relatively dense consistency of wheat meal. The delayed gastric emptying of wheat meal had a profound impact on glycemic response and antral contraction wave frequency. Thus, in delaying gastric emptying, a meal's nutrient content has a more significant effect than viscosity	112
Two whey protein emulsion gels (soft and hard gels), each with a different structure	To replicate the *in vitro* gastric digestion of simulated gel boluses, a dynamic stomach model [human gastric simulator (HGS)] was employed	The gastric emptying pattern of gel boluses was evaluated with solid retention (gel retention) in HGS. Gel disintegration was measured using wet sieve analysis (particle size distribution)	Disintegration kinetics and particle size	The particle size distribution of gel boluses and gastric disintegration kinetics, which are controlled by food structure, bolus characteristics, and metabolic effects, play major roles in the gastric emptying of both hard and soft gels.	113

				Pepsin had a significant acceleration effect on the disintegration of both gels and gastric emptying; however, the harder gels always exhibited slower disintegration governed by abrasion compared with soft gels (abrasion and fragmentation), irrespective of pepsin. Owing to improved disintegration in the presence of pepsin, soft gels showed faster gastric emptying than hard gels after 120 min	
Two liquid meals with the same volume but different calorific values (low-calorie soup and high-calorie drink)	The randomized study was conducted among 32 healthy volunteers (age 19–35 years) to explore the influence of calorie content on gastric emptying	Ultrasonography was used to measure the cross-sectional area of the antrum of the stomach before and after ingestion to investigate gastric emptying	Caloric content	Ultrasonography aids in evaluating gastric emptying by measuring the antral area. The antral area after the ingestion of a low-calorie liquid meal returned nearly to the fasting size at a faster rate (within 30 min), signifying rapid gastric emptying. However, in the case of a high-calorie liquid meal the antral area remained fairly constant in each postprandial measurement until 30 min, indicating slower gastric emptying compared with the low-calorie liquid meal	91

Table 2.2 (*Continued*)

Meal/food	Study methodology	Characterization technique	Key factors	Significance of the study	Ref.
Two drinks containing whey protein isolates, (i) 30 g (120 kcal) and (ii) 70 g (280 kcal), and a flavored water as control drink	The study was conducted among eight healthy young men (age 25 ± 2 years and BMI 23 ± 1 kg m^{-2}) and eight women (age 23 ± 1 years and BMI: 24 ± 0.4 kg m^{-2}), to investigate the effects of whey protein ingestion on appetite, gastric emptying, energy intake, and hormone responses	Three-dimensional ultrasonography was used to measure gastric volume. Plasma total insulin [enzyme-linked immunosorbent assay (ELISA)]. Perceptions of appetite were assessed using VAS. Blood glucose and plasma gut hormone concentrations (glucose oxidase method). Plasma glucagon, total ghrelin, CCK-8, total GIP, and total GLP-1 concentrations were determined by radioimmunoassay	Gender	Compared with men, healthy young women who consumed whey protein drinks showed delayed gastric emptying. Women also reported less control of their appetite and energy intake. In addition, the levels of several hormones (plasma glucagon, CCK, and GLP-1) secreted were lower in women than in men	114

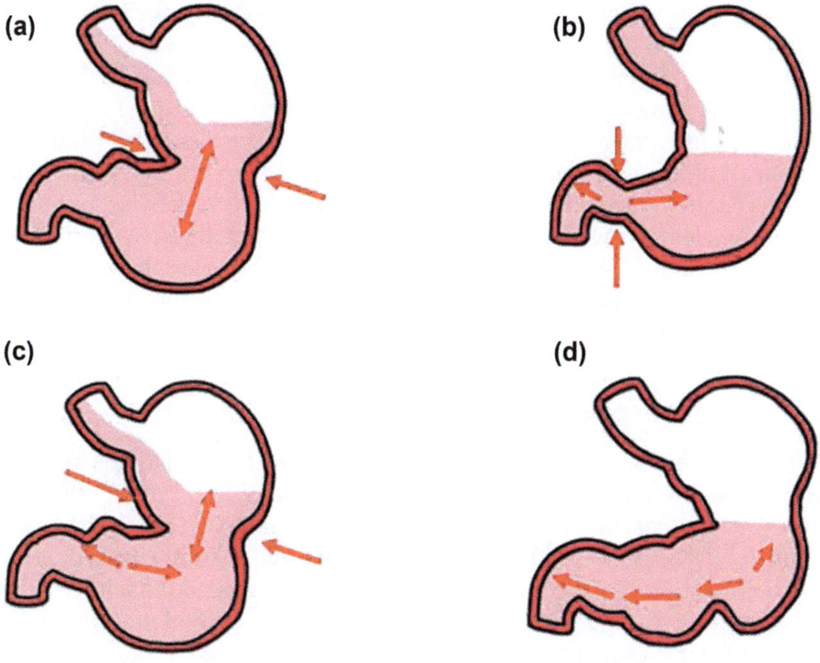

Figure 2.4 The process of gastric emptying: (a) tonic contractions, (b) peristaltic contractions, (c) retropulsion, and (d) emptying. Reproduced from ref. 19 with permission from Taylor & Francis, Copyright 2021.

retropulsion along with the influence of enzymes and acids results in a decrease in particle size. Finally, the gastric emptying of small particles (<1–2 mm) occurs along with fluids flowing from the stomach to the duodenum continuously once the pylorus has partly opened.[19] In addition, the duodenum of the small intestine works in coordination with the pyloric complex to regulate gastric emptying. "Antroduodenal coordination" is a phenomenon in which the duodenum relaxes during antral contractions. Further, the duodenal bulb contracts upon receiving the chyme to aid the steady flow of chyme towards the subsequent portion of the duodenum.[17]

2.6.2 Regulation of the Gastric Emptying Process

The gastric emptying rate is precisely regulated by the intestinal feedback mechanism mediated through a range of gastrointestinal hormones. Accordingly, the gastrointestinal hormones are classified as gastric emptying "braking" hormones [cholecystokinin (CCK), GLP-1, and leptin] and "accelerating" hormones (ghrelin and motilin), as they are responsible for preventing and accelerating the gastric emptying process.[82,83] Moreover, during the digestive period hormones responsible for gastric acceleration are not released. In addition, gastric emptying is facilitated by two parallel neural circuits, namely the gastric excitatory vagal motor circuit and the gastric inhibitory vagal motor circuit, which mediate the gastric excitation and inhibition.[17,84] Further, both motor circuits are controlled differentially by unique sets of neurons in the solitary tract nucleus and vagal afferents.

Gastric emptying is inhibited by the hormones CCK and GLP-1 through a gastric inhibitory vagal motor circuit during the digestive period and the hormones ghrelin and motilin accelerate gastric emptying by exciting the gastric excitatory vagal motor circuit during the inner digestive period.[17] Thus, gastric emptying during the digestive and inter-digestive periods is differently regulated. Moreover, the gastric emptying process is also associated with appetite, satiety, and hunger and their hedonic aspects along with chronic food intake and energy homeostasis.

2.6.3 Mechanism of Gastric Emptying

The gastric emptying rates are unique for different types of foods such as solids (digestible and indigestible), liquids (high and low calorie content), and water. The gastric emptying of liquids is remarkably faster than that of solid foods because solid foods need to undergo a sufficient reduction in particle size and viscosity. In addition, the average stomach empties *ca.* 1–4 kcal min^{-1}. Figure 2.5 illustrates the emptying of food materials with different physical properties and chemical compositions.

2.6.3.1 *Liquid Foods*

Gastric emptying of liquids (water, non-calorie/low-calorie liquids, equimolar solutions, salt solutions, hot inert solutions, and neutral solutions) starts immediately after ingestion and proceeds exponentially. However, a linear emptying curve is observed in the case of a nutrient-rich solution, owing to

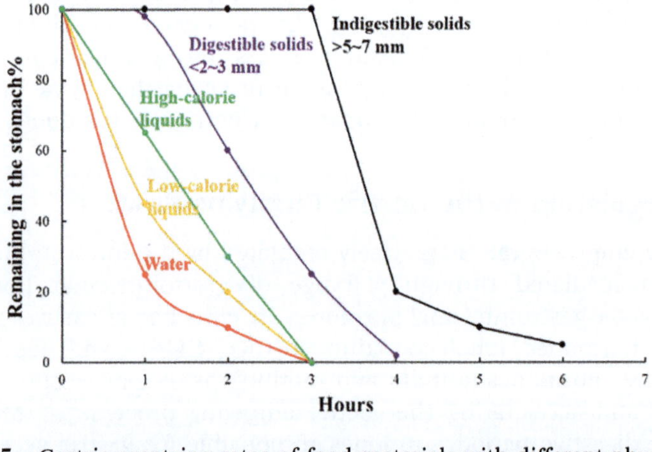

Figure 2.5 Gastric emptying rates of food materials with different physical properties and chemical compositions. Reproduced from ref. 19 with permission from Taylor & Francis, Copyright 2021.

the existence of a lag phase during gastric emptying. The gastric emptying process of liquids lasts for 2–3 h depending on the volume, caloric density, and viscosity of the liquid.[19] The major reason for the difference in the emptying patterns of liquids with a high calorie content is the nutrient-sensing mechanism (mucosal receptors) present in the small intestine, which controls the number of nutrients entering the duodenum by limiting the rate of gastric emptying.[85] In general, for most liquid foods, based on their energy content the gastric emptying half-time ($t_{\frac{1}{2}}$) usually ranges between 10 and 120 min. For instance, after 2 h of ingestion 50% of the high-calorie liquid remained in the stomach, whereas after 10 min *ca.* 50% of the water had left the stomach.[17] Further, the tonic contractions of the proximal stomach act as a driving force for the gastric emptying of liquids by regulating the proximal stomach's pressure in accordance with the duodenum to control the gastric emptying process of liquid foods.[17]

2.6.3.2 Solid Foods

The gastric emptying mechanism of solid foods follows a biphasic pattern.[28] The initial phase is a delayed phase comprising gastric mixing and antral grinding, leading to distribution and particle size reduction (2–3 mm), respectively, to form chyme. During the initial lag period, there is almost no gastric emptying of contents. Further, during the latter stage, gastric emptying occurs through the pylorus linearly.[19] Thus, in short, solid foods exhibit an initial lag period before the initiation of gastric emptying. Gastric emptying lasts *ca.* 2–3 h for digestible solids after ingestion of food. Nevertheless, during the inter-digestive period, the indigestible solids that are resistant to the gastric digestion process are forcefully dumped into the duodenum of the small intestine.[17]

2.6.4 Factors That Influence Gastric Emptying

The gastric emptying rate is affected by various factors that can be classified into four categories, namely food-related factors (food composition, specific ingredients, processing methods, calories, and viscosity), biological factors (age and gender), gastrointestinal features (hormones, gastric motility, and feedback effect of nutrients) and medical conditions (diabetes, gastric infections, nervous system diseases, and surgery). Figure 2.6 shows various key factors that influence the gastric emptying of foods.

2.6.4.1 Food-related Factors

Several research studies have established that various characteristics of food, such as composition (carbohydrates, sugars, proteins, and fat), energy density (high- and low-calorie content), viscosity (high and low viscosity), processing methods, and the presence of specific ingredients, affect the gastric digestion of food, which significantly impacts the rate of gastric emptying.

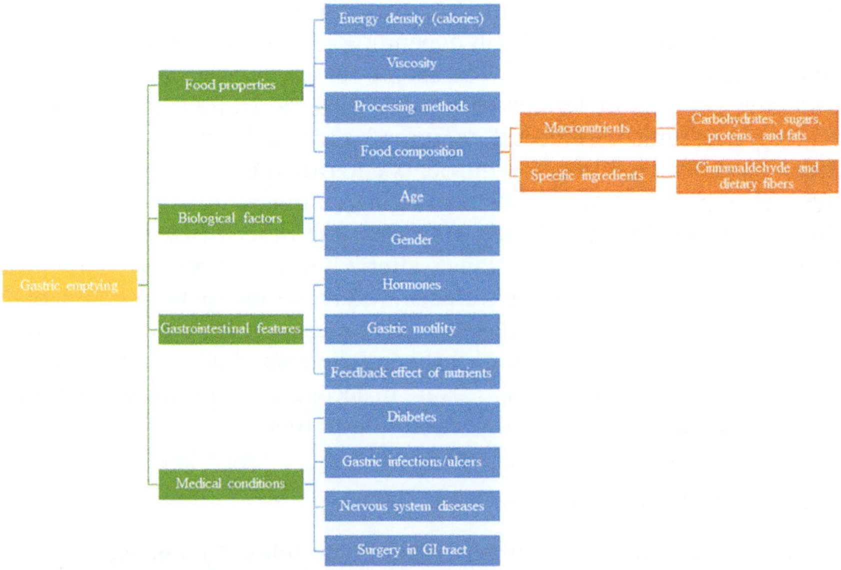

Figure 2.6 Key factors that influence the gastric emptying of foods.

Gastric emptying rates of sugars are fastest, followed by proteins, and are slowest in the case of lipids.[86,87] In addition, the gastric emptying rates of carbohydrates and sugars are of great importance as they could interfere with the management of diabetes and the release of intestinal hormones if there is faster and slower gastric emptying, respectively. Further, various studies have indicated that the gastric emptying rates of high-viscosity foods are slower than those of low-viscosity foods. In addition, highly viscous foods can provide prolonged satiety and a regulated sugar response, owing to extended gastric emptying.[88,89] Nevertheless, the energy density of the food consumed is the single most important factor for regulating the gastric emptying rate.[90] Correspondingly, foods with a high calorific value empty more slowly than low-calorie foods.[91] Further, the quantity of food (meal volume) consumed also has a minimal effect on the gastric emptying process. Since satiation can be achieved only by consuming nutritious food and not by drinking large volumes of water.[92]

The processing of foods also affects the digestion and in turn the gastric emptying process as it modifies the internal matrix structure of the food. The gastric emptying of roasted almonds was faster than that of raw almonds because the roasting process resulted in structural modification and disruption of proteins.[93] In another study, gastric digestion in pigs was explored using white and brown rice, and delayed gastric emptying ($t_{\frac{1}{2}} = 309$ min) was noted for brown rice proteins compared with white rice proteins ($t_{\frac{1}{2}} = 237$ min), which could be due to the accumulation of bran layers in the distal stomach.[94]

In a study conducted using traditional African foods prepared from millet and sorghum, they exhibited remarkably slower gastric emptying compared with non-traditional foods (rice, potato, or pasta), regardless of the viscosity of the meal.[88] The ability of the millet and sorghum meals to provide the sensations of satiety, fullness, and sustained energy was believed to be the reason for delayed gastric emptying.[88] Similar observations were reported in another study that compared different types of breakfast such as cornflakes (no fiber) and oatmeal (containing fiber), where the high-fiber oatmeal exhibited higher scores for fullness and lower scores for hunger, which was related to its slower gastric emptying.[95] Further, the presence of specific functional ingredients such as cinnamaldehyde and dietary fibers (konjac glucomannan, arabinoxylan, and β-glucan) was found to delay gastric emptying owing to activation of transient receptor potential ankyrin 1 (TRPA1), which decreases ghrelin secretion and provides enhanced satiety in addition to imparting viscosity.[96–98]

2.6.4.2 Biological Factors

Biological factors such as the age and gender of individuals play an extremely important role in gastric emptying. Several critical gastrointestinal factors, namely the pH of gastric secretion, gastric motility, and digestive enzymes, are considerably different for individuals in different age groups. Usually, newborns, particularly premature babies with aplasia of neonatal GI development, show slower gastric emptying than older children.[19] In contrast, an investigative study was conducted to determine the effects of age and other covariates on the rate of gastric emptying by analyzing 49 published research reports relating to 1457 persons from 28 weeks of gestation to adulthood.[99] It was concluded that age is not an important covariate and that diet type plays a significant role in gastric emptying.[99] Further, elderly men exhibited delayed gastric emptying at the half-time compared with young men.[100]

Several studies indicated that there is no direct correlation between gastric emptying and gender.[101] However, in another study, gastric emptying was slower for women in the premenopausal stage than for women in the postmenopausal stage and for men, because of the presence of estrogen.[102] Nevertheless, the impact of age and gender on gastric emptying is not understood completely and further research is required.

2.6.4.3 Gastrointestinal Features

Gastrointestinal hormones and gastric motility are modulated according to the feedback signals triggered by the ingested meal characteristics and nutrients present in the food. The nutrients in the meal deliver several signals to the brain that carry information about content and quantity. As a consequence, hormones are secreted in the GI tract to lessen meal consumption and promote satiety, to adjust the food intake. In addition, the endocrine cells in the gut produce a unique mix of satiety peptides according to the type (carbohydrates, fats, and proteins) of macronutrients in the food.[83,103] Further, the satiety signals from the gut reach the hindbrain through the gut–brain connection; as the signals intensify to a sufficient threshold, they inhibit the ingestion of food. Hence these sequence feedback mechanisms are linked to human gastric emptying.[104,105]

Various gastrointestinal hormones are released in response to the gut contents and also their time and degree of release are associated with several food-related gastric and intestinal signals such as macronutrient composition and calorie content of the meal. For instance, foods rich in fats or proteins signal the type I enteroendocrine cells to secrete CCK. The key role of CCK is to hinder food intake and gastric emptying by stimulating pancreatic enzyme secretion and gallbladder contractions.[106] Likewise, GLP-1 also plays a prominent role in hindering the gastric emptying process.[17]

In contrast, another gastrointestinal hormone, ghrelin (hunger hormone), is also connected with gastric emptying as it initiates hunger and hence meal consumption. Ghrelin (plasma) usually peaks before a normal meal and then falls during consumption. In addition, ghrelin also impacts gastrointestinal motility like many other gastrointestinal hormones and it also competes against the actions of leptin (inhibits hunger by regulating energy).[28] However, further research is required to understand more clearly the precise relationships between gastric signals and gastrointestinal hormone release in relation to satiety, hunger, and subsequently gastric emptying.

2.6.4.4 Medical Conditions

Gastroparesis means partial paralysis of the stomach, and is a medical condition that is also commonly known as delayed gastric emptying.[59] A person with gastroparesis cannot empty ingested food in a normal fashion as it affects and reduces normal gastric motility, which results in slower and delayed gastric emptying of food. The reasons for the occurrence of gastroparesis are not well known, but it is believed that it is caused as a result of damage to the vagus nerve (a nerve that controls the stomach muscles).[53] Gastroparesis is common in individuals with prolonged diabetes, gastric infections, hypothyroidism, nervous system diseases (Parkinson's disease and multiple sclerosis), and those who have undergone surgery in the stomach, esophagus, and small intestine.[59,107] Symptoms of gastroparesis include vomiting, stomach pain, abdominal bloating, nausea, acid reflux, lack of appetite, weight loss, and malnutrition.[108] It is evident that proper gastric emptying of foods is required for proper digestion and nutrient absorption.

2.7 Gastric Digestion Behavior of Foods

In general, all foods have unique hierarchal length scale-dependent structural features that impact the release, transport, and utilization of macro-

and micronutrients in the human GI tract.[109] The gastric digestion of food facilitates destructuring (gastric mixing, antral grinding, swelling, acid and enzyme hydrolysis) and restructuring (biopolymer interactions induced by pH and ionic strength changes and phase separation of gastric digesta) of foods.[14] Further, food destructuring and restructuring significantly affect the gastric digestion and nutrient absorption process by modulating gastric emptying and nutrient release. Raw foods have a nature-designed sophisticated structure, whereas processed foods undergo various structural modifications during processing. In addition, several processing techniques, such as roasting, boiling, frying, coagulation (acid and rennet), and heating, alter the structure of the foods significantly, which in turn impacts their gastric digestion and also nutrient bioavailability and bioaccessibility after digestion.[1,48] Moreover, understanding the effects of food structure on food destructuring and restructuring, transport, and nutrient release and absorption within the body will support the development of new products with enhanced functionalities. Table 2.3 surveys recent studies on the gastric digestion behavior of foods by highlighting the impact of processing methods and the significance of food structure in gastric digestion.

2.8 Conclusion and Future Perspectives

In recent times, there has been an upsurge in research to understand the food destructing and restructuring processes that occur during food digestion. In this chapter, we have tried to summarize the current knowledge and understanding regarding gastric food digestion and the impact of anatomical and physiological features of the stomach on food digestion. The gastric food disintegration process is highly complicated as it involves several different physicochemical, physical, and biochemical mechanisms. All the mechanisms are dependent on the food material properties and gastric physiological factors, which may vary depending on individual differences such as age, gender, and existing medical conditions. However, our understating of how the food-related factors (composition, structure, and processing method), gastric factors, and other individual differences affect the gastric digestion process remains incomplete. Moving forward, we need to develop more sophisticated approaches to understand the relationship between food material properties (such as composition and structure) and gastric food digestion, to quantify the mobility of water, enzymes, and acids during gastric digestion, the effect of gastric digestion on probiotic viability and nutrient availability, and also the impact of specific disease conditions on the gastric digestion process. Thus, precise and better knowledge regarding the gastric digestion of food and nutrient release mechanisms can be utilized in the future to develop foods with personalized nutritional profiles and improved functionality for targeted populations.

Table 2.3 Recent studies on the gastric digestion behavior of foods.

Food type	Scope of the study	Processing methods	Digestibility study	Other parameters studied	Significant interferences	Ref.
Meat protein (pig muscle proteins)	To evaluate the digestion rate and nutritional quality of pig muscle proteins in relation to different meat processing methods	Aging (days 1 and 4), mincing, and cooking at different temperatures (70, 100, 140 °C, to represent cooked ham, boiled meat, and grilled meat)	*In vitro* pepsin, trypsin, and α-chymotrypsin digestion. Initial slope, maximum degradation (rate and time), and half-life time	Antioxidant enzyme activities, glycolytic potential, biochemical parameters (carbonyls, myofibrillar protein hydrophobicity, lipid oxidation), granulometry parameters (protein aggregation), and cooking loss	Cooking temperature had a significant impact on the meat protein digestion rate and degradation potential among other processing methods. Heat treatment at 70 °C denatured the meat proteins, which in turn improved the enzyme accessibility and increased the rate of pepsin digestion. However, cooking above 100 °C resulted in enhanced overall meat protein digestibility but hindered pepsin digestion due to oxidation-related protein aggregation	115
Milk protein (caseins and β-lactoglobulin) Acid and rennet gels	To investigate the kinetics of milk protein digestion and amino acid absorption in gelled dairy matrices with different coagulation modes (acidification/renneting) but the	Two different modes of coagulation: acid coagulation (glucono-D-lactone 3% w/w) and rennet coagulation (rennet 0.3% v/w)	*In vivo* animal experiment (six multi-cannulated mini-pigs). Dynamic *in vitro* gastric acidification (visual monitoring)	Chemical analysis (effluents and plasma sampling to quantify amino acids, chromium, CCK, and ghrelin concentrations). Duodenal and jejunal samples were used to	In comparison with acid gels, rennet gels exhibited lower levels of residual milk proteins (caseins and β-lactoglobulin) and essential amino acids in the duodenum and plasma, respectively. This difference in digestion behavior was attributed to the higher resistance to	116

	same composition, and also similar rheological and structural features			quantify casein and β-lactoglobulin. Rheology (loss tangent, viscous modulus, and elastic modulus). Image analysis by confocal microscopy (gel microstructure)	hydrolysis of the rennet gel due to greater protein network shrinkage (when exposed to an acidic gastric environment), restricted accessibility of gastric enzymes to the proteins, prolonged gastric retention, and slower gastric emptying of the rennet gel. Additionally, the plasma concentrations of CCK and ghrelin demonstrated that the rennet gel had a greater satiating effect than the acid gel	
Almonds	To determine the effects of gastric protein hydrolysis of raw and roasted almonds on the rate of gastric emptying and subsequent small intestinal protein digestibility	Diced raw and hot-air roasted almonds (~130–154 °C)	*In vivo* animal experiment (male pigs)	Tricine: SDS-PAGE, peptide band analysis and gastric hydrolysis (%)	Raw and roasted almonds showed an increasing trend of gastric protein hydrolysis with increase in gastric digestion time. However, raw almonds showed significantly higher gastric protein hydrolysis and in turn a faster rate of gastric protein emptying compared with roasted almonds irrespective of the difference in the biochemical environment (pH and enzyme activity) in various regions of the stomach (proximal and distal regions).	93 and 117

Table 2.3 (*Continued*)

Food type	Scope of the study	Processing methods	Digestibility study	Other parameters studied	Significant interferences	Ref.
					Hence the observed variation in the gastric digestion behavior of almonds is due to the structural modification and disruption of protein (across various structural length scales) as a result of the roasting process (high temperature). Further, modifications of the almond macro-, micro-, and protein structure lower the penetration of gastric secretions into the almond matrix and hinder the hydrolysis of almond proteins	
Bread (six varieties: almond wheat, barley, rye, wheat, sourdough, and white bread)	To determine the key features (flour type, oral and gastric conditions) responsible for the disintegration of bread boluses in the gastric environment	Various ingredients (flour and others) were kneaded into dough and baked for 65 min at 125 °C, then cooled (at room temperature for 1 h) to prepare bread	*In vitro* digestion: simulated oral processing (saliva level and α-amylase) and gastric digestion (static and agitated soaking in gastric secretions under various pH and pepsin conditions)	Bread property analysis (moisture content, firmness, and water holding capacity). Bolus property analysis (moisture content, salivary impregnation and cohesive force). Mass retention test	The disintegration kinetics of various bread boluses were significantly impacted by the bread type (flour type, bread structure, and moisture content) during *in vitro* digestion (oral and gastric conditions). Further, the mass retention kinetics of bread boluses followed three distinct gastric	118

disintegration profiles: exponential (almond wheat, barley, and wheat bread), sigmoidal (white bread), and delayed sigmoidal (rye and sourdough bread). Hence the key factor that influenced the disintegration profile is the variation in the water absorption of the bread boluses, which occurred due to the difference in bread structure, water holding capacity, and moisture content. Moreover, oral processing conditions such as the presence of α-amylase (starch digestion) and an increase in the saliva level decreased the bolus cohesive force, which in turn increased the bolus disintegration rate. In the case of gastric conditions, the disintegration kinetics were greatly influenced by the hydrodynamic forces and pepsin, but the pH conditions (addition of acids) had only a limited effect

Table 2.3 (*Continued*)

Food type	Scope of the study	Processing methods	Digestibility study	Other parameters studied	Significant interferences	Ref.
Hazelnut particles (1–2 mm), oil bodies (both raw and roasted), defatted hazelnut particles, and hazelnut oil	To explore the effect of cell walls on the rate and extent of lipid hydrolysis during *in vitro* digestion of raw and roasted hazelnuts	Hazelnuts were roasted at 140 °C for 20 min. Further, both raw and roasted nuts were mechanically dry-grained and sieved to obtain particles of 1–2 mm. Oil body cream (oleosomes) was extracted from both raw and roasted hazelnuts after a rigorous extraction and purification process	*In vitro* digestion using a slightly modified version of the harmonized INFOGEST method	Characterization of hazelnuts (lipid content, moisture content, and protein content and composition). Hazelnut oil body characterization [average droplet size (light scattering technique), surface charge (zeta potential), refractive index ratio, mean particle diameter]. Cryo-scanning electron microscopy (cryo-SEM) analysis: microstructural changes	This study investigated the barrier role of cell structural factors (physical encapsulation of nutrients by the plant cell wall) and the impact of heat processing (roasting) on lipid hydrolysis and its bioaccessibility in hazelnuts during *in vitro* digestion of raw and roasted hazelnut particles and isolated oil bodies. Additionally, compared with other oil bodies, the oil bodies from roasted hazelnuts displayed a much higher rate of lipid hydrolysis. Further, the oil bodies extracted from roasted hazelnuts showed a significantly higher lipid hydrolysis rate than others. Thus, in the case of isolated oil bodies, the heat treatment resulted in destabilization of the interfacial protein membrane, which facilitated the proteolysis of oil bodies under gastric conditions and also enhanced fatty acid	119

					release during intestinal digestion. Furthermore, the structure of the hazelnut cell walls was considerably modified during the roasting process, which led to more effective diffusion of gastric secretions (digestive enzymes and fluids) into the hazelnut cells	
Cheese (stabilized Camembert, smear cheese, young Cheddar, aged Cheddar, and string mozzarella)	To investigate the digestion profiles, matrix degradation, and nutrient release of various commercial cheeses that differ in texture	The processing/manufacturing conditions and techniques varied for each variety of cheese, which in turn resulted in diverse textures and compositions	The *in vitro* digestion model (maintained at 37 ± 1 °C) consists of the oral phase (2 min, pH 6.8, amylase), gastric phase (10 min, pH 2–3, pepsin, 200 rpm), and duodenal phase (180 min, pH 6–7, pancreatin and lipase, 200 rpm)	Mass balance determination, cheese disintegration, and protein/peptide profile (SDS-PAGE) and release	The great variations in cheese texture and composition had a significant influence on cheese disintegration and protein digestion patterns during gastric digestion. The cheeses with higher initial hardness, cohesiveness, and chewiness (elastic cheese: mozzarella) showed relatively lower matrix disintegration at the end of gastric digestion compared with ripened and soft cheeses (aged Cheddar and Camembert, respectively). Further, the extent of proteolysis, nitrogen release, and fat content also affected the cheese's textural characteristics, which in	120

Table 2.3 (*Continued*)

Food type	Scope of the study	Processing methods	Digestibility study	Other parameters studied	Significant interferences	Ref.
					turn accelerated the disintegration of the cheese matrix. Thus, the structure and texture of the cheese matrix can be controlled by adjusting the cheese processing parameters and composition, which will help in modulating the digestion of cheese proteins	
Milk protein matrices (unheated and heated skim milk and their rennet gels)	To investigate the impact of the structural modification of dairy products using different technological processes on the digestion kinetics of the main milk proteins (caseins and β-lactoglobulin), without altering the biochemical composition of various matrices	Thermal treatment (heating for 10 min at 90 °C) and a gelation process (rennet, 24 h at 20 °C)	*In vivo* animal experiment (six female adult mini-pigs)	Chemical analyses: total nitrogen, non-protein nitrogen, and non-casein nitrogen (Kjeldahl method) Rheological measurements (elastic modulus, viscous modulus, and loss tangent) and compression test for rennet milk gels. Image analysis (gel microstructure) Protein profiling and mean retention test	Heat treatment considerably modified the microstructure, which had a significant influence on milk protein hydrolysis and also affected the mean retention time of meals in the stomach. In the case of unheated matrices, casein was sensitive and β-lactoglobulin was resistant to gastric hydrolysis. However, both caseins and β-lactoglobulin (heat-induced susceptibility) displayed similar digestion behavior with the heated matrices.	121

					Further, variations of the matrix structure at the macro level (*i.e.* liquid/gelled matrices) had a significant impact on gastric emptying, amino acid absorption, and bioavailability. The heated rennet milk gels had a higher mean retention time, delayed gastric outflow, increased amino acid absorption, and lower bioavailability compared with other matrices, since the combination of the stomach's acidity and the rennet caused it to form a coagulum with a high degree of stiffness	
Protein gels (soy protein, pea protein, albumin from chicken egg white, and whey protein isolate)	To study the influence of the protein source, state (liquid and gel), and gel microstructure acquired at various temperatures (heat-induced) on gastric digestibility (*in vitro*)	The gels were prepared from various protein sources at three temperatures (90, 120, and 140 °C) for 30 min	Simulated *in vitro* gastric digestion (pH 2)	Gel microstructure (SEM). Texture analysis: compression tests (resistance to penetration and hardness). Molecular characterization/protein profiling (SDS-PAGE). High-performance size-exclusion chromatography (HPSEC).	The protein source, state, gelation temperature, gel microstructure, and mechanical strength had a significant impact on the *in vitro* gastric protein digestibility. Further, the gels induced at higher temperature (140 °C) were digested faster than gels induced at lower temperatures. However, the albumin gels were slow hydrolyzers regardless of the gelation temperature,	122

Table 2.3 (*Continued*)

Food type	Scope of the study	Processing methods	Digestibility study	Other parameters studied	Significant interferences	Ref.
				Degree of hydrolysis (free amino groups: *o*-phthaldialdehyde assay method)	even though gels formed at higher temperatures showed faster hydrolysis. Moreover, both soy and pea proteins were digested faster in the solution state compared with their gelled counterparts. However, pea protein gels were hydrolyzed faster in gastric conditions than soy protein gels, and this difference in digestion behavior is associated with the mechanical strength, porosity, and microstructures of the gels	
Milk (unheated fresh milk, pasteurized and UHT homogenized milk)	To explore the impact of different milk processing techniques on the physicochemical behavior and the curd microstructures	Unheated fresh milk (homogenized at 20/4 MPa and 50 °C using a two-stage valve homogenizer), commercial pasteurized milk	*In vitro* dynamic human gastric simulator and *in vivo* rat stomach	pH analysis of samples (milk and curds) and curd analysis (weight, fat content, protein hydrolysis, and microstructure)	This study showed that the coagulation of protein, curd structure, rate of protein breakdown, and release of fat globules from the stomach are all significantly affected by	123

formed during gastric digestion	(85 °C for 15 s), and UHT milk (140 °C for 4 s)		both homogenization and heat treatment of milk. Whereas unheated milk formed a firm clot, heat-treated milk (pasteurized and UHT) created a soft curd with a loose structure and larger and more numerous voids within the curd. The UHT-treated milk showed higher rates of protein hydrolysis (digestion) and release of fat globules during digestion in comparison with others (unheated fresh and pasteurized milk) due to its excessively fragmented and crumbled curd structure (casein micelle aggregation was sterically prevented by β-lactoglobulin binding with the micelles under gastric conditions), which generated a larger surface area and enhanced rapid diffusion of pepsin into the curd

Table 2.3 (*Continued*)

Food type	Scope of the study	Processing methods	Digestibility study	Other parameters studied	Significant interferences	Ref.
Whey protein gels	To explore the effect of swelling of solid foods during gastric digestion on digestion kinetics	Whey protein isolate gels were prepared with water (15 wt%) and various concentrations of NaCl (0, 0.01, 0.05, and 0.1 M) at neutral pH	*In vitro* gastric digestion (37 °C and stirring at 100 rpm)	Rheological characteristics of gels (Young's modulus). Free amino acid groups, mass change, swelling ratio (%), acid uptake, and pepsin (green fluorescent protein transport)	This study emphasized the significance of swelling on the gastric digestion behavior of protein gels. The digestion rate of the gels decreased as the NaCl concentration increased (0–0.1 M) due to the reduction in swelling and increased hardness (Young's modulus). Further, the swelling behavior of food in the gastric environment affected the rate of the digestion process by enhancing the acid diffusion and by controlling the partitioning property of pepsin at the food–gastric fluid interface (*i.e.* increasing the concentration of pepsin in the food particles). Hence the swelling properties of food particles play a crucial role in the gastric digestion of foods	124

References

1. H. Singh, A. Ye and M. J. Ferrua, Aspects of food structures in the digestive tract, *Curr. Opin. Food Sci.*, 2015, **3**, 85–93.
2. G. M. Bornhorst and R. Paul Singh, Gastric digestion *in vivo* and *in vitro*: how the structural aspects of food influence the digestion process, *Annu. Rev. Food Sci. Technol.*, 2014, **5**, 111–132.
3. H. W. Davenport, Gastrointestinal physiology, 1895–1975: motility, *Compr. Physiol.*, 2010, 1–101.
4. J. J. Feher, *Quantitative human physiology: an introduction*, Academic Press, 2017.
5. P. Sethupathy, J. A. Moses and C. Anandharamakrishnan, Food Oral Processing and Tribology: Instrumental Approaches and Emerging Applications, *Food Rev. Int.*, 2020, 1–34.
6. J. Chen, Food oral processing: Mechanisms and implications of food oral destruction, *Trends Food Sci. Technol.*, 2015, **45**(2), 222–228.
7. O. Gouseti, G. M. Bornhorst, S. Bakalis and A. Mackie, *Interdisciplinary Approaches to Food Digestion*, Springer, 2019.
8. I. R. Daniels and W. H. Allum, The anatomy and physiology of the stomach, in *Upper Gastrointestinal Surgery*, Springer, 2005, pp. 17–37.
9. M. J. Ferrua and R. P. Singh, Modeling the fluid dynamics in a human stomach to gain insight of food digestion, *J. Food Sci.*, 2010, **75**(7), R151–R162.
10. G. M. Bornhorst, O. Gouseti, M. S. J. Wickham and S. Bakalis, Engineering digestion: multiscale processes of food digestion, *J. Food Sci.*, 2016, **81**(3), R534–R543.
11. C. Li, W. Yu, P. Wu and X. D. Chen, Current *in vitro* digestion systems for understanding food digestion in human upper gastrointestinal tract, *Trends Food Sci. Technol.*, 2020, **96**, 114–126.
12. J.-R. Malagelada and F. Azpiroz, Determinants of gastric emptying and transit in the small intestine, *Compr. Physiol.*, 2010, 909–937.
13. W. Schwizer, A. Steingoetter and M. Fox, Magnetic resonance imaging for the assessment of gastrointestinal function, *Scand. J. Gastroenterol.*, 2006, **41**(11), 1245–1260.
14. Q. Guo, A. Ye, H. Singh and D. Rousseau, Destructuring and restructuring of foods during gastric digestion, *Compr. Rev. Food Sci. Food Saf.*, 2020, **19**(4), 1658–1679.
15. D. I. Soybel, Anatomy and physiology of the stomach, *Surg. Clin.*, 2005, **85**(5), 875–894.
16. R. L. Wilson and C. E. Stevenson, Anatomy and physiology of the stomach, in *Shackelford's Surgery of the Alimentary Tract, 2 Volume Set*, Elsevier, 2019, pp. 634–646.
17. R. K. Goyal, Y. Guo and H. Mashimo, Advances in the physiology of gastric emptying, *Neurogastroenterol. Motil.*, 2019, **31**(4), e13546.
18. J. Punkkinen, I. Konkka, O. Punkkinen, T. Korppi-Tommola, M. Färkkilä and J. Koskenpato, Measuring gastric emptying: comparison of

13 C-octanoic acid breath test and scintigraphy, *Dig. Dis. Sci.*, 2006, **51**(2), 262–267.
19. W. Liu, Y. Jin, P. J. Wilde, Y. Hou, Y. Wang and J. Han, Mechanisms, physiology, and recent research progress of gastric emptying, *Crit. Rev. Food Sci. Nutr.*, 2020, 1–14.
20. G. J. Tortora and B. H. Derrickson, *Principles of Anatomy and Physiology*, John Wiley & Sons, 2018.
21. G. M. Bornhorst and R. P. Singh, Bolus formation and disintegration during digestion of food carbohydrates, *Compr. Rev. Food Sci. Food Saf.*, 2012, **11**(2), 101–118.
22. G. Somaratne, F. Nau, M. J. Ferrua, J. Singh, A. Ye and D. Dupont, *et al.*, Characterization of egg white gel microstructure and its relationship with pepsin diffusivity, *Food Hydrocolloids*, 2020, **98**, 105258.
23. K. A. Kelly, Gastric emptying of liquids and solids: roles of proximal and distal stomach, *Am. J. Physiol.: Gastrointest. Liver Physiol.*, 1980, **239**(2), G71–G76.
24. P. M. Treuting, S. M. Dintzis and K. S. Montine, *Comparative Anatomy and Histology: A Mouse, Rat, and Human Atlas*, Academic Press, 2017.
25. G. M. Bornhorst, Gastric mixing during food digestion: Mechanisms and applications, *Annu. Rev. Food Sci. Technol.*, 2017, **8**, 523–542.
26. D. O. Castell, J. A. Murray, R. Tutuian, R. C. Orlando and R. Arnold, The pathophysiology of gastro-oesophageal reflux disease-oesophageal manifestations, *Aliment. Pharmacol. Ther.*, 2004, **20**, 14–25.
27. K. Schulze, Imaging and modelling of digestion in the stomach and the duodenum, *Neurogastroenterol. Motil.*, 2006, **18**(3), 172–183.
28. P. M. Hellström, P. Grybäck and H. Jacobsson, The physiology of gastric emptying, *Best Pract. Res. Clin. Anaesthesiol.*, 2006, **20**(3), 397–407.
29. R. Wyllie, J. S. Hyams and M. Kay, *Pediatric Gastrointestinal and Liver Disease E-Book*, Elsevier, 2020.
30. P. Greaves, Digestive System, in *Histopathology of Preclinical Toxicity Studies*, ed. P. Greaves, Academic Press, Boston, 2012, ch. 8, pp. 325–431.
31. Y. Li, L. Fortner and F. Kong, Development of a Gastric Simulation Model (GSM) incorporating gastric geometry and peristalsis for food digestion study, *Food Res. Int.*, 2019, **125**, 108598.
32. S. Ishida, T. Miyagawa, G. O'Grady, L. K. Cheng and Y. Imai, Quantification of gastric emptying caused by impaired coordination of pyloric closure with antral contraction: a simulation study, *J. R. Soc., Interface*, 2019, **16**(157), 20190266.
33. M. Camilleri and M. I. Vazquez Roque, Gastrointestinal Functions, in *Encyclopedia of the Neurological Sciences*, ed. M. J. Aminoff and R. B. Daroff, Academic Press, Oxford, 2nd edn, 2014, pp. 411–416. Available from: http://www.sciencedirect.com/science/article/pii/B9780123851574005017
34. B. M. Carlson, The Digestive System, in *The Human Body*, ed. B. M. Carlson, Academic Press, 2019, ch. 8, pp. 321–355. Available from: http://www.sciencedirect.com/science/article/pii/B9780128042540000120

35. E. K. Ulleberg, I. Comi, H. Holm, E. B. Herud, M. Jacobsen and G. E. Vegarud, Human gastrointestinal juices intended for use in *in vitro* digestion models, *Food Dig.*, 2011, **2**(1–3), 52–61.
36. W. D. W. Rees, V. L. W. Go and J.-R. Malagelada, Antroduodenal motor response to solid-liquid and homogenized meals, *Gastroenterology*, 1979, **76**(6), 1438–1442.
37. L. Marciani, P. A. Gowland, R. C. Spiller, P. Manoj, R. J. Moore and P. Young, *et al.*, Effect of meal viscosity and nutrients on satiety, intragastric dilution, and emptying assessed by MRI, *Am. J. Physiol.: Gastrointest. Liver Physiol.*, 2001, **280**(6), G1227–G1233.
38. S. Mattioli, V. Felice, V. Pilotti, M. L. Bacchi, M. Pastina and G. Gozzetti, Indications for 24-hour gastric pH monitoring with single and multiple probes in clinical research and practice, *Dig. Dis. Sci.*, 1992, **37**(12), 1793–1801.
39. A. Lanas and B. T. Htrschowitz, Influence of smoking on basal and on vagally and maximally stimulated gastric acid and pepsin secretion, *Scand. J. Gastroenterol.*, 1992, **27**(3), 208–212.
40. M. Feldman, B. Cryer, K. E. McArthur, B. A. Huet and E. Lee, Effects of aging and gastritis on gastric acid and pepsin secretion in humans: a prospective study, *Gastroenterology*, 1996, **110**(4), 1043–1052.
41. O. Oektedalen, I. Guldvog, P. K. Opstad, A. Berstad, D. Gedde-Dahl and R. Jorde, The effect of physical stress on gastric secretion and pancreatic polypeptide levels in man, *Scand. J. Gastroenterol.*, 1984, **19**(6), 770–778.
42. M. Rubach, R. Lang, G. Bytof, H. Stiebitz, I. Lantz and T. Hofmann, *et al.*, A dark brown roast coffee blend is less effective at stimulating gastric acid secretion in healthy volunteers compared to a medium roast market blend, *Mol. Nutr. Food Res.*, 2014, **58**(6), 1370–1373.
43. H. T. Debas, M. M. Cohen, I. B. Holubitsky and R. C. Harrison, Caffeine-stimulated acid and pepsin secretion: dose-response studies, *Scand. J. Gastroenterol.*, 1971, **6**(5), 453–457.
44. J. G. Moore, P. E. Christian, J. A. Brown, C. Brophy, F. Datz and A. Taylor, *et al.*, Influence of meal weight and caloric content on gastric emptying of meals in man, *Dig. Dis. Sci.*, 1984, **29**(6), 513–519.
45. M. Camilleri, Integrated upper gastrointestinal response to food intake, *Gastroenterology*, 2006, **131**(2), 640–658.
46. J. Keller, Gastrointestinal Digestion and Absorption, in *Encyclopedia of Biological Chemistry*, ed. W. J. Lennarz and M. D. Lane, Academic Press, Waltham, 2nd edn, 2013, pp. 354–359. Available from: http://www.sciencedirect.com/science/article/pii/B9780123786302001067
47. P.-J. Lu, P.-I. Hsu, C.-H. Chen, M. Hsiao, W.-C. Chang and H.-H. Tseng, *et al.*, Gastric juice acidity in upper gastrointestinal diseases, *World J. Gastroenterol.: WJG*, 2010, **16**(43), 5496.
48. G. Somaratne, M. J. Ferrua, A. Ye, F. Nau, J. Floury and D. Dupont, *et al.*, Food material properties as determining factors in nutrient release during human gastric digestion: a review, *Crit. Rev. Food Sci. Nutr.*, 2020, 1–17.

49. M. Minekus, M. Alminger, P. Alvito, S. Ballance, T. Bohn and C. Bourlieu, *et al.*, A standardised static *in vitro* digestion method suitable for food–an international consensus, *Food Funct.*, 2014, **5**(6), 1113–1124.
50. R. M. Arin, A. Gorostidi, H. Navarro-Imaz, Y. Rueda, O. Fresnedo and B. Ochoa, Adenosine: direct and indirect actions on gastric acid secretion, *Front. Physiol.*, 2017, **8**, 737.
51. S. J. Czinn and S. S. Blanchard, 25 - Developmental Anatomy and Physiology of the Stomach, in *Pediatric Gastrointestinal and Liver Disease*, ed. R. Wyllie and J. S. Hyams, W.B. Saunders, Saint Louis, 4th edn, 2011. pp. 262–268.e1. Available from: http://www.sciencedirect.com/science/article/pii/B9781437707748100259
52. S. Brandstaeter, S. L. Fuchs, R. C. Aydin and C. J. Cyron, Mechanics of the stomach: A review of an emerging field of biomechanics, *GAMM-Mitteilungen*, 2019, **42**(3), e201900001.
53. K. E. Jarrett and R. E. Glasgow, Normal Gastric Motility, in *Gastroparesis*, Springer, 2020, pp. 3–20.
54. A. Husainy Yusuf, L. Adamu, M. Babakura, H. A. Salami, S. Nuhu and U. K. Sandabe, Advances in the Measurement of Gastric Motility: A Review, *IOSR Journal of Dental and Medical Sciences Ver VIII January*, 2018, **17**, 1–14. Available fromwww.iosrjournals.org.
55. J. Feher, 8.5 Digestion and Absorption of the Macronutrients, *Quantitative Human Physiology*, Academic Press, Boston, 2017.
56. A. Vella and M. Camilleri, The gastrointestinal tract as an integrator of mechanical and hormonal response to nutrient ingestion, *Diabetes*, 2017, **66**(11), 2729–2737.
57. M. Camilleri and M. I. Vazquez-Roque, Gastric dysmotility at the organ level in gastroparesis, in *Gastroparesis*, Springer, 2012, pp. 37–46.
58. E. W. Englander and G. H. Greeley, Postpyloric Gastrointestinal Peptides, in *Physiology of the Gastrointestinal Tract*, ed. L. R. Johnson, Academic Press, Burlington, 4th edn, 2006, ch. 5, pp. 121–159. Available from: http://www.sciencedirect.com/science/article/pii/B9780120883943500088
59. A. Ibele and J. Gould, *Gastroparesis: A Comprehensive Approach to Evaluation and Management*, Springer, 2019.
60. C. Feinle-Bisset, D. O'Donovan and M. Horowitz, Gastric Motility, in *Encyclopedia of Gastroenterology*, ed. L. R. Johnson, Elsevier, New York, 2004, pp. 132–138. Available from: http://www.sciencedirect.com/science/article/pii/B012386860200304X
61. G. W. Hennig and N. J. Spencer, Physiology of Gastric Motility Patterns, in *Physiology of the Gastrointestinal Tract*, ed. H. M. Said, Academic Press, 6th edn, 2018, ch. 21, pp. 469–484. Available from: http://www.sciencedirect.com/science/article/pii/B9780128099544000219
62. H. P. Parkman, P. Orthey and A. H. Maurer, Gastric Motility, in *Encyclopedia of Gastroenterology*, ed. E. J. Kuipers, Academic Press, Oxford, 2nd edn, 2020, pp. 598–612. Available from: http://www.sciencedirect.com/science/article/pii/B978012801238311092X

63. K. Indireshkumar, J. G. Brasseur, H. Faas, G. S. Hebbard, P. Kunz and J. Dent, *et al.*, Relative contributions of "pressure pump" and "peristaltic pump" to gastric emptying, *Am. J. Physiol.: Gastrointest. Liver Physiol.*, 2000, **278**(4), G604–G616.
64. L. Marciani, P. A. Gowland, A. Fillery-Travis, P. Manoj, J. Wright and A. Smith, *et al.*, Assessment of antral grinding of a model solid meal with echo-planar imaging, *Am. J. Physiol.: Gastrointest. Liver Physiol.*, 2001, **280**(5), G844–G849.
65. Y. A. Mennah-Govela, G. M. Bornhorst and R. P. Singh, Acid diffusion into rice boluses is influenced by rice type, variety, and presence of \$α\$-amylase, *J. Food Sci.*, 2015, **80**(2), E316–E325.
66. M. J. Ferrua, Z. Xue and R. P. Singh, On the kinematics and efficiency of advective mixing during gastric digestion–A numerical analysis, *J. Biomech.*, 2014, **47**(15), 3664–3673.
67. A. Pal, K. Indireshkumar, W. Schwizer, B. Abrahamsson, M. Fried and J. G. Brasseur, Gastric flow and mixing studied using computer simulation, *Proc. R. Soc. London, Ser. B*, 2004, **271**(1557), 2587–2594.
68. S. M. Harrison, P. W. Cleary and M. D. Sinnott, Investigating mixing and emptying for aqueous liquid content from the stomach using a coupled biomechanical-SPH model, *Food Funct.*, 2018, **9**(6), 3202–3219.
69. A. Steingoetter, M. Sauter, J. Curcic, D. Liu, D. Menne and M. Fried, *et al.*, Volume, distribution and acidity of gastric secretion on and off proton pump inhibitor treatment: a randomized double-blind controlled study in patients with gastro-esophageal reflux disease (GERD) and healthy subjects, *BMC Gastroenterol.*, 2015, **15**(1), 111.
70. G. M. Bornhorst, S. M. Rutherfurd, M. J. Roman, B. J. Burri, P. J. Moughan and R. P. Singh, Gastric pH distribution and mixing of soft and rigid food particles in the stomach using a dual-marker technique, *Food Biophys.*, 2014, **9**(3), 292–300.
71. A. Mößeler, S. Köttendorf, V. G. Liesner and J. Kamphues, Impact of diets' physical form (particle size; meal/pelleted) on the stomach content (dry matter content, pH, chloride concentration) of pigs, *Livestock Sci.*, 2010, **134**(1–3), 146–148.
72. A. K. Mößeler, M. F. Wintermann, M. Beyerbach and J. Kamphues, Effects of grinding intensity and pelleting of the diet–fed either dry or liquid–on intragastric milieu, gastric lesions and performance of swine, *Anim. Feed Sci. Technol.*, 2014, **194**, 113–120.
73. S. Kumar, H. J. Kim and A. Beskok, Numerical simulations of peristaltic mixing, *J. Fluids Eng.*, 2007, **129**(11), 1361–1371.
74. F. Marra, M. J. Ferrua and R. P. Singh, Experimental characterization of the fluid dynamics in an *in-vitro* system simulating the peristaltic movement of the stomach wall, *Proc. Food Sci.*, 2011, **1**, 1473–1478.
75. C. V. Maxwell, E. M. Reimann, W. G. Hoekstra, T. Kowalczyk, N. J. Benevenga and R. H. Grummer, Use of tritiated water to assess, *in vivo*, the effect of dietary particle size on the mixing of stomach contents of swine, *J. Anim. Sci.*, 1972, **34**(2), 212–216.

76. K. D. Ange, J. H. Eisemann, R. A. Argenzio, G. W. Almond and A. T. Blikslager, Effects of feed physical form and buffering solutes on water disappearance and proximal stomach pH in swine, *J. Anim. Sci.*, 2000, **78**(9), 2344–2352.
77. C. Rayner, G. Hebbard and M. Horowitz, *Physiology of the Antral Pump and Gastric Emptying*, Elsevier, 2012.
78. M. L. Schubert and D. A. Peura, Control of gastric acid secretion in health and disease, *Gastroenterology*, 2008, **134**(7), 1842–1860.
79. A. Mitra and F. Kesisoglou, Impaired drug absorption due to high stomach pH: a review of strategies for mitigation of such effect to enable pharmaceutical product development, *Mol. Pharmaceutics*, 2013, **10**(11), 3970–3979.
80. J. K. Aronson, Inhibiting the proton pump: mechanisms, benefits, harms, and questions, *BMC Med.*, 2016, **14**(1), 1–4.
81. S. Bellmann, J. Lelieveld, T. Gorissen, M. Minekus and R. Havenaar, Development of an advanced *in vitro* model of the stomach and its evaluation *versus* human gastric physiology, *Food Res. Int.*, 2016, **88**, 191–198.
82. R. K. Goyal, V. Cristofaro and M. P. Sullivan, Rapid gastric emptying in diabetes mellitus: Pathophysiology and clinical importance, *J. Diabetes Complications*, 2019, **33**(11), 107414.
83. M. Camilleri, Gastrointestinal hormones and regulation of gastric emptying, *Curr. Opin. Endocrinol., Diabetes Obes.*, 2019, **26**(1), 3.
84. H. Y. Chang, H. Mashimo and R. K. Goyal, IV. Current concepts of vagal efferent projections to the gut, *Am. J. Physiol.: Gastrointest. Liver Physiol.*, 2003, **284**(3), G357–G366.
85. A. H. Maurer, Advancing gastric emptying studies: standardization and new parameters to assess gastric motility and function, in *Seminars in Nuclear Medicine*, 2012, pp. 101–112.
86. L. Lundin, M. Golding and T. J. Wooster, Understanding food structure and function in developing food for appetite control, *Nutr. Diet.*, 2008, **65**, S79–S85.
87. A. R. Mackie, H. Rafiee, P. Malcolm, L. Salt and G. van Aken, Specific food structures supress appetite through reduced gastric emptying rate, *Am. J. Physiol.: Gastrointest. Liver Physiol.*, 2013, **304**(11), G1038–G1043.
88. F. Cisse, D. P. Erickson, A. M. R. Hayes, A. R. Opekun, B. L. Nichols and B. R. Hamaker, Traditional malian solid foods made from sorghum and millet have markedly slower gastric emptying than rice, potato, or pasta, *Nutrients*, 2018, **10**(2), 124.
89. Y. Zhu, W. H. Hsu and J. H. Hollis, The impact of food viscosity on eating rate, subjective appetite, glycemic response and gastric emptying rate, *PLoS One*, 2013, **8**(6), e67482.
90. G. Camps, M. Mars, C. De Graaf and P. A. M. Smeets, Empty calories and phantom fullness: a randomized trial studying the relative effects of energy density and viscosity on gastric emptying determined by MRI and satiety, *Am. J. Clin. Nutr.*, 2016, **104**(1), 73–80.

91. T. Mazzawi, E. Bartsch, S. Benammi, R. M. C. Ferro, E. Nikitina and N. Nimer, *et al.*, Gastric emptying of low-and high-caloric liquid meals measured using ultrasonography in healthy volunteers, *Ultrasound Int. Open*, 2019, **5**(1), E27.
92. M. A. Kwiatek, D. Menne, A. Steingoetter, O. Goetze, Z. Forras-Kaufman and E. Kaufman, *et al.*, Effect of meal volume and calorie load on postprandial gastric function and emptying: studies under physiological conditions by combined fiber-optic pressure measurement and MRI, *Am. J. Physiol.: Gastrointest. Liver Physiol.*, 2009, **297**(5), G894–G901.
93. G. M. Bornhorst, K. C. Drechsler, C. A. Montoya, S. M. Rutherfurd, P. J. Moughan and R. P. Singh, Gastric protein hydrolysis of raw and roasted almonds in the growing pig, *Food Chem.*, 2016, **211**, 502–508.
94. G. M. Bornhorst, L. Q. Chang, S. M. Rutherfurd, P. J. Moughan and R. P. Singh, Gastric emptying rate and chyme characteristics for cooked brown and white rice meals *in vivo*, *J. Sci. Food Agric.*, 2013, **93**(12), 2900–2908.
95. A. Geliebter, C. L. Grillot, R. Aviram-Friedman, S. Haq, E. Yahav and S. A. Hashim, Effects of oatmeal and corn flakes cereal breakfasts on satiety, gastric emptying, glucose, and appetite-related hormones, *Ann. Nutr. Metab.*, 2015, **66**(2–3), 93–103.
96. D. Y. Low, A. M. Pluschke, W. J. J. Gerrits, D. Zhang, K. J. Shelat and M. J. Gidley, *et al.*, Cereal dietary fibres influence retention time of digesta solid and liquid phases along the gastrointestinal tract, *Food Hydrocolloids*, 2020, **104**, 105739.
97. L. Shang, Y. Wang, Y. Ren, T. Ai, P. Zhou and L. Hu, *et al.*, *In vitro* gastric emptying characteristics of konjac glucomannan with different viscosity and its effects on appetite regulation, *Food Funct.*, 2020, **11**(9), 7596–7610.
98. S. Camacho, S. Michlig, C. de Senarclens-Bezençon, J. Meylan, J. Meystre and M. Pezzoli, *et al.*, Anti-obesity and anti-hyperglycemic effects of cinnamaldehyde *via* altered ghrelin secretion and functional impact on food intake and gastric emptying, *Sci. Rep.*, 2015, **5**, 7919.
99. J. J. Bonner, P. Vajjah, K. Abduljalil, M. Jamei, A. Rostami-Hodjegan and G. T. Tucker, *et al.*, Does age affect gastric emptying time? A model-based meta-analysis of data from premature neonates through to adults, *Biopharm. Drug Dispos.*, 2015, **36**(4), 245–257.
100. A. Brogna, M. Loreno, F. Catalano, A. M. Bucceri, M. Malaguarnera and L. A. Muratore, *et al.*, Radioisotopic assessment of gastric emptying of solids in elderly subjects, *Aging: Clin. Exp. Res.*, 2006, **18**(6), 493–496.
101. S. Hellmig, F. Von Schöning, C. Gadow, S. Katsoulis, J. Hedderich and U. R. Fölsch, *et al.*, Gastric emptying time of fluids and solids in healthy subjects determined by 13C breath tests: influence of age, sex and body mass index, *J. Gastroenterol. Hepatol.*, 2006, **21**(12), 1832–1838.
102. Y. T. Wang, S. D. Mohammed, A. D. Farmer, D. Wang, N. Zarate and A. R. Hobson, *et al.*, Regional gastrointestinal transit and pH studied in

215 healthy volunteers using the wireless motility capsule: influence of age, gender, study country and testing protocol, *Aliment. Pharmacol. Ther.*, 2015, **42**(6), 761–772.
103. M. Clegg and A. Shafat, Energy and macronutrient composition of breakfast affect gastric emptying of lunch and subsequent food intake, satiety and satiation, *Appetite*, 2010, **54**(3), 517–523.
104. S. C. Woods, Gastrointestinal satiety signals I. An overview of gastrointestinal signals that influence food intake, *Am. J. Physiol.: Gastrointest. Liver Physiol.*, 2004, **286**(1), G7–G13.
105. S. C. Woods, Signals that influence food intake and body weight, *Physiol. Behav.*, 2005, **86**(5), 709–716.
106. G. J. Dockray, Cholecystokinin and gut–brain signalling, *Regul. Pept.*, 2009, **155**(1–3), 6–10.
107. H. M. Said, *Physiology of the Gastrointestinal Tract, Two Volume Set*, Academic Press, 2012.
108. M. Camilleri, V. Chedid, A. C. Ford, K. Haruma, M. Horowitz and K. L. Jones, *et al.*, Gastroparesis, *Nat. Rev. Dis. Primers*, 2018, **4**(1), 1–19.
109. M. Boland, M. Golding and H. Singh, *Food structures, digestion and health*, Academic Press, 2014.
110. P. Wu, R. Deng, X. Wu, Y. Wang, Z. Dong and S. Dhital, *et al.*, In vitro gastric digestion of cooked white and brown rice using a dynamic rat stomach model, *Food Chem.*, 2017, **237**, 1065–1072.
111. O. Ménard, M.-H. Famelart, A. Deglaire, Y. Le Gouar, S. Guérin and C.-H. Malbert, *et al.*, Gastric emptying and dynamic *in vitro* digestion of drinkable yogurts: effect of viscosity and composition, *Nutrients*, 2018, **10**(9), 1308.
112. R. Gopirajah, K. P. Raichurkar, R. Wadhwa and C. Anandharamakrishnan, The glycemic response to fibre rich foods and their relationship with gastric emptying and motor functions: An MRI study, *Food Funct.*, 2016, **7**(9), 3964–3972.
113. Q. Guo, A. Ye, M. Lad, M. Ferrua, D. Dalgleish and H. Singh, Disintegration kinetics of food gels during gastric digestion and its role on gastric emptying: An *in vitro* analysis, *Food Funct.*, 2015, **6**(3), 756–764.
114. C. Giezenaar, N. D. Luscombe-Marsh, A. T. Hutchison, K. Lange, T. Hausken and K. L. Jones, *et al.*, Effect of gender on the acute effects of whey protein ingestion on energy intake, appetite, gastric emptying and gut hormone responses in healthy young adults, *Nutr. Diabetes*, 2018, **8**(1), 1–12.
115. M.-L. Bax, L. Aubry, C. Ferreira, J.-D. Daudin, P. Gatellier and D. Rémond, *et al.*, Cooking temperature is a key determinant of *in vitro* meat protein digestion rate: investigation of underlying mechanisms, *J. Agric. Food Chem.*, 2012, **60**(10), 2569–2576.
116. F. Barbé, O. Ménard, Y. Le Gouar, C. Buffière, M.-H. Famelart and B. Laroche, *et al.*, Acid and rennet gels exhibit strong differences in the kinetics of milk protein digestion and amino acid bioavailability, *Food Chem.*, 2014, **143**, 1–8.

117. G. M. Bornhorst, M. J. Roman, K. C. Dreschler and R. P. Singh, Physical property changes in raw and roasted almonds during gastric digestion *in vivo* and *in vitro*, *Food Biophys.*, 2014, **9**(1), 39–48.
118. G. M. Bornhorst and R. P. Singh, Kinetics of *in vitro* bread bolus digestion with varying oral and gastric digestion parameters, *Food Biophys.*, 2013, **8**(1), 50–59.
119. E. Capuano, N. Pellegrini, E. Ntone and C. V. Nikiforidis, *In vitro* lipid digestion in raw and roasted hazelnut particles and oil bodies, *Food Funct.*, 2018, **9**(4), 2508–2516.
120. X. Fang, L.-E. Rioux, S. Labrie and S. L. Turgeon, Commercial cheeses with different texture have different disintegration and protein/peptide release rates during simulated *in vitro* digestion, *Int. Dairy J.*, 2016, **56**, 169–178.
121. F. Barbé, O. Ménard, Y. Le Gouar, C. Buffière, M.-H. Famelart and B. Laroche, *et al.*, The heat treatment and the gelation are strong determinants of the kinetics of milk proteins digestion and of the peripheral availability of amino acids, *Food Chem.*, 2013, **136**(3–4), 1203–1212.
122. M. Opazo-Navarrete, M. D. Altenburg, R. M. Boom and A. E. M. Janssen, The effect of gel microstructure on simulated gastric digestion of protein gels, *Food Biophys.*, 2018, **13**(2), 124–138.
123. A. Ye, W. Liu, J. Cui, X. Kong, D. Roy and Y. Kong, *et al.*, Coagulation behaviour of milk under gastric digestion: Effect of pasteurization and ultra-high temperature treatment, *Food Chem.*, 2019, **286**, 216–225.
124. R. Deng, M. Mars, R. G. M. Van Der Sman, P. A. M. Smeets and A. E. M. Janssen, The importance of swelling for *in vitro* gastric digestion of whey protein gels, *Food Chem.*, 2020, 127182.

CHAPTER 3

Small Intestine, Digestion, and Nutrient Absorption

3.1 Introduction

The small intestine is a convoluted hollow tube-like structure that starts at the pylorus of the stomach and terminates at the ileocecal valve, which separates the small intestine's ileum from the large intestine's cecum.[1] It is the longest part of the gastrointestinal (GI) tract, has a highly specialized morphology, and is capable of performing two important roles, namely the digestion of macromolecules and absorption of water and nutrients.[2] The anatomical features of the small intestine can be broadly classified into two categories, microscopic and macroscopic anatomy. The macroscopic anatomy of the small intestine deals with gross anatomical features, namely the duodenum, jejunum, ileum, and mesentery.[3]

The microscopic anatomy of the small intestine (also known as histology) elucidates the microscopic structural features such as different layers of the intestinal wall, villi, and the major cell types of the intestinal epithelium. The wall of the small intestine has four layers, namely the serosa (adventitia), muscularis propria, submucosa, and mucosa.[4] The mucosa is the innermost layer that faces the small intestinal lumen. Further, the mucosa has a highly folded structure due to the presence of plicae circulares and, along with the existence of a huge number of villi and microvilli, it provides an enormous surface area for the absorption of nutrients and water.[5] The submucosa, containing a rich network of blood vessels and lymphatics, supports the mucosa in its specific function of electrolyte, nutrient, and fluid absorption. In addition, the lumen of the small intestine is made up of several unique structures that are arranged in a complex manner to support the nutrient absorption process.[6] All these specialized structures increase

Food Chemistry, Function and Analysis No. 42
Food Digestion and Absorption: Its Role in Food Product Development
By C. Anandharamakrishnan, Jeyan Arthur Moses and S. Priyanka

Published by the Royal Society of Chemistry, www.rsc.org

the total absorptive area of the small intestine to 250–400 m^2.[4] In addition, epithelial cells present on the luminal surface continuously proliferate and become differentiated into one of the seven different specialized cell types, namely absorptive enterocytes, goblet cells, enteroendocrine cells, M cells, tuft cells, cup cells, and Paneth cells.[5,7]

The small intestinal secretions such as mucus, bicarbonates, water, and other electrolytes have a significant role in the protection of the small intestine against microbes and acidic chyme, maintain an alkaline environment for the proper functioning of digestive enzymes, maintain the passage of luminal contents, and also aid in the digestion and absorption of nutrients.[8] The small intestine is a very complex organ with numerous folds and it lies between the stomach and large intestine, which consists of a small amount of microbiota (*ca.* 10^4 CFU mL^{-1}) and an abundant quantity of microbiota.[9] Further, the three major functions of the small intestinal mucosal immune system are (i) to safeguard the intestinal mucosal surface against colonization or invasion by harmful microbes, (ii) to act as a defense against undigested foreign antigens that are ingested with food and those that are secreted by non-pathogenic commensal flora, and (iii) to avoid the generation of immune responses against antigens and microbes that might be dangerous to the host itself.[10]

The motility of the small intestine is a very crucial and significant process that ensures the thorough mixing of the intestinal secretions and pancreatic and liver secretions along with the gastric chyme that enters the small intestine. Further, it is a key responsibility of intestinal motility to transport the above-mentioned mixture slowly along the whole length of the small intestine, so that there is sufficient time for digestion and absorption of nutrients.[11] In addition, intestinal motility also aids in the excretion of unwanted, undigested, excess, non-absorbed, and toxic contents from the gastrointestinal system by dumping them into the cecum of the large intestine, which collects and excretes the waste. The muscularis propria, which consists of longitudinal muscle (outer layer, thinner) and circular muscle (inner layer, thicker), lengthens the small intestine to permit the transport of the intestinal content in a single direction (caudally towards the colon), and allows the mixing of intestinal contents, breaks down the larger food particles, and also prevents the movement of the intestinal content in the reverse direction.[4] Further, the small intestinal motility is faster at the proximal end and comparatively slower at the distal end. Thus, it takes around 3–5 h for the chyme to transit from the stomach's pyloric region to the ileocecal junction.[3] There are various patterns of small intestinal motility, namely spontaneous contractions, segmentation contractions, peristalsis, migrating motor complex or migrating motility complex (MMC), and reverse peristalsis.[7] Further, a combination of factors, namely myogenic, neural, and hormonal factors, supports the regulation of small intestinal motility, of which the myogenic factors are of prime importance and the hormonal and neural factors aid in adjusting the myogenic-initiated motor patterns.[4]

The small intestine plays a crucial and significant role in the digestion of macronutrients along with the aid of three additional digestive organs, the liver, gallbladder, and pancreas.[6,12] The food, after undergoing various digestive phases (oral, gastric, and pancreatic digestion), enters the small intestine with the chyme, which consists principally of various short glucose oligomers, branched carbohydrate oligomers, α-dextrin, short peptide oligomers and amino acids, mixed micelles consisting of free fatty acids and monoglycerides (lipids), and indigestible foodstuff (fibers).[6] The macronutrients present in food have to undergo enzymatic digestion, acid denaturation, and enzyme hydrolysis in the oral and gastric phases, followed by pancreatic enzyme digestion in the small intestine before being acted upon by the brush border digestive enzymes.[13,14] The small intestine's brush border digestive enzymes are launched through the microvilli and account for the terminal phase of macronutrient breakdown (luminal digestion) into their subsequent monomers before absorption.[15,16]

In addition, the small intestine is the key location for the absorption of digested macronutrients, micronutrients, water, and electrolytes. These have specialized transport mechanisms and routes, namely active transport, passive diffusion (pore route and lipid route), cotransport (or secondary active transport), facilitated diffusion, and endocytosis for the absorption of various macro- and micronutrients.[6] Further, the specific transport proteins present in the epithelial membrane carry out optimal transport of the oligomeric and monomeric subunits of macronutrients from the lumen into enterocytes.[4] Likewise, the absorption of micronutrients (vitamins, minerals), electrolytes, and water occurs in the small intestine and these also have specific transport mechanisms.[17,18] Further, the size of the substance and its solubility properties are the two specific criteria that decide whether the specific substance is absorbed into the blood or the lymph.[19] Usually, small water-soluble substances are absorbed into the blood and lipid-soluble substances, large molecules, and particles are transported through the lymphatic system.[20]

In this chapter, we discuss in detail the anatomy and physiology of the small intestine. Further, special emphasis is given to the digestion and absorption of various nutrients such as carbohydrates, proteins, lipids, minerals, and vitamins.

3.2 Anatomy and Histology of the Small Intestine

3.2.1 Macroscopic Anatomy of the Small Intestine

The small intestine has an average length and diameter of about 6–7 m and 3.5 cm, respectively.[6,21] Commonly, it can be divided into three segments, namely the duodenum, jejunum, and ileum, which are located in the central and lower parts of the stomach cavity. In addition, the mesentery, which was recently recognized as an organ, is accommodated in the small intestine.[22] The mesentery is a contiguous structure made up of

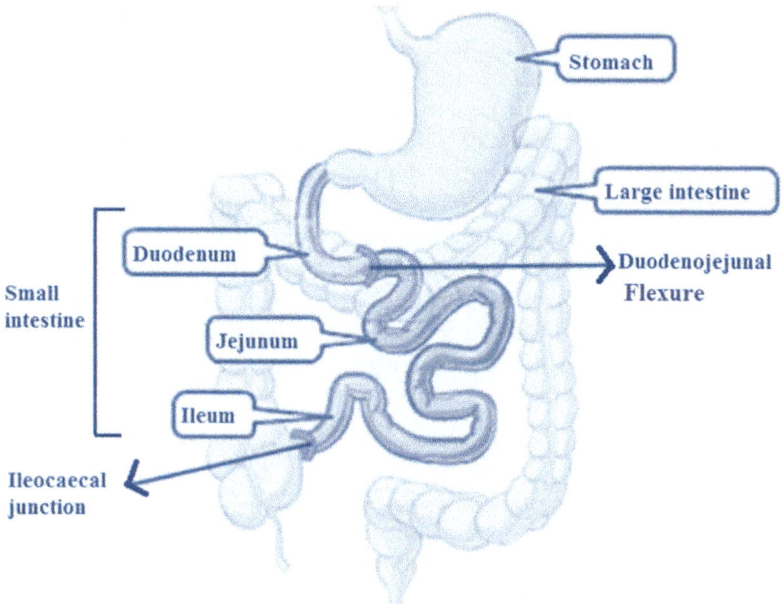

Figure 3.1 Macroscopic anatomy of the small intestine.

a double layer of visceral peritoneum and it aids in securing the small intestine to the posterior abdominal wall, which prevents the sagging of the small intestine into the pelvis when the body is in a vertical position.[3,23] Further, the crescent-shaped folds present in the luminal surface of the small intestine are called plicae circulares, which increase the surface area of the small intestine threefold and support the absorption of nutrients.[6,24] Plicae circulares are most prominently present in the distal duodenum and proximal jejunum. They become smaller in size and number as one moves distally in the small intestine, and at the terminal ileum they are absent.[4] Figure 3.1 illustrates the macroscopic anatomy of the small intestine.

3.2.1.1 *Duodenum*

The initial smallest portion of the small intestine commences at the pylorus and finishes at the duodenojejunal flexure (ligament of Treitz), marking the transition from the duodenum to the jejunum. In adults, it measures *ca.* 25 cm in length.[4] The duodenum is further divided into four sections: (1) duodenal bulb (~5 cm), (2) descending duodenum (~7.5 cm), (3) transverse duodenum (~12–13 cm) and (4) ascending duodenum (~2.5 cm).[25] Further, the retroperitoneal region consists of a major portion of the duodenum, and the intraperitoneal region consists of the jejunum and ileum. In addition, the duodenal region does not have mesentery, unlike the remaining small intestine.[26] However, the first section of the duodenum

(2–5 cm) is not retroperitoneal and is occasionally supported on a short mesentery and also it is attached to the liver through a hepatoduodenal ligament.[4] In the descending portion of the duodenum, the major and minor duodenal papillae are present. The major duodenal papillae consist of an opening called the ampulla of Vater through which the common bile duct and the main pancreatic duct (the duct of Wirsung) jointly empty the bile juice and pancreatic juice, respectively, into the duodenum.[27,28] Further, the duct of Santorini (or the minor pancreatic duct) also empties its contents through the minor duodenal papillae into the duodenum.[29] The main function of the duodenum is to mix the digestive secretions from the liver and pancreas (bile and pancreatic juice) and also the intestinal secretions from its wall along with the chyme emptied from the stomach.

3.2.1.2 Jejunum and Ileum

The jejunum starts at the ligament of Treitz and the ileum terminates at the ileocecal valve, which serves as a barrier to prevent the backward flow of colonic contents into the small intestine. The jejunum and ileum measure *ca.* 2.5 and 3.5 m in length, respectively.[6] The jejunum is located in the upper abdominal cavity and the ileum is positioned within the hypogastric region (lower abdominal cavity) and pelvic cavity.[4] The jejunum and ileum are bound to the posterior abdominal wall by the mesentery and they are loosely suspended in the peritoneal cavity, which permits mobility and helps to accommodate the shape and position variations brought on by propulsive peristaltic contractions.[24] Since there is no distinct histological demarcation or gross anatomical marker that divides the jejunum from the ileum, the transition between the two is arbitrary. However, a few anatomical characteristics progressively differentiate the proximal jejunum from the distal ileum. In comparison with the ileum, the jejunum has the largest luminal diameter, a thicker mucosal lining, more numerous and prominent mucosal folding (plicae circulares), and less fatty mesentery. Further, as we progress distally towards the ileum, the intestinal wall thickness, mucosal fold count, and luminal diameter gradually decrease.[6] In addition, the absence of mucosal folds in the terminal ileum and the gradual decrease in villi size and number with distance from the duodenum result in a fourfold decrease in the surface area of the small intestine from the distal duodenum and jejunum to the ileum. Further, the intestines consist of a copious amount of small aggregates of lymphoid tissues called Peyer's patches, especially in the ileum region from the mid-ileum to the ileocecal valve.[4]

3.2.2 Microscopic Anatomy of the Small Intestine

3.2.2.1 Structure of the Intestinal Wall

3.2.2.1.1 Mucosa. The mucosal layer of the small intestine is subdivided into three layers, namely the epithelium, lamina propria, and muscularis mucosae. Further, in the proximal region of the small intestine the mucosa is highly vascular and thick, and in the distal portion it is thin and less vascular.[4] The main function of the mucosa is absorption because of its large surface area of about 200 m^2.[3] Further, the presence of unique morphological and physiological features such as the microvilli, villi, and plicae circulares on the epithelial surface of mucosa provides 600-fold amplification of the absorptive surface area in the small intestine.[24] The lamina propria acts as a structural support to the epithelium and as a base for the connective tissues that extends into the villous protrusions and also contains the crypts. Further, the lamina propria contains an abundant amount of blood capillaries (arterioles and veins), lymph capillaries, lacteals, nerve fibrils, collagen, and fibroblasts.[30] In addition, it also contains lymphocytes, macrophages, eosinophils, mast cells, and neutrophils. Thus, it acts as an immune-protective tissue in the body that aids in protecting the tissues against bacteria that cross the epithelial membrane.[31] The deepest layer of the mucosa, the muscularis mucosae, is located close to the submucosa. It comprises two thin layers of smooth muscles, namely the outer longitudinal layer and inner circular layer of disposed fibers with some elastic tissue. Further, it also plays a small part in intestinal peristalsis.[32]

3.2.2.1.2 Submucosa. This is a connective tissue layer that lies next to the mucosa and it provides support and strength to the intestinal wall. It also contains Peyer's patches and Brunner's glands along with fibroblasts and mast cells.[3] Brunner's glands are exclusively present in the submucosal layer and secrete mucus, which aids in lubricating and protecting the delicate epithelial surface of the mucosa from the acidic chyme of peptic digestion. Moreover, Brunner's glands secrete a mucus-like substance that is alkaline in nature and which, in addition to its lubricating action, provides a conducive alkaline luminal environment that is vital for the activation of enzymes secreted from the pancreas and aids in the contraction of the gallbladder.[33] Peyer's patches, containing B lymphocytes, T lymphocytes, and M cells (microfold cells), play a significant role in mucosal immunity by identifying and processing antigens.[4] Further, the extensive nerve network of the submucosa works through the enteric nervous system (ENS) to guarantee appropriate intestinal motility and hormone secretion and offers necessary control over effective digestion and absorption.[11] In addition, Meissner's plexus (or the submucosal plexus), which is an essential constituent of the ENS, is accountable for secretions in the mucosal layer and the regulation of bowel motility.[32]

3.2.2.1.3 Muscularis Propria. The muscularis propria, which lies just below the serosa, consists of two layers of smooth muscles, namely a thicker inner circular layer and a thinner outer longitudinal layer.[3] Further, the myenteric plexus (or Auerbach plexus), which is located between the two muscle layers, is composed of parasympathetic and postganglionic sympathetic fibers. The Auerbach plexus is responsible for managing the peristaltic function of the small intestine and also helps in controlling bowel motility and secretions.[34] Interestingly, it is the interstitial cells of Cajal (ICCs), also called the pacemakers of intestinal contractile activity, that regulate the intestinal tone, and any irregularities in the ICCs can result in intestinal motility disorders.[11]

3.2.2.1.4 Serosa. This is the outermost layer of the wall of the small intestine and it is composed of one thin layer of flattened mesothelial cells that are held by the adventitia (a tiny portion of loose connective tissue structure).[7] All other parts of the whole of the small intestine are fully vested in a serosal coat, except for the retroperitoneal portions (the duodenum and the very terminal portion of the ileum).[33] Figure 3.2 illustrates the microscopic anatomy of the small intestine.

3.2.2.2 Villi

Villi are tiny hair-like projections that are located throughout the whole length of the small intestine's mucosa and it gives a velvety appearance and feel.[35] Further, the villi are differently shaped in different regions of the small intestine, namely ridge-shaped and broader (proximal duodenum), predominantly leaf-shaped (distal duodenum and proximal jejunum), and finger-shaped (distal jejunum and ileum).[33] Based on the location, the length of the villi varies between 0.5 and 1.5 mm. The villi are smaller and fewer in the distal area of the ileum and longer and denser in the duodenum.[36] They are considered to be the unit of absorption and they increase the absorptive area 10-fold.[4] Each villus consists of blood vessels, such as arteries, veins, and blood capillary networks, and lymphatic vessels (lacteals), which absorb the chylomicrons containing lipid particles and lipid-soluble vitamins.[18] The villi are covered by a single layer of simple columnar epithelial cells that are well known as enterocytes (absorptive cells).[6] Further, the enterocytes have abundant cytoplasmic extensions recognized as microvilli, which are tiny projections of the luminal surface of the intestine that make up the "brush border". The microvilli act as a protective barrier layer and aid in nutrient absorption since they are coated with a thick glycocalyx coating. Further, the millions of microvilli on the luminal surface of the intestine increase its surface area a further 20-fold.[4] In addition, within the microvillus layer, numerous brush border enzymes such as nucleosidases, peptidases, and disaccharidases are released and they aid in the intestinal digestion and absorption process.[37] The villus core encompasses macrophages, fibroblasts, eosinophils, lymphocytes, plasma cells, mast cells, and small nerve fibers.[33] Additionally, the cascade capillary bed at the tips of the villi enables the rapid absorption and transport of nutrients, liquids, and electrolytes into the systemic circulation.[38]

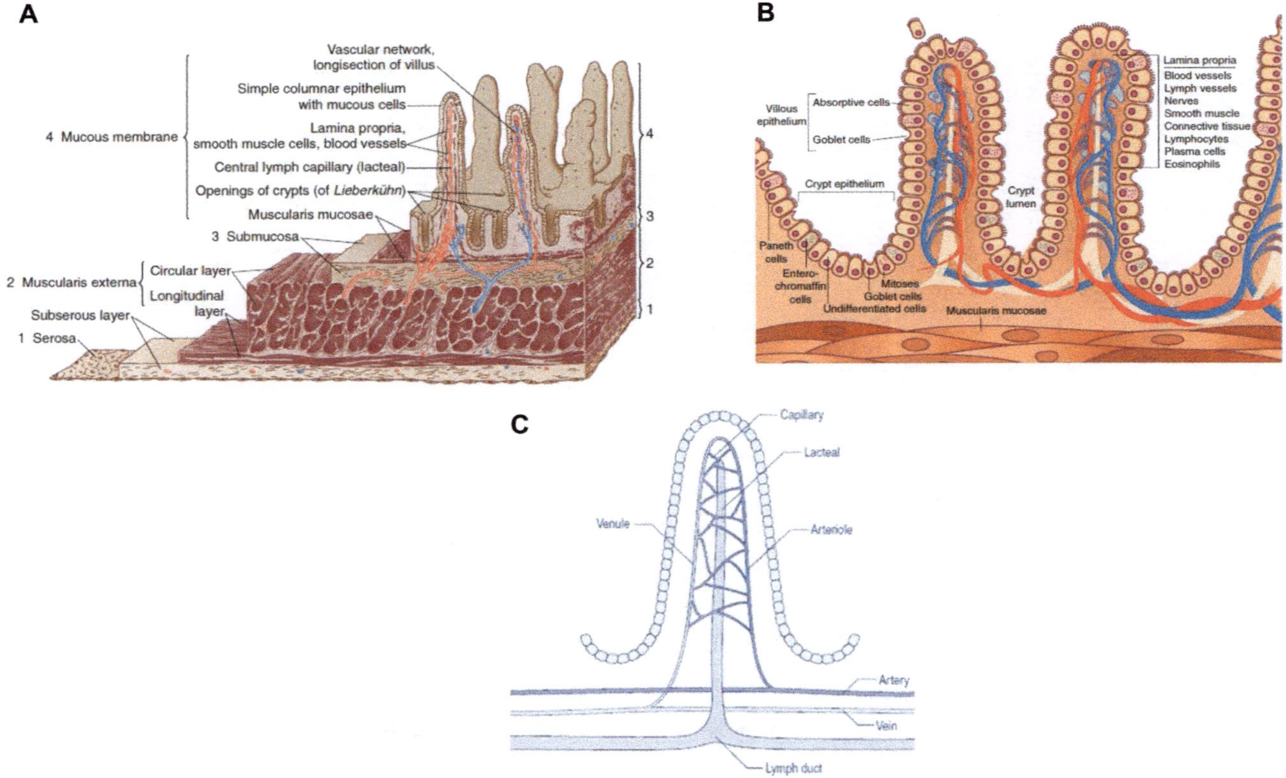

Figure 3.2 Microscopic anatomy of the small intestine. (A) Four main layers of the wall of the small intestine; (B) small intestinal mucosa with different intestinal cells and brush border; (C) structural features of the villus. (A and B) Reproduced from ref. 4 with permission from Elsevier, Copyright 2019. (C) Reproduced from ref. 6 with permission from Elsevier, Copyright 2010.

3.2.2.3 *Intestinal Cell Types and Function*

The intestinal crypts or crypts of Lieberkühn are 0.3–0.5 mm pit-like structures that are present at the base of each villus, extending all the way down through the lamina propria to the muscularis mucosa.[4] Each monoclonal crypt includes only one stem cell, which divides into two daughter cells by mitosis. Further, one daughter cell remains secured in the crypt for further mitosis while the other daughter cell undergoes a division process as it migrates through the sides of the crypts towards the villus. Furthermore, on the villus, the daughter cells go through a differentiation process where they are differentiated into one of seven different specialized cell types, namely enteroendocrine cells, absorptive enterocytes, Paneth cells, cup cells, tuft cells, M cells, and goblet cells.[7] These differentiated cells perform several specialized functions, which are summarized in Table 3.1. Secretory epithelial cells, which are less developed and younger, border the crypts. In addition, the epithelial cells present on the villi surface are replenished every 5–7 days and also epithelial cells of both the crypts and the villi are readily stocked by stem cells present in the crypts.[6] Moreover, the epithelium of the villi consists predominantly of absorptive cells called enterocytes and the luminal surface of each enterocyte consists of about 3000 microvilli.[3] The goblets cells secrete mucin, which lubricates and protects the intestinal wall located between the enterocytes.[4] Further, there are specialized immune cells such as M cells and cells of Paneth that provide immunity and protect the luminal surface from harmful agents such as bacteria and viruses.[31]

3.3 Small Intestinal Secretions

The alkaline intestinal fluid secreted throughout the small intestine comprises electrolytes, mucus, and water. However, the composition and control mechanisms of the secretions vary among different regions of the small intestine.[4] In addition, small intestinal secretions are strictly regulated and controlled through hormones (such as neurotensin, histamine, gastrin, prostaglandins, serotonin, and others), paracrine factors, and the nervous system.[6] Further, the secretions are secreted through either active or passive transport. Nevertheless, any disturbance in the intestinal secretions could lead to various gastrointestinal diseases.[32]

3.3.1 Mucus and Bicarbonates

Goblet cells secrete mucus, which protects the mucosal lining of the small intestine against damage by acids and pepsin from chyme along with bicarbonates. The mucus layer of the duodenum is 80–280 μm thick, which establishes a pH gradient and acts as an antimicrobial barrier against bacteria.[21] The largest amount of bicarbonate per unit area is secreted by the duodenal epithelium, which helps to neutralize roughly 40% of the total postprandial acid load.[4] The secretion of bicarbonate by the duodenum

Table 3.1 Different types of cells located in the small intestine and their functions.

Type of cell	Location in the mucosa	Function	Structure	Other significant features
Absorptive cells	Predominately present in the epithelium of villi and intestinal glands (crypt of Lieberkühn)	Produce digestive enzymes that support digestion and also aid in the absorption of nutrients in chyme	Tall columnar cells containing a clear, oval-shaped nucleus and several nucleoli at the base. The apical side of the cells is studded with densely packed (3000 per cell) finger-like, cylindrical projections called microvilli	Also known as enterocytes. These enterocytes, which are firmly attached to the basal lamina, are strongly connected to the adjacent absorptive cells by complex intracellular tight junctions at the apical pole
Cup cells	Present in very meagre amounts in the crypts and the villi, specifically limited to the ileum region of the small intestine	They are believed to have significant and unique functions that are still unknown at present	Cup cells are tall and columnar with a characteristic cup-like depression of the apical surface but they are narrower compared with surrounding enterocytes	A distinct feature of the cup cell is the shorter microvilli, which have a thickened and extensive glycocalyx coating. The cup cells also have higher cholesterol levels on their microvillus membranes compared with other enterocytes
Goblet cells	Present in a scattered pattern throughout the epithelium and intestinal glands of the small intestine. They are found in higher concentrations in the ileum region of the small intestine and in the large intestine	Goblet cells secrets mucin, which provides lubrication against shear stress and protects the mucosal linings of the small intestine from the acidic chyme and other chemical damage. It is also expected to bind with surface antigens and inhibit their attachment to the epithelial surfaces	Goblet cells have a characteristic wine-glass shape (goblet shaped) with a narrow base and expanded mucin-secreting granules located apically in the cell which appear as an oval-shaped apical portion. Goblet cells also appear cylindrically shaped if special care is taken during tissue fixation	Goblet cells have a distinct polarized morphology, in which the nucleus and Golgi apparatus of the cells are basally situated. Mucin secreted from the goblet cells is composed of highly glycosylated proteins that are suspended in an electrolyte solution

Table 3.1 (*Continued*)

Type of cell	Location in the mucosa	Function	Structure	Other significant features
Microfold cells (M cells)	These are spread over the surface of the Peyer's patches. They are also found in other parts of the body where there is an interface between the mucosal and the external environment such as the tonsils, adenoids, airways and ocular mucosa	M cells are the crucial antigen and luminal bacteria sampling sites of the small intestine (from the lumen to the Peyer's patches) where the endocytosis bacteria and viruses are present. The main function of M cells is transcellular transport from the intestinal lumen across the epithelial barrier to the basal membrane, where interactions with immune cells take place	M cells are epithelial cells, which are structurally distinct in the absence of brush borders and also contain numerous apical endocytic vesicles. Further, they usually have an oval or globular configuration, with a widened base and narrowed apex	M cells transport foreign antigens and microorganisms through the epithelial layer and deliver them to the Peyer's patches (mucosal lymphoid tissue) for recognition and handling; this is currently being exploited in vaccine production
Paneth cells	These are present in meagre amounts exclusively at the base of the crypts of Lieberkühn (intestinal glands)	Paneth cells secrete lysozymes and phospholipase A2, both of which have antimicrobial activity and act as a defense mechanism against bacteria. They are also involved in the phagocytosis process and also secrete α-defensins and adenosine monophosphates in response to cholinergic stimulation	Paneth cells are usually large and pyramidal in shape with the widest portion at the base where the nucleus is located. Further, the cytoplasm is rich in eosinophilic granules present at the apical end, which contains lysozyme and other glycoproteins	The secretions of the Paneth cells prevent the colonization of potentially pathogenic microbes and prevent enteric infection by maintaining and regulating the gastrointestinal antimicrobial barrier and intestinal flora

Tuft cells	Tuft cells are a rare cell type that comprise around 0.4% of the intestinal epithelium	Aid in chemoreception and potentially support parasitic defense	The morphological features are very distinct, containing a wide base and narrow apex with a tuft of long microvilli projecting from the apical surface protruding into the gastrointestinal lumen	The microvilli are attached through a core of long microfilaments passing deep into the apical cytoplasm. In addition, the morphology of the tuft cells is highly differentiated, similarly to cells specialized for absorption
Enteroendocrine cells	Primarily present in both crypts and villi but can also be found at other locations within the epithelium	The different enteroendocrine cells aid in the local regulation of intestinal activity by secreting various products such as secretin, grehlin, gastrin, vasoactive intestinal peptide (VIP), cholecystokinin (CCK), somatostatin, substance P, pancreatic polypeptide, enteroglucagon, gastric inhibitory polypeptide (GIP), and serotonin	They appear tall and columnar with the apical surface studded with microvilli. They have a large nucleus that is basally located with the Golgi apparatus situated above the nucleus	The most unique feature of the gut endocrine cells is the prominent cytoplasmic secretory granules, which are distributed mainly in the basal region of the cell rather than apical end of the cell
(a) APUD cells (amine precursor uptake and in gastrointestinal epithelium mucosa and in digestive glands	APUD cells are a type of enteroendocrine cells that produce peptide hormones that aid in regulating secretion and motility of the GI tract, liver, and pancreas			

Table 3.1 (*Continued*)

Type of cell	Location in the mucosa	Function	Structure	Other significant features
(b) G cells	Situated in the intestinal glands of the duodenum	Involved in the secretion of the hormone gastrin		
(c) I cells	Found in the intestinal glands of the duodenum	Aid in the secretion of the hormone cholecystokinin, which stimulates the release of pancreatic juices and bile		
(d) K cells	Commonly located in the intestinal glands	Secrete glucose-dependent insulinotropic peptide hormone, which stimulates the release of insulin		
(e) S cells	Found in the intestinal glands	Secrete the hormone secretin		
(f) M cells	Situated in the intestinal glands of the duodenum and jejunum	Secrete the hormone motilin, which accelerates gastric emptying, stimulates intestinal peristalsis, and also stimulates the production of pepsin		

increases significantly when there is a heightened presence of acid in the duodenum, which is mediated by both prostaglandin activity and neural reflexes [neural effectors: VIP (vasoactive intestinal peptide) and acetylcholine].[25] Additionally, an apical anion-conductive channel and the cystic fibrosis transmembrane conductance regulator (CFTR) govern the secretion of bicarbonate from the duodenum.[39] Nevertheless, the enzyme carbonic anhydrase and energy are required for the effective transportation of bicarbonate.[40]

3.3.2 Electrolytes

In the case of electrolytes released in the small intestine, in addition to bicarbonates, transcellular secretion of chloride by intestinal crypt cells plays a prominent role in the secretion of fluids in the small intestine.[6] The chloride is secreted throughout the small intestine through three specific cell membrane transporters, the basolateral membrane K^+ channel, the apical anion channel, and basolateral membrane electroneutral $Na^+/K^+/2Cl^-$ cotransporters, which are activated by cyclic nucleotides such as cyclic AMP (cAMP) *via* protein kinase signaling.[32] In addition, there are three apical anion membrane channels, namely (1) calcium-activated Cl^- channels, (2) the CFTR, and (3) type 2 chloride channels, all of which are identified as being responsible for the secretion of Cl^- into the intestinal lumen.[33] The chloride is usually absorbed in the large intestine. Further, to maintain cellular electroneutrality among luminal and serosal compartments, the Na^+ follows chloride passively by paracellular flux through tight junctions and potassium channels in the basolateral membrane permit charge compensation through K^+ efflux.[4] The secretion of chloride can be stimulated by the increase in certain substances, namely cyclic AMP, cyclic GMP, and calcium, and also by substances that increase the secretion of cyclic AMP, such as prostaglandins and VIP.[41,42]

3.3.3 Water

Water is secreted in the small intestinal lumen in reaction to the net movement of osmolytes (NaCl) into the lumen. Further, chloride is the major electrolyte along with Na^+ following passively *via* the paracellular space, and also the bicarbonates to a lesser extent are responsible for creating the osmotic gradient (established by CFTR), which stimulates the secretion of water into the small intestine.[43] The secretion of water by the small intestine aids in the solubilization and dilution of intraluminal nutrients and also supports maintaining intraluminal fluidity.[4] Further, the luminal chlorine draws out the sodium and water *via* the paracellular pathway, which results in net fluid and ion secretion. Water movement within the small intestine is both paracellular and transcellular *via* intercellular tight junctions and water channels, known as aquaporins, respectively.[44,45] In addition, the secretion of water is regulated by a multitude of effectors, hormonal factors (hormonal effectors: serotonin and VIP), and signaling by the ENS.[46]

3.4 Small Intestinal Mucosal Immunity and Barrier Function

The mucosal immunity of the small intestine acts as a barricade to many microbes present in the large intestine and the bacterial load of the small intestine increases as we move distally from about 10^2 in the proximal regions to 10^8 in the distal regions.[47] Further, in the distal ileum, the concentration of Gram-positive bacteria was less than that of Gram-negative bacteria.[48] It is noteworthy that in a healthy human, about 80% of all immune cells belong to the mucosa-associated lymphoid tissues (MALTs), and of these a major portion, *i.e.* around 70% of immune cells, belongs to the gut-associated lymphoid tissues (GALTs). In general, except for the mucosal surface, which is always surrounded by a foreign material environment, all other parts of the human body that are under the surveillance of the immune system are completely sterile.[49] In addition, various bacterial symbionts that aid in metabolizing the food components also support maintaining the integrity of the epithelial membrane and the development and maturation of mucosa by providing a defense against pathogenic bacteria.[4] Nevertheless, various pathogenic bacteria, such as *Helicobacter pylori*, *Salmonella typhimurium*, *Vibrio cholerae*, and their detrimental toxins can impair the body's immune responses and result in numerous inflammatory diseases and disorders.[3]

The small intestine experiences frequent attacks from bacteria, viruses, and bacterial toxins, and it also constantly comes into contact with foreign proteins and harmful chemical compounds through ingested food. Therefore, to protect against these harmful microbes and foreign antigens, the human body has various immune accompaniments and mechanisms, such as the mucosal immune system, expression of innate immune receptors, adaptive immune response, production of antimicrobial proteins, secretion of immunoglobulin A (IgA), and autophagy of intracellular bacteria.[50] Further, the mucosal immune responses are diverse; apart from innate immunity, adaptive immunity, and IgA secretion, they also involve various specialized antigen-presenting cells, mucosal macrophages, specialized B and T cells, and secretion of antibacterial proteins from the epithelium.[51,52] Consequently, these mechanisms and immune responses aid in the maintenance of the homeostasis of the small intestinal milieu by keeping the pathogenic and commensal bacteria under control.

3.5 Motility of the Small Intestine

3.5.1 Types of Intestinal Motility

There are four different patterns of motility of the small intestine, namely spontaneous contractions of smooth muscles, segmentation, peristalsis, and migrating motor complex or migrating motility complex (MMC). In addition, there are two different phases of intestinal motility, namely the fed state, which comprises segmentation and peristalsis, and the fasting/inter-digestive period, which includes the spontaneous contractions of smooth muscles and the MMC.[6] Spontaneous contractions of the smooth muscles of the small intestine occur even during the fasting period. Further, these spontaneous contractions increase in strength and frequency after the ingestion of food and are transformed into segmented contractions, which aid in mixing the contents of the lumen.[11]

The peristalsis contraction patterns aid in the propulsion of the intestinal contents (a mixture of chyme and secretions from the intestine, pancreas, and liver) towards the colon.[7] In addition, similarly to the stomach, the MMC that occurs during the inter-digestive period (fasting state) plays a significant role in the motility of the small intestine, and it consists of four phases that occur over a period of 85–115 min.[8,53] The MMC is discussed in detail in Chapter 2. It also plays a critical part in the motility of the small intestine as the strong contractions during phase III propagate caudally and completely occlude the intestinal lumen. Thus, the peristaltic and segmental contractions are aided by longitudinal and circular muscles, respectively.[11] Segmentation and peristalsis are the two fundamental intestinal contractions during the fed state that aid in the mixing of the contents of the lumen and propulsion of intestinal contents, respectively. Figure 3.3 illustrates the motility patterns in the small intestine

3.5.1.1 Spontaneous Contractions

The spontaneous contractile activity of smooth muscles even in the absence of food in the GI tract is due to the tonus activity, which occurs even with the non-existence of hormones, stretch, or nervous activity.[6] The rhythmic and slow wave activity of the smooth muscle is generated by specialized cells, the

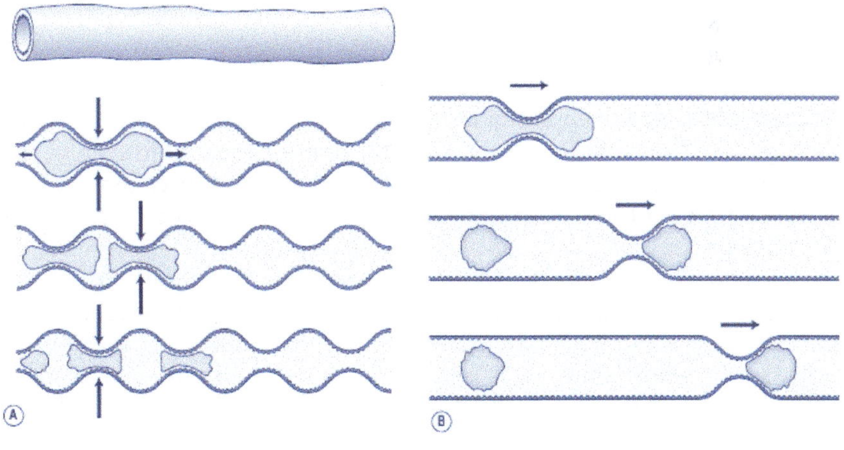

Figure 3.3 Motility in the small intestine: (A) segmentation and (B) peristalsis. Reproduced from ref. 6 with permission from Elsevier, Copyright 2010.

interstitial cells of Cajal (ICCs), situated in the muscle layer of the GI tract, and they are amplified due to the release of acetylcholine from the nerves in the neighborhood.[8] The slow waves generated result in contractions of smooth muscles, which first spread circumferentially and then travel along the small intestine in an aboral direction. The contraction frequency of the small intestine is higher at 11–12 min^{-1} in the duodenum and 7–8 min^{-1} in the terminal ileum compared with only 3 min^{-1} in the stomach.[7,54]

3.5.1.2 Segmentation: Mixing of Intestinal Content

Segmentation motility is the main pattern of pressure waves that occurs after the ingestion of a meal. When the nearby circular smooth muscle segments of the small intestine's local areas contract and relax, segmentation contractions take place. Segmentation contractions result in continuous rhythmic motion and sub-partitioning of the intestinal content, which moves in both forward and backward directions, aiding in chyme mixing and improving chyme's interaction with villous surfaces.[55] However, given that segmentation occurs more often and strongly in the proximal duodenum region (about 12 contractions min^{-1}) compared with the distal ileum region (about 8 min^{-1}), the likelihood that the food digesta will be pushed towards the large intestine is higher than that towards the stomach.[6] This is desirable as there is a need to mix the incoming acidic chyme from the stomach with the intestinal secretions, pancreatic juice, and bile mixture in order to neutralize the acidity and provide the appropriate alkaline environment for further digestion, micelle formation, and subsequent absorption. The segmentation pattern in the jejunum is known as the minute rhythm, where the bursts of contractions occur for about 1 min followed by a break with weak or no contractions. Moreover, apart from mixing the chyme, the segmentation contractions are also responsible for moving the chyme towards the large intestine.[56]

3.5.1.3 Peristalsis: Propulsion of Intestinal Content

Peristalsis is a wave of contractions that move the intestinal content towards the large intestine and it occurs as a result of progressive contractions of the adjacent rings of circular smooth muscle in the aboral direction due to excitatory motor neurons followed by the relaxation of rings of circular muscle by inhibitory neurons and contractions of the longitudinal muscle.[57] The peristalsis waves are usually short and of low strength, so they transport the intestinal content only a short distance of about 10 cm.[6] However, peristaltic rushes (long waves), which are waves of incredibly strong contractions that can span the whole small intestine, can occasionally occur. As a result, the chyme is propelled through the small intestine and towards the large intestine by both long and short peristalsis waves.[38] Further, the small and large intestines experience peristaltic reflex, a reflex that is neurally driven and results in the caudad propulsion of chyme.[11]

3.5.2 Control of Intestinal Motility and Factors that Affect Intestinal Motility

The motility of the small intestinal can be controlled by several physiological factors, namely stretch, circulating hormones, extrinsic autonomic nerves, intrinsic nerves of the intramural plexi, and paracrine factors.[4] However, the fundamental patterns of contractile activity are programmed into the neural circuitry of the intrinsic nerve plexi (myenteric plexus), hence even with complete obstruction of neural signals intestinal motor activity persists. Further, the reflex contraction of smooth muscles, known as the myenteric reflex, occurs because of the stimulation of pressure receptors as a response to the muscle stretch even in the absence of hormones or nerve involvement since it is an intrinsic property.[11] Even in the absence of extrinsic nerves and hormones, intrinsic nerve activity in the intramural plexi can regulate the basal electrical rhythm and consequently control segmentation and peristalsis. Parasympathetic and sympathetic nerves make up the ENS, which controls motility. Further, the stimulation of parasympathetic nerves and sympathetic nerves increases and inhibits the segmentation and peristalsis, respectively.[6]

In addition, various transmitters such as acetylcholine, catecholamines, VIP, somatostatin, substance P, nitric oxide, and opioids can influence intestinal motility.[58] Numerous hormones have a crucial role in controlling and modifying the motility of the intestine through electrical signals and contractile patterns. The stomach and small intestine activate pressure receptors and chemoreceptors that trigger hormone secretions.[4] Gastrin is released into the blood due to the presence of peptides in the stomach, while cholecystokinin (CCK) and secretin are released into the intestinal chyme due to the presence of fat and acid, respectively. Further, when the intestinal chyme develops alkalinity, the walls of the duodenum and jejunum secrete motilin (peptide) into the blood. Injecting hormones such as gastrin, secretin, CCK, and motilin increases intestinal motility and peristalsis.[59] Enteroglucagon is a peptide hormone released by the small intestine in response to glucose and fat in the chyme when food is present in the intestine. It aids in inhibiting peristalsis to allow more time for the absorption process before the chyme is propelled towards the large intestine. In addition, prostaglandin and ghrelin also speed up gastrointestinal motility.[60] Figure 3.4 illustrates the hormonal control of small intestinal motility.

Various drugs, medications, and disease conditions greatly affect the small intestinal motility patterns, which significantly compromises nutrient digestion and absorption. Further, the incidence of bacterial growth, connective tissue disorders (*e.g.* scleroderma), non-alcoholic steatohepatitis (NASH), chronic renal failure, colonic distention, obesity, infiltrative disorders, circadian rhythm disruption, GI disorders and diseases, long-term steroid use, and chronic intestinal pseudo-obstruction (inherited conditions) significantly affect intestinal motility.[4,61,62]

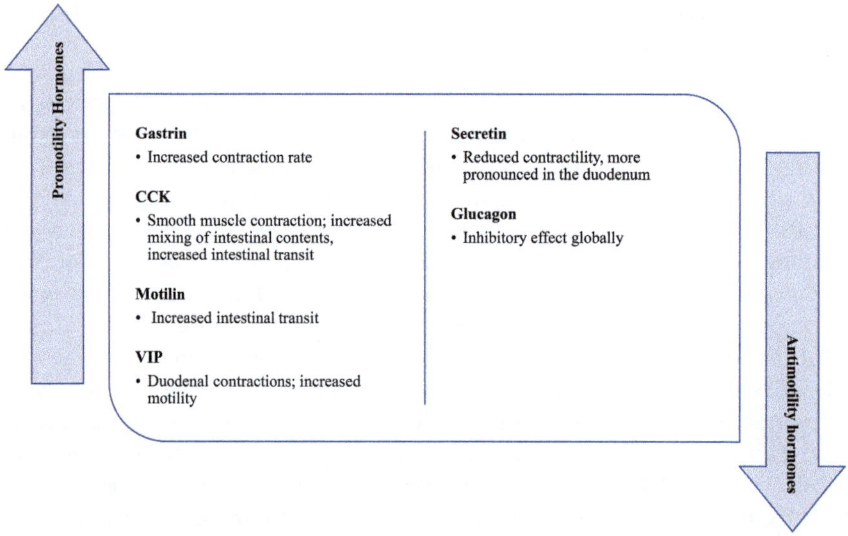

Figure 3.4 Hormonal control of small intestinal motility.

3.6 Accessory Digestive Organs: Liver, Gallbladder and Pancreas

3.6.1 Role of the Liver and Gallbladder in Intestinal Digestion

3.6.1.1 Liver

The liver is an important and the largest gland in the human body.[63] It performs various key functions, *e.g.* it absorbs toxins and drugs, metabolizes drugs, detoxifies chemicals, produces certain blood plasma proteins, acts a reservoir for glucose storage and release, regulates blood clotting, and aids in fighting infections by producing immune factors and by removal of microorganisms, in addition to its significant role in food digestion and nutrient absorption.[64] Further, the liver acts as an auxiliary digestive organ that plays a significant role in food digestion and nutrient absorption through the secretion of bile, cholesterol, and specific carrier proteins, all of which support fat digestion in the small intestine.[65] The liver secretes about 0.5–1.0 L of bile per day.[66]

3.6.1.2 Gallbladder

The gallbladder, which is a small sac-like structure made up of smooth muscle, stores the bile produced by the liver. A gallbladder's storage capacity ranges from 15 to 60 mL, and it is situated directly above the duodenum.[67] During the inner digestive period, the mucosa of the gall bladder absorbs water and ions from the bile and concentrates it up to 10-fold. Additionally, during the fed state, the common bile duct at the ampulla of Vater allows the gallbladder to discharge the concentrated bile into the duodenal region of the small intestine.[68] When the chyme enters the duodenum, the presence of fat digestion products and amino acids triggers the release of the hormone CCK, which in turn stimulates the gallbladder contraction.[69]

3.6.1.3 Bile

Alkaline bile solution, with a yellow-brown/green color and a pH 7.6–8.6, helps with lipid emulsification in the small intestine. Bile is made up of various constituents, such as bile acids, bile salts, bile pigments (bilirubin and biliverdin), water, electrolytes, phospholipids (lecithin), cholesterol, mucin, triglycerides, and xeno-chemicals.[67] Various constituents of bile play specific roles in the digestion and absorption processes, *e.g.* bile acid, bile salts, and phospholipids act as emulsifying agents and aid in fat digestion and absorption in the small intestine. In addition, bile salts and bile acids play a crucial role in the small intestinal absorption of vitamins A, D, E, and K (fat-soluble vitamins).[6]

The bile duct epithelial lining is protected from the damaging effects of bile acids by mucin present in bile, the binding of phospholipids and cholesterol to the bile acids protects the cell from the cytotoxic detergent effects of bile acids, and the electrolytes present in bile provide the osmotic pressure required for the flow of bile.[70] Nevertheless, the micelles formed due to the aggregation of bile acids, phospholipids, and cholesterol lessen their osmotic activity.[71] In the small intestine, the conjugated bile acids demonstrate antimicrobial properties and also promote an innate defense against bacteria through regulating gene expression.[72] Further, it is interesting that when chyme rich in fat arrives at the duodenum, bile production by hepatocytes surges and thereby triggers secretin (gut hormone) secretion.[6] Most of the bile constituents, such as bile acids (0.6 g) and pigments, are excreted in the feces. However, the enterohepatic circulation of the liver allows the recovery of about 95% of the bile acids through absorption *via* the apical sodium-dependent bile acid transporter situated at the brush boundary membrane of the ileum.[73]

3.6.2 Role of the Pancreas in Intestinal Digestion

3.6.2.1 Pancreas

The pancreas is a soft, oblong, glandular organ that has both exocrine and endocrine functions to perform such as secreting digestive enzymes and pancreatic juice though ducts into the GI tract and releasing the hormones into the blood directly (ductless), respectively.[74] It is located in the retroperitoneum region behind the stomach in a transverse position. Additionally, the pancreatic head area nestles in the duodenum's "C-shaped" curvature, with its body extending 15.2 cm to the left and its tapered tail portion terminating in the hilum of the spleen.[6] The pancreas consists of small clusters of endocrine cells knowns as the islets of Langerhans, which are dispersed among the abundantly present exocrine cells, which are small grape-like cell clusters known as acinar cells. The islets of Langerhans secrete various hormones, such as insulin, glucagon, pancreatic polypeptide, and somatostatin, which aid metabolism regulation and decide the fate of nutrients absorbed.[8] The acinar cells of the exocrine pancreas are hollow spheroids made up of 20–50 pyramidal cells organized around a central lumen. Each acinus develops lobules that are divided by loose connective tissue.[75] Enzyme-rich pancreatic juice secreted by acinar cells is drained into an intralobular duct, which along with other ducts forms interlobular ducts that progress until they reach the two dominant pancreatic ducts, specifically the primary pancreatic duct (duct of Wirsung) and the auxiliary pancreatic duct (duct of Santorini).[6]

3.6.2.2 Pancreatic Juice

The pancreas secretes 1–2 L of pancreatic juice daily, with a pH range of 7.1–8.2. The arrival of acidic chyme and the presence of proteins and fats in the duodenum stimulate the release of secretin and the secretion of CCK, respectively, which in turn signals the acinar cells of the pancreas to release bicarbonate-rich pancreatic juice and enzyme-rich pancreatic juice, respectively.[74] Various protein enzymes and enzyme precursors that are later activated in the small intestine to aid the digestion of various macronutrients (carbohydrates, lipids, and proteins) are secreted by acinar cells of the pancreas.[75] Further, along with enzymes they also secrete a colorless, watery, alkaline solution (rich in sodium bicarbonate with some salts) that carries the pancreatic enzymes forwards towards the small intestine and aids in neutralizing the acidity of the chyme entering the duodenum to offer a perfect setting for the small intestine's pH-sensitive digesting enzymes to function.[6]

3.6.2.3 Pancreatic Enzymes

Four classes of enzymes (protein in nature), namely the proteolytic, amylolytic, lipolytic, and nucleolytic enzymes, are secreted by the pancreas. Owing to the variable rate of protein secretion and the watery solution, the protein concentration of the pancreatic secretion ranges between 0.7 and 10%.[76] Additionally, the protein secretion of the pancreas is mostly composed of digestive enzymes, in addition to trypsin inhibitors, plasma proteins, and mucus. Further, the postprandial secretion of pancreatic enzymes is regulated through the cephalic phase (accounting for around 50% of pancreatic enzyme secretion), the gastric phase (mediated by the vasovagal reflex), and the intestinal phase (nutrient delivery into the intestine strongly triggers pancreatic enzyme secretion).[6] The pancreas produces a starch-digesting enzyme (pancreatic α-amylase), fat-digesting enzymes (phospholipase A_2, cholesterol esterase, pancreatic lipase, and colipase), and nucleic acid-digesting enzymes (pancreatic ribonuclease and pancreatic deoxyribonuclease) in their active forms.[75] However, protein-digesting enzymes (procarboxypeptidase A, trypsinogen, proelastase, chymotrypsinogen, and procarboxypeptidase B) are secreted in their inactive forms by the pancreas as they have the potential to digest the pancreas if they were produced in their active form.[18] The trypsinogen in the duodenum is first activated into trypsin by the intestinal brush border enzyme enteropeptidase (enterokinase). Consecutively, the trypsin then self-activates the remaining trypsin and other pancreatic enzymes into their active forms, namely procarboxypeptidase into carboxypeptidase and chymotrypsinogen into chymotrypsin.[4] The roles of various pancreatic digestive enzymes, their sources, substrates, and digestion products are presented in Table 3.2. Figure 3.5 illustrates the accessory digestive organs, namely the liver, gallbladder, and pancreas.

3.7 Roles of Brush Border Enzymes in Small Intestinal Digestion

The roles of various digestive enzymes of the oral cavity, stomach, and small intestine (pancreatic and brush border enzymes) are presented in detail in Table 3.2.

3.7.1 Brush Border Digestive Enzymes (BBDEs)

The brush border membrane vesicles or the enterocyte's apical membrane serve as the attachment points of BBDEs, and their catalytic sites protrude extracellularly into the intestinal lumen.[15] Further, BBDEs are liberated from the BBMV by pancreatic proteolysis or through disruption of the brush border vesicle membrane and the interior of enterocytes *via* lysis. The BBDEs will act on the nutrient oligomers present in the chyme while they are attached to the brush border membrane vesicles or apical membrane of the enterocytes and after their release from the brush border membrane vesicles. The main purpose of the BBDEs is to break down nutrient oligomers and dimers into their component monomers so that they can be properly absorbed.[78,79] However, only a very small number of short di- and tripeptides are directly carried across the apical membrane by peptide transporters, where they are then intracellularly digested by cytosolic enzymes such as aminotripeptidase and dipeptidases.[80]

3.7.2 Oligosaccharidase Consortium

Carbohydrates are hydrolyzed by the salivary α-amylase and pancreatic α-amylase at the α-(1,4)-linked glucose moieties to produce various

Table 3.2 Comprehensive overview of the chemical/enzymatic digestion of macronutrients in the oral cavity, stomach, and small intestine.

Digestion site	Macronutrient category	Enzymes responsible for digestion	Source of enzymes	Substrate for the enzymes	End products of digestion
Oral cavity	Carbohydrates	Salivary α-amylase (endosaccharidases)	Salivary glands	Polysaccharides are cleaved at the α-(1,4)-glycosidic bonds	Maltose, maltotriose, and dextrins
	Lipids	Lingual lipase	Lingual glands	Acts on triglycerides	Free fatty acids and mono- and diglycerides
Stomach	Proteins	Pepsin (endopeptidases)	Chief cells	Acts on protein peptide bonds and cleaves at the amino acids tyrosine, phenylalanine, leucine, and aspartic acid	Denatured (acid) and partially digested large peptide fragments and free amino acids
	Lipids	Gastric lipase	Chief cells	Acts on triglycerides	Free fatty acids and mono- and diglycerides
Small intestine	Carbohydrates	Pancreatic α-amylase (endosaccharidases)	Acinar cells of the pancreas	Digests starch by cleaving starch molecules at the α-(1,4)-glycosidic bonds	Maltose and maltose oligosaccharides
		Isomaltase (α-dextrinase)	Brush border enzymes of the small intestine	Maltose is cleaved at its α-(1,4) bond and limit dextrins at their α-(1,6) bond	Glucose
		Maltase glucoamylase (it has two subunits)	Brush border enzymes of the small intestine	Oligosaccharides: they are cleaved at the α-(1,4)-glycosidic bonds at their non-reducing end	Releases the terminal glucose
		Sucrase	Brush border enzymes of the small intestine	Cleaves the glycosidic linkage of sucrose to separate its subunits	Glucose and fructose
		Lactase	Brush border enzymes of the small intestine	Cleaves the β-(1,4)-glycosidic linkage of the lactose to separate its subunits	Glucose and galactose
		Trehalase	Brush border enzymes of the small intestine	Trehalose is an α-(1,1)-linked dimer of glucose	Glucose

	Trypsin (endopeptidases)	Acinar cells of the pancreas	Proteins are cleaved at the peptide bonds in which basic amino acids (lysine and arginine) contribute the carboxyl group	Oligopeptides and some free amino acids
	Chymotrypsin (endopeptidases)	Acinar cells of the pancreas	Proteins are cleaved at those peptide bonds in which aromatic amino acids (tyrosine, phenylalanine, methionine, leucine, and tryptophan) contribute the carboxyl group	Oligopeptides and some free amino acids
	Elastase (endopeptidases)	Acinar cells of the pancreas	Cleave elastin and peptide bonds adjacent to the amino acids such as alanine, glycine, and serine	Oligopeptides (2–6 amino acids)
Proteins	Carboxypeptidase A (exopeptidases)	Acinar cells of the pancreas	Cleave off single aromatic or branched-chain amino acids from the free carboxyl ends of proteins	Free amino acids and a mixture of oligopeptides
	Carboxypeptidase B (exopeptidases)	Acinar cells of the pancreas	Cleave off single basic amino acids from the free carboxyl ends of proteins	Free amino acids and a mixture of oligopeptides
	Endopeptidases	Brush border enzymes of the small intestine	Cleave/hydrolyze non-terminal amino acids in oligopeptides with 30 or fewer amino acid residues	Oligopeptides and dipeptides
	Aminopeptidase N	Brush border enzymes of the small intestine	Cleaves off one neutral amino acid	Free neutral amino acid
	Aminopeptidase A	Brush border enzymes of the small intestine	Cleaves off one anionic amino acid (such as glutamate, aspartate) from the amino terminus	Free anionic amino acid
	Aminopeptidase P	Brush border enzymes of the small intestine	Cleaves off one amino acid that is bound to a penultimate proline	Free amino acid

Table 3.2 (*Continued*)

Digestion site	Macronutrient category	Enzymes responsible for digestion	Source of enzymes	Substrate for the enzymes	End products of digestion
		Aminopeptidase W	Brush border enzymes of the small intestine	Cleaves off one amino acid that is bound to a penultimate tryptophan	Free amino acid
		Carboxypeptidase P	Brush border enzymes of the small intestine	Cleaves a single amino acid preferentially bound to a penultimate proline	Free amino acid
		Dipeptidylcarboxypeptidases	Brush border enzymes of the small intestine	Cleave off dipeptide fragments from the carboxyl end	Dipeptides
		Intracellular proteases/cytosolic peptidases such as aminotripeptidase and various dipeptidases	Cytoplasm of the enterocytes in the intestinal epithelium	Cleave the amino acids off the NH_2 terminus of tripeptides and specific dipeptides containing specific amino acids	Free amino acids and dipeptides and specific amino acids, respectively
Lipids	Pancreatic colipase	Acinar cells of the pancreas	Bind to the triglyceride surface with the help of the bile acids and aid in the anchoring of pancreatic lipase for lipolysis	—	
	Pancreatic lipase	Acinar cells of the pancreas	Hydrolyze positions 1 and 3 of triglycerides	Free fatty acids and 2-monoglyceride	
	Pancreatic phospholipase A_2	Acinar cells of the pancreas	Hydrolyze the fatty acyl ester bond at position 2 in phospholipids	Free fatty acid and lysophosphatidylcholine or lysolecithin	
	Pancreatic cholesterolesterase	Acinar cells of the pancreas	Cleave cholesterol esters	Cholesterol and free fatty acid	
Nucleic acids	Pancreatic ribonuclease (RNase) and pancreatic deoxyribonuclease (DNase)	Acinar cells of the pancreas	Degrade nucleic acids (RNA and DNA) in the ingested food	Nucleotides	

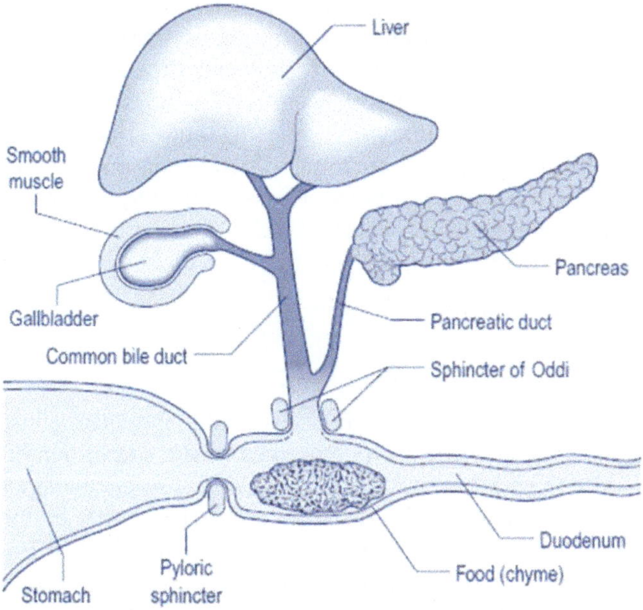

Figure 3.5 Diagrammatic representation of accessory digestive organs: the liver, gallbladder and pancreas. Reproduced from ref. 6 with permission from Elsevier, Copyright 2010.

oligosaccharide end products such as maltose, maltotriose, and α-dextrins.[4] Further, both the salivary α-amylase and pancreatic α-amylase are incapable of hydrolyzing amylopectin at or close to its α-(1,6)-linkages.[18] Thus, brush border digestive enzymes are critical for the digestion of the oligosaccharides into their component monosaccharides, which can be absorbed later. The brush border digestive enzymes consist of a suite of oligosaccharides such as maltase, glucoamylase, sucrase, isomaltase, lactase, and trehalase, which play a crucial role in breaking down various oligosaccharides and disaccharides. A group of oligosaccharides such as sucrase, isomaltase, maltase, and glucoamylase work together to break the α-(1,4)-glycosidic bonds of these resultant oligomers (maltose, maltotriose, and α-dextrins) of oral and pancreatic digestion. Further, sucrase, lactase, and trehalase break down the α_1,β_2-glycosidic linkages of sucrose, β-(1,4)-glycosidic linkages of the lactose, and α_1,α_1-glycosidic linkages of the trehalose. In addition, isomaltase and glucoamylase (to a comparatively lesser extent) have a vital role in cleaving the α-(1,6)-glycosidic linkages of carbohydrates.[15]

3.7.3 Oligopeptidase Consortium

In the case of proteins, the gastric pepsin and acid and pancreatic proteases produce denatured large peptides, a high proportion of small peptides (oligopeptides), and some free amino acids.[13] A conglomerate of brush border oligopeptidases with various specificities is essential for oligopeptide digestion, such as a larger group of N-terminal exopeptidases that includes aminopeptidases (specifically cleave N-terminal dipeptides and/or amino acids), a smaller group of exopeptidases that includes carboxypeptidase P (specifically cleaves C-terminal amino acids) and angiotensin-converting enzyme (specifically cleaves C-terminal dipeptides), and a group of endopeptidases that includes enteropeptidase.[8] Further, the congenital deficiency of brush border enteropeptidase causes gross protein malabsorption, nutrient deprivation, diarrhea, and edema as it plays a very crucial role in activating the pancreatic trypsin.[81]

3.7.4 Lipolytic Consortium

Lipids are efficiently digested into their absorbable fatty acid constituents by the full range of pancreatic lipolytic enzymes within the lumen of the small intestine. Hence there is not much literature on the presence of lipolytic brush border enzymes in the case of humans.[15] Nevertheless, other mammals, such as rats and mice, contain a substantial amount of some lipolytic brush border enzymes such as alkaline sphingomyelinase and neutral ceramidase that are responsible for the hydrolysis of phospholipids and sphingolipids, which are abundantly present in cheese, soybeans, cream, and eggs.[15]

3.8 Digestion and Absorption in the Small Intestine

Table 3.2 provides a comprehensive overview of the chemical/enzymatic digestion of macronutrients in the oral cavity, stomach, and small intestine. Table 3.3 presents details on various mechanisms and specific carriers involved in the absorption of digested macronutrients and their post-absorptive fate.

3.8.1 Digestion and Absorption of Macronutrients

Digestion of carbohydrates and proteins has already begun in the oral cavity (salivary amylase) and the stomach (acid and pepsin), respectively, before the food reaches the small intestine, whereas fat digestion occurs completely within the small intestine. Additionally, the small intestine is where the majority of the chemical digestion process (the breakdown of macronutrients into subsequent components) occurs. Chemical digestion is a complex interplay of various secretions, such as pancreatic enzymes (amylases, lipases, and proteases), bicarbonate ions, bile, and the brush border enzymes. Figure 3.6 illustrates absorbable forms of various macronutrients.

3.8.1.1 Carbohydrates

Carbohydrates are major constituents of human diets. In general, 40–50% of the daily energy intake comes from carbohydrates in comparison with other nutrients such as proteins (15–20%) and fats (30–40%).[18]

Table 3.3 Mechanism and specific carriers involved in the absorption of digested macronutrients and their post-absorptive fate.

Macronutrient category	End products of digestion	Mechanism of absorption into the cell	Specific carriers/ transport systems	Absorption into the blood/lymph	Final destination	Fate of absorbed macronutrients
Carbohydrates	Glucose	Secondary active transport	SGLT1 (sodium glucose-linked transporter 1) carries glucose into the cell using Na^+ as the cotransporter ion	Monosaccharides leave the cells by penetrating the basolateral membranes using a facilitated diffusion mechanism and GLUT2 (glucose transporter) carrier. They are then absorbed through capillary blood vessels in the villi	Liver *via* the hepatic portal vein	The liver releases the glucose directly into the bloodstream
	Galactose	Secondary active transport	SGLT1 carries galactose into the cell using Na^+ as the cotransporter ion		Liver *via* the hepatic portal vein	Galactose is converted to glucose in the liver and then released into the bloodstream
	Fructose	Facilitated diffusion	GLUT5 (glucose transporter 5) specifically transports fructose from the intestinal lumen into the enterocytes		Liver *via* the hepatic portal vein	Fructose is converted to glucose in the liver and then released into the bloodstream
Proteins	Neutral amino acids[a]	Secondary active transport. It requires ACE2 for its activity	The carrier B^0AT1 is a symport that uses the Na^+ gradient to drive the inward flux of amino acids. The ACE2 may feed substrate amino acid to the carrier	Absorbed through the capillary blood vessels in the villi	The absorbed amino acids either end up in the liver *via* the hepatic portal vein or they are taken up to the muscles and other tissues	In the liver the amino acids are converted to keto acids, which are oxidized *via* the citric acid cycle and other pathways, and the keto acids act as the liver's main source of energy during the absorptive state. The amino acids taken up by other tissues such as muscles are used for the synthesis of proteins
	Anionic (acidic) amino acids, aspartic acid, and glutamic acid	Secondary active transport	The carrier X_{AG}^- cotransports 3 Na^+ and 1 H^+ and counter transports 1 K^+ for carrying a single anionic amino acid into the cell			

Table 3.3 (*Continued*)

Macronutrient category	End products of digestion	Mechanism of absorption into the cell	Specific carriers/ transport systems	Absorption into the blood/lymph	Final destination	Fate of absorbed macronutrients
	Cationic (basic) amino acids (arginine, lysine, and cystine)	Facilitated diffusion	The heterodimer carrier rBAT/$b^{0,+}$ transports cationic amino acids and cystine in exchange for neutral amino acids			
	Imino amino acids (proline and hydroxyproline)	Secondary active	The carrier IMINO transports proline and hydroxyproline across the brush border membrane, and their transport is coupled to the Na^+ electrochemical gradient			
	Proline, glycine, and alanine	Secondary active	The carrier PAT1 along with Na^+ cotransports proline, glycine, and alanine into the enterocytes			
	Di- and tripeptides	Secondary active	The carrier PepT1 carries oligopeptides across the brush border membrane, coupled to H^+ cotransport			

Lipids	Glycerol	Simple diffusion	—	Absorbed through the capillary blood vessels in the villi	Liver *via* the hepatic portal vein	The glycerol produced is either taken up by the adipose tissue cells or transported to other tissues
	Free fatty acids and 2-monoglycerides	Diffusion or by endocytosis mediated	The transport is mediated through caveolin to form CEVs (caveolae endocytotic vesicles). Further, the transport of lipolytic products may involve various lipid-binding proteins including FABP pm (fatty acid binding protein, plasma membrane), FATP4 (fatty acid transport protein), CD36 (cluster of differentiation 36), cytosolic FABP (fatty acid binding protein), and other proteins such as Cav-1 (caveolin-1) and clathrin	In the Golgi apparatus of the enterocytes the re-esterified triglycerides, phospholipid, and cholesterol are packaged into lipoproteins by microsomal triglyceride transfer protein (MTP). Various lipoproteins are produced by the small intestine such as chylomicrons, very low-density lipoprotein (VLDL), and high-density lipoprotein (HDL). The lipoproteins are released at the basolateral membrane to travel	Systemic circulation *via* lymph entering thoracic duct	They are stored in the adipose tissue present in the body in various regions such as perirenal, subcutaneous, mesenteric, and other regions

Table 3.3 (*Continued*)

Macronutrient category	End products of digestion	Mechanism of absorption into the cell	Specific carriers/ transport systems	Absorption into the blood/lymph	Final destination	Fate of absorbed macronutrients
	Cholesterol	Not established	The transportation involves NPC1-L1 (Nieman5Pick C1-like 1) protein and it is specifically inhibited by some drugs (*e.g.* ezetimibe). Further, free cholesterol is secreted out of the enterocytes by ABCG5/ABCG8, a heterodimeric ABC transport complex (ATP-binding cassette transport proteins)	in the lymph or in the portal blood vessels that perfuse the intestine. The chylomicrons are transported solely in the lacteals of the villi (lymphatic system)		

[a] Neutral amino acids include methionine, leucine, isoleucine, valine, glutamine, asparagine, phenylalanine, cysteine, alanine, glycine, serine, tyrosine, threonine, histidine, proline, lysine, and tryptophan.

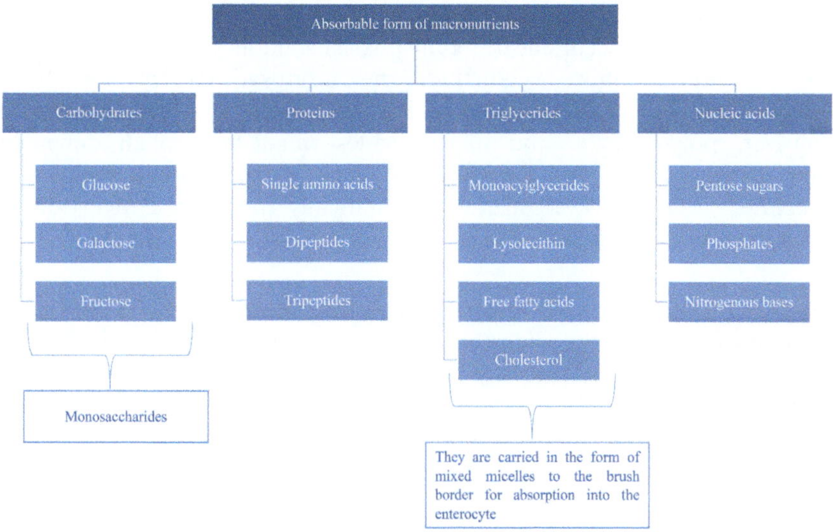

Figure 3.6 Absorbable forms of various macronutrients.

3.8.1.1.1 Carbohydrates in Food. Dietary carbohydrates can be classified into two categories based on their digestibility: (1) digestible carbohydrates, namely sugars (mono- and disaccharides), oligosaccharides, and complex polysaccharides (starches and glycogen), and (2) non-digestible carbohydrates such as insoluble cellulose and soluble dietary fiber such as gums, pectin, and hemicellulose. The human gastrointestinal system does not have appropriate enzymes that can break down the β-(1,3)- or β-(1,4)-glycosidic linkages of these complex carbohydrates, which serve as a source of dietary fiber.[82] Hence they all reach the colon intact and serve as a source of nutrition to the colonic microflora, which digest them and produce relatively water-soluble short-chain fatty acids that are readily absorbed and locally metabolized by colonocytes.[83] Further, the carbohydrates in the diet are mostly from plant sources such as starches and sugars, except for lactose (milk sugar) and glycogen (animal carbohydrate from meat).

Glucose and fructose are the common monosaccharides found in foods. Also, disaccharides such as sucrose, maltose, and lactose are frequently seen in human diets. Additionally, polysaccharides are significantly longer and heavily branched, whereas oligosaccharides are made up of 2–10 monosaccharide units. Nevertheless, starch, which exists in the forms of amylose and amylopectin and is mostly found in plant-derived foods such as grains (wheat and rice) and vegetables (such as potatoes), accounts for around 50% of the calories that come from carbohydrates.[84]

3.8.1.1.2 Digestion and Absorption of Carbohydrates. Digestion and absorption of carbohydrates constitute a multi-step process beginning in the oral cavity with salivary amylase, followed by pancreatic amylase in the small intestine and finally digestion by the brush border enzymes before absorption, which occurs at different phases such as the membrane phase and a luminal phase.[18] Further, the monosaccharides are directly absorbed in the small intestine and do not need any further digestion. However, poly-, oligo-, and disaccharides must be broken down into their corresponding monosaccharides before being absorbed in the small intestine.[85] Figure 3.7 illustrates schematically the overall digestion and absorption of carbohydrates.

3.8.1.1.2.1 Luminal and Brush Border Membrane Phase. The luminal phase encompasses the digestion of carbohydrates due to the activity of α-amylase, which is released by the pancreas (small intestine) and salivary glands (oral cavity). Since salivary amylase has only limited access to digestible carbohydrates as the food remains in the oral cavity for a much

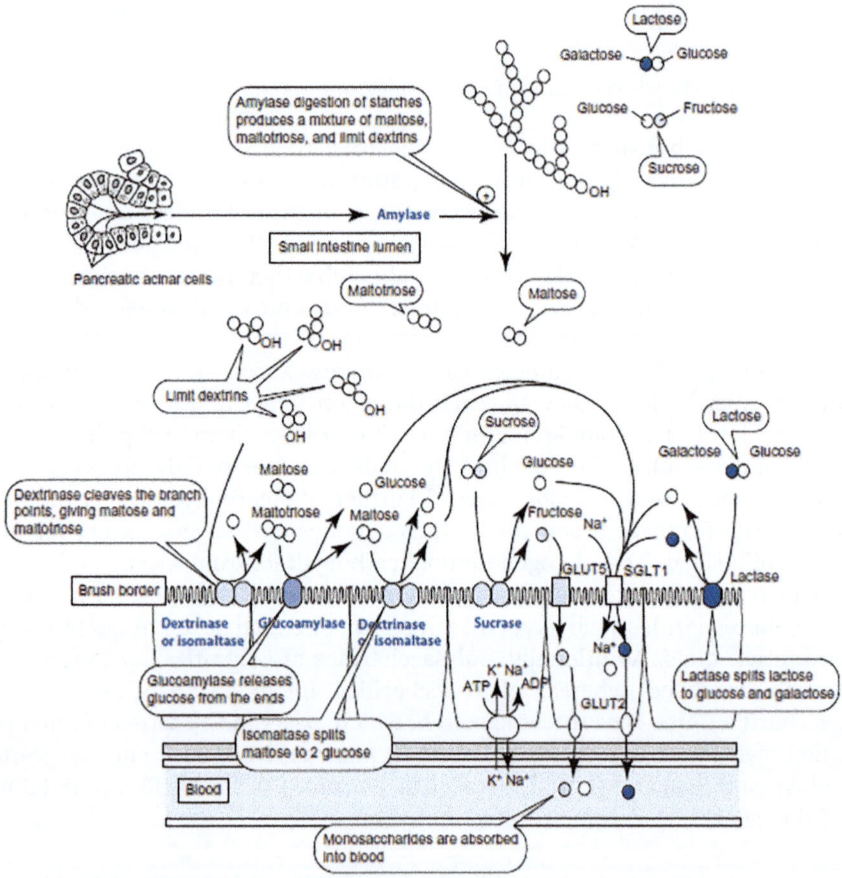

Figure 3.7 Schematic representation of overall digestion and absorption of carbohydrates. Reproduced from ref. 18 with permission from Elsevier, Copyright 2017.

shorter time (30 s–1 min).[86] Further, when the bolus/food reaches the stomach, the carbohydrates are inactivated due to the acidic pH. Thus, pancreatic amylase plays a very crucial and dominant role in the digestion of carbohydrates. Both the amylases digest the starches and other oligosaccharides by breaking down the α-(1,4)-glycosidic bonds and yield maltose, limit dextrins, and maltotriose. The amylases cannot cleave the α-(1,6)-glycosidic bonds of the amylopectin.[7] Additionally, there are several brush border enzymes, such as sucrase–isomaltase, trehalase, lactase, and maltase–glucoamylase, in the apical membrane of the duodenum and jejunal enterocytes that digest the luminal digestion by-products to create monosaccharides such as galactose, fructose, and glucose.[17]

3.8.1.1.2.2 Carrier-mediated Absorption of Monosaccharides. The Na^+/K^+-ATPase pump powers the sodium–glucose cotransporter (SGLT1), a carrier-mediated cotransporter system that is located in the brush border of enterocytes and actively transports glucose and galactose across the apical and basolateral membranes of the enterocytes.[84] However, fructose enters the enterocytes *via* facilitated diffusion mediated by a selective carrier protein, GLUT5, rather than an Na^+ cotransporter system. Thus, facilitated diffusion, which is driven by the glucose transporter GLUT2, permits the monosaccharides (glucose, galactose, and fructose) to leave the basolateral membrane of the cell.[87] Transferring the monosaccharides into the bloodstream is the last step in their absorption process.

3.8.1.2 Proteins

Three important sources of protein undergo the digestion process, namely dietary protein sources (∼70–100 g), endogenous protein secretions (35–200 g), and desquamated cells.[88] Thus, a range of proteolytic enzymes with specificity are secreted along the GI tract to digest food proteins completely into small peptides (di- and tripeptides) and amino acids, so that the small intestine can efficiently absorb them. In addition, all the dietary proteins are digested and completely absorbed in a healthy human.[89] The jejunum portion of the small intestine is where the majority of protein absorption takes place. The protein excreted in the fecal matter is derived from the colonic bacteria and colonic epithelial cells. Further, oral salivary secretion does not have any proteolytic enzymes.[6] Figure 3.8 illustrates schematically significant events in protein digestion and absorption.

3.8.1.2.1 Digestion of Proteins. Protein digestion has two phases, namely the initial gastric phase and the final intestinal phase.

3.8.1.2.1.1 Gastric Phase. Protein digestion commences in the stomach due to the pepsin's activity and it cleaves peptide bonds of the polypeptide chain at various cleavage points to produce several larger polypeptide fragments and some amino acids, as it is an endopeptidase. Further, pepsin will

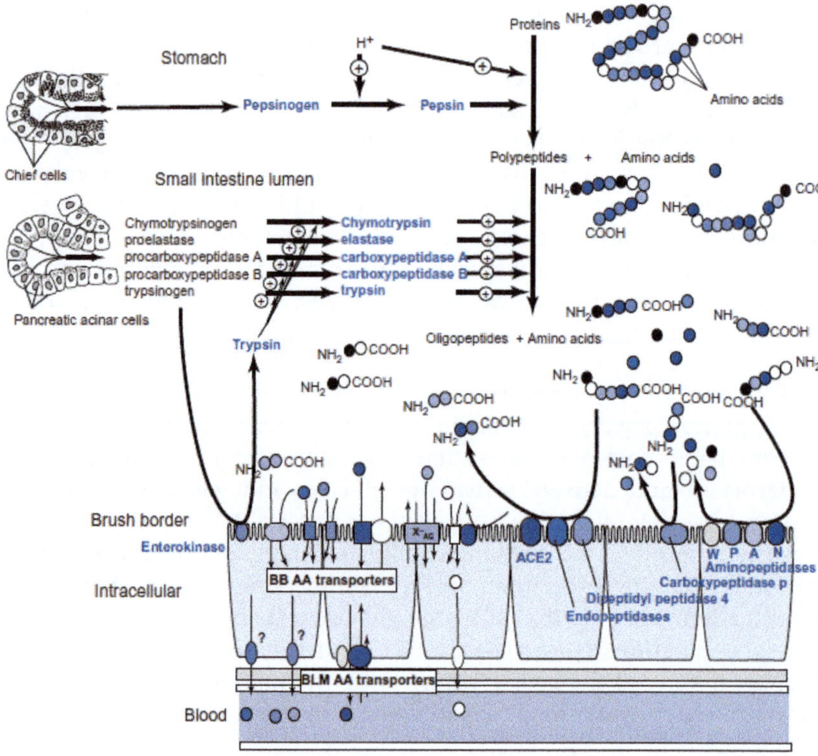

Figure 3.8 Schematic representation of significant events in protein digestion and absorption. Reproduced from ref. 18 with permission from Elsevier, Copyright 2017.

be active only at acidic pH (<4).[90] Once the gastric chyme has reached the duodenal region, the pepsin becomes inactivated due to the alkaline nature of the small intestine contributed by the bicarbonates of the biliary and pancreatic secretions.[91] The gastric phase of protein digestion is not critical, as persons with complete gastrectomies fully digest dietary protein.[84]

3.8.1.2.1.2 Intestinal Phase. Three crucial stages of protein digestion take place in the small intestine, namely the luminal phase, the brush border, and the intracellular phase. In the luminal phase, the endopeptidases such as elastase, chymotrypsin, and trypsin cleave the proteins at the middle of their chain, and the exopeptidases (such as carboxypeptidase A and B) cleave single amino acids off the free carboxyl ends of proteins with specificity.[84] Consequently, 30% of the final products of pancreatic and stomach protease digestion are free amino acids and 70% are oligopeptides (2–8 amino acids). The brush border phase consists of a variety of aminopeptidases, dipeptidylcarboxypeptidase, and carboxypeptidase P, which are found on the membranes of the brush border.

To liberate amino acids and di- and tripeptides, the brush border enzymes further hydrolyze the oligopeptides with greater specificity.[18]

3.8.1.2.2 Absorption of Proteins

3.8.1.2.2.1 Amino Acid- and Peptide-specific Transportation Systems Across the Brush Border Membrane. Proteins are hydrolyzed by gastric, pancreatic, and brush border proteases into a mixture of amino acids and small peptides of 2–6 amino acids. These di- and tripeptides and free amino acids are moved through the brush border membrane with the aid of six major transport systems that exhibit a variety of specificities and transport mechanisms such as facilitated diffusion and secondary active transport.[92] Further, the amino acids are transported into enterocytes through the electrochemical gradients facilitated by Na^+ and in some cases Cl^-. However, the Na^+/H^+ exchanger expressed on the luminal enterocyte membrane produces an H^+ concentration gradient that drives the electrochemical gradient in the case of dipeptide and tripeptide transit.[84] Finally, the intracellular phase of the protein digestion occurs inside the enterocytes where the intracellular/cytosolic peptidases such as aminotripeptidase and a variety of dipeptidases split the small peptides and specific dipeptides containing specific amino acids, respectively, into their components. Nevertheless, certain peptides are resistant to proteolysis and are not broken down.[8]

3.8.1.2.2.2 Specific Carriers Transporting Amino Acids Across the Basolateral Membrane. Single amino acids are transported *via* specific carriers over the enterocyte's basolateral membrane and into the portal blood circulation. The primary transporters and mechanisms that control the efflux of amino acids across the basolateral membrane remain unidentified to date.[93] Nevertheless, three transporters partially explain amino acid efflux and facilitate the exit of cytosolic amino acids from the cells into the blood, namely the y^+L system (facilitates an obligatory exchange of amino acids, with the influx of amino acids linked to Na^+ cotransport), T system (mediates the transport of aromatic amino acids), and L system (facilitates an obligatory exchange of amino acids).[88] It is crucial to understand that basolateral amino acid transporters are bidirectional and have overlying specificities for amino acids. The A and ASC systems are two additional transporters that let amino acids enter cells from the blood for cellular nourishment. They are presumably Na^+-dependent cotransporters.[18]

3.8.1.3 Lipids

Lipids (fats and oils) are important constituents of our daily diet and their daily consumption on average will be 90–100 g. The major forms of dietary lipids are triglycerides (~90%), phospholipids (~5%), cholesterol, and

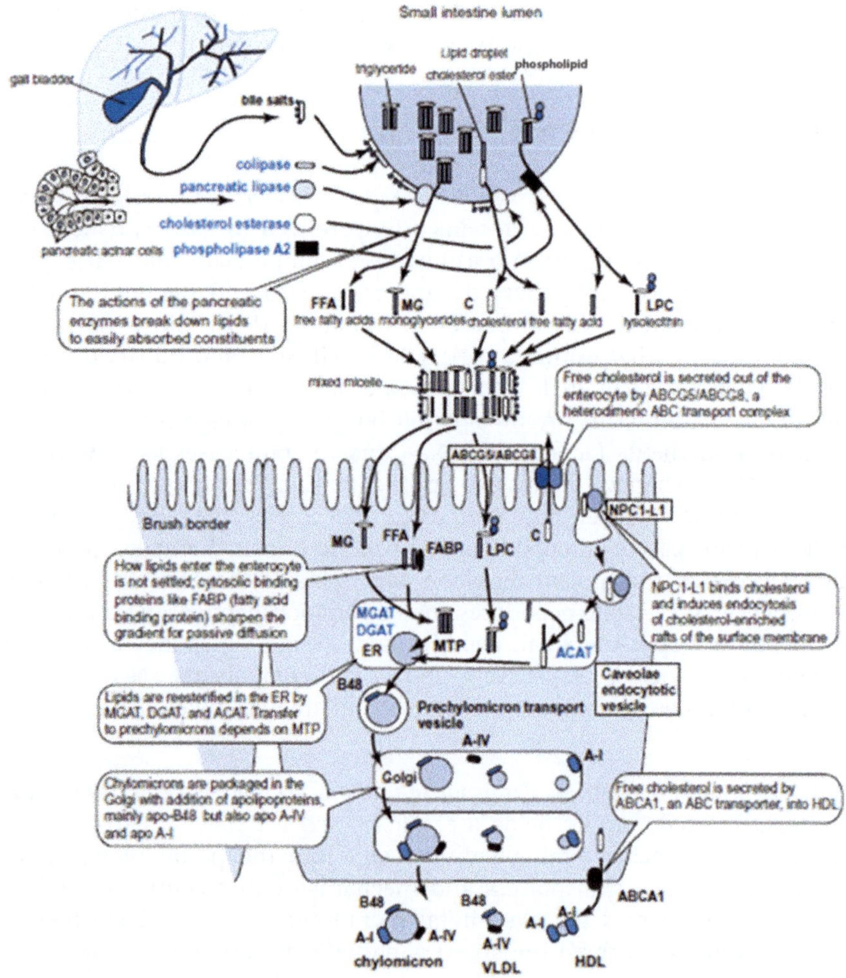

Figure 3.9 Schematic representation of overall processing of lipids in the small intestine. Reproduced from ref. 18 with permission from Elsevier, Copyright 2017.

cholesterol esters. Figure 3.9 illustrates schematically the overall processing of lipids in the small intestine.

3.8.1.3.1 Role of Emulsification in Lipid Digestion. Emulsification of the dietary lipids is crucial and necessary for their digestion owing to the insolubility of lipids in water. Emulsification of lipids produces a stable dispersion of lipids with a larger surface area, which in turn enhances the accessibility for lipase enzymes (as the lipase enzymes are water soluble).[6] Further, various factors such as cooking, chewing during oral processing and churning, and peristalsis of the stomach play a crucial role in the emulsification process.[94] In the stomach, a crude emulsion with larger lipid droplets is formed, where most of the lipid is present in the lipid core and only a very small amount is present on the surface. Hence the lipase cannot fully access its substrate and the digestion can occur only on the surface.[95]

However, the pancreatic lipase and bile salts in the small intestine play an important role in the digestion of ingested lipids (~70–80%). Bile salts and the by-products of lipid digestion (phospholipids and cholesterol) operate as significant emulsifiers in the small intestine to produce a finer dispersion of lipids with a higher surface area, increasing the likelihood of lipase activity.[96]

3.8.1.3.2 Digestion of Lipids. The digestion of lipids starts in the oral cavity by the activity of lingual lipases during the oral processing of foods.[97] Further, in the stomach, the chief cells in the fundus secrete gastric lipase, which aids in the hydrolysis of triglycerides (~15–20%) by cleaving preferentially at position 3 of the fatty acyl ester bond, generating fatty acid and diglycerides.[13] Thus, various lipolytic enzymes secreted from the pancreas, such as pancreatic lipase (along with pancreatic colipase), pancreatic cholesterolesterase, and pancreatic phospholipase A_2, digest the triglycerides (into component fatty acid and 2-monoglycerides), cholesterol esters (into component cholesterol and fatty acid), and phospholipids (into component head groups and fatty acid), respectively, in the small intestine.[98]

3.8.1.3.3 Mixed Micelle Formation and Absorption of Lipid Digestion Products. All of the lipid digestion products, such as lysophospholipids, cholesterol, fatty acids, and monoglycerides, are separated from the large lipid droplets and solubilized by bile salts and phospholipids to form complexes called mixed micelles. The mixed micelles are tiny disk-shaped aggregates, stabilized by the unique topology of bile acids, and they facilitate the transfer of lipolytic substances to the brush border membranes of the enterocytes that are present throughout the undisturbed water layer.[6] Furthermore, lipid micelles diffuse more easily across the undisturbed water layer than free fatty acids as they are more soluble in the aqueous phase and fuse with the enterocytes of the villi upon contact with the cell membrane. Hence the formation of micelles is as important to lipid absorption as digestion.[99] Although the exact transport processes used for transporting lipolytic products across the brush border membrane are still unknown, several lipid-binding proteins may be involved. However, currently, it is believed that short- and long-chain fatty acids and 2-monoglycerides enter the enterocytes by simple diffusion, and either by diffusion or by endocytosis facilitated through caveolin to form caveolae endocytotic vesicles (CEVs), respectively.[18]

3.8.1.3.4 Post-absorptive Fate of Lipids. Triglycerides are resynthesized in the enterocytes from fatty acids and monoglycerides. Further, to generate chylomicrons, high-density lipoprotein (HDL), and very low-density lipoprotein (VLDL), the triglycerides will combine with phospholipids, cholesterol, and apoproteins in the Golgi. Later, they are packed into secretory vesicles within the cell and are released from the basolateral membrane into either the blood vessels or lymph that perfuse the intestine.[100] However, the chylomicrons are solely released into the lacteals (lymphatic circulation) *via* exocytosis and then they are transported through the intestinal lymphatics to drain into the thoracic duct and finally reach the systemic circulation.[101] Additionally, more short- and medium-chain fatty acids are absorbed straight into the portal blood circulation instead of the lymph. Additionally, the cholesterol is taken up by the lymphatic system after being absorbed as VLDL in the small intestine.[102]

3.8.2 Absorption of Micronutrients

Vitamins and minerals are together classified as micronutrients, and they are also the two fundamental nutrient groups apart from the three macronutrients (carbohydrates, proteins, and fat) that are essential for good health and sustenance of life.[103] The micronutrients have a wide range of therapeutic functions and physiological roles, such as serving as coenzymes, cofactors, biochemical substrates, and hormones.[104,105] Deficiencies of micronutrients will lead to serious clinical effects, diseases, and adverse medical conditions. There is also evidence that some vitamins and minerals in excessive concentrations or overdosage will also cause certain ailments and disorders.[106]

Micronutrient absorption occurs in three stages: (i) the intraluminal stage, which includes the chemical reactions in the intestine and stomach dominated by the composition of the food entering the stomach and the pH of the digested luminal content, (ii) the translocation stage, which comprises transport or diffusion of the micronutrients across the intestinal cell membrane *via* various unique mechanisms such as simple diffusion, facilitated diffusion, or active transport, and (iii) the mobilization stage, which involves the transport of nutrients across the serosal surface of the intestinal cells into the circulation.[6,107]

Several factors affect the efficient absorption of micronutrients and lead to nutrient malabsorption, such as gastrointestinal diseases/disorders (diarrhea, hepatic disorders, celiac disease, sprue, Crohn's disease, pancreatic insufficiency, cystic fibrosis, biliary atresia, cholestasis, *etc.*), antinutrient or inhibiting factors present in the diet, age, interactions among micronutrients, excessive alcohol consumption, genetic disorders related to transport molecules, and interactions with medications/drugs.[108,109] Figure 3.10 depicts the classification of different micronutrient categories.

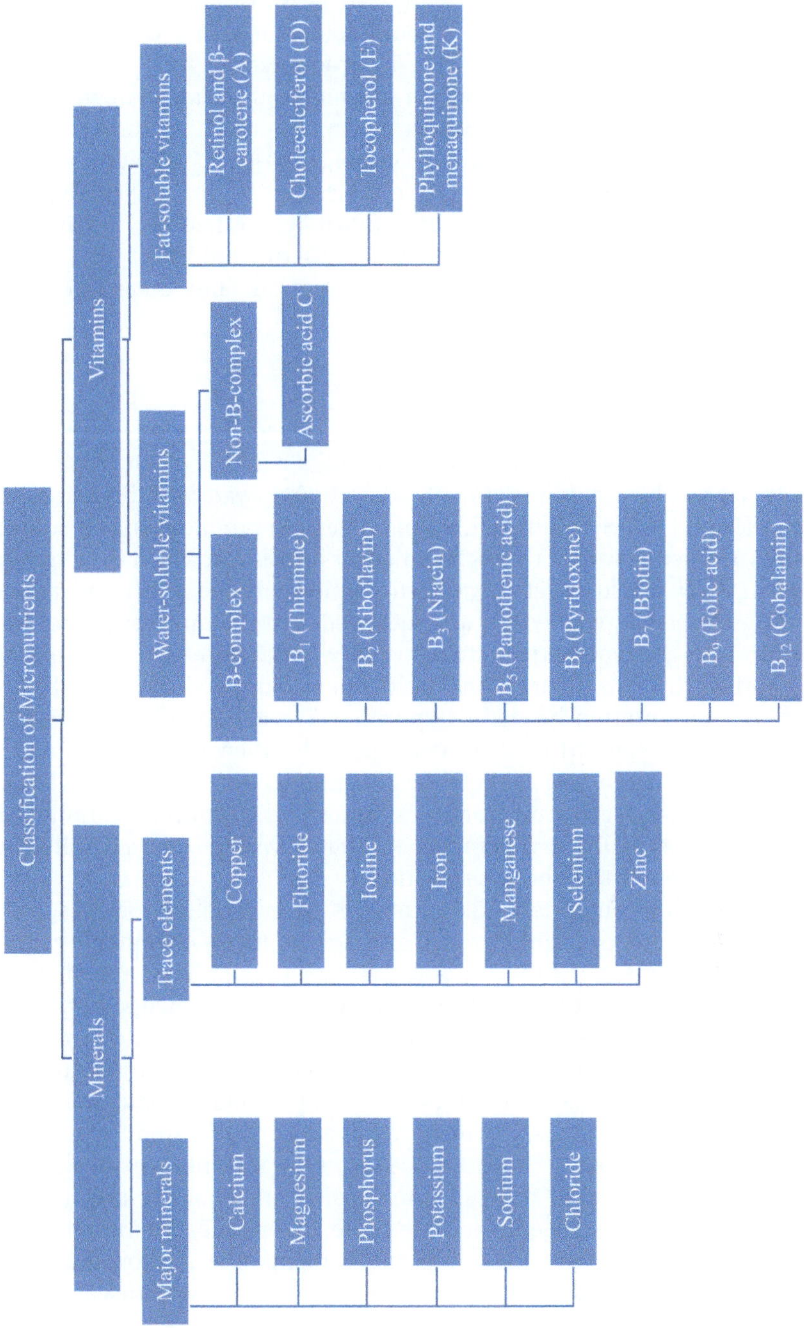

Figure 3.10 Classification of micronutrients.

3.8.2.1 Mineral Absorption

Minerals required by the human body can be further classified into two categories based on the quantities required, namely macro-minerals and trace elements/micro-minerals. The macro-minerals such as calcium, phosphorus, magnesium, and sodium are required in larger quantities as they are responsible for various major roles in physiology.[110] Trace elements such as copper, iron, zinc, selenium, iodine, fluorine, chromium, *etc.*, are needed only in small quantities; nevertheless, their insufficiency causes various serious life-threatening conditions if left untreated.[105] Table 3.4 provides a comprehensive overview of various mineral absorption mechanisms, sites, factors that affect their absorption, and their significant functions and deficiency diseases.

3.8.2.2 Vitamin Absorption

Vitamins are broadly categorized into two kinds based on their solubilities, namely fat-soluble vitamins (A, D, E, K) and water-soluble vitamins (B complex and C).[6] Similarly to the lipid digestion products, the fat-soluble vitamins are also absorbed through a micellar process, in which the vitamins are taken up by mixed micelles with other lipids and bile acids, transported into the enterocytes, where they are packaged into chylomicrons, and then taken up into the lymphatic system.[85] Further, the water-soluble vitamins are absorbed through the intestinal epithelia *via* simple diffusion, followed by active transportation by vitamin-specific Na^+ vitamin cotransporters.[111] Vitamin B_{12} (cobalamin) is absorbed especially when the intrinsic factor produced by stomach parietal cells first combines with vitamin B_{12} before the complex attaches to a membrane receptor at the terminal ileum and is absorbed. Additionally, the complex separates within the cell, allowing vitamin B_{12} to enter the portal circulation.[17]

Vitamins play an important role in numerous metabolic processes, making them necessary for healthy growth and development. As humans cannot synthesize vitamins or synthesize only inadequate quantities, the vitamins must be obtained from the diet or through medical supplements. In general, most water-soluble vitamins are available for intestinal absorption from two sources, namely the diet and produced by bacteria in the large intestine. Dual-origin vitamins include biotin, thiamine, pantothenic acid, folic acid, and riboflavin. Additionally, the body can produce niacin from tryptophan, but it may also absorb it from food sources in the intestines.[6] Further, vitamins play very crucial roles in intermediary metabolism and the specialized metabolism of specific organs. Table 3.5 provides a comprehensive overview of various vitamin absorption mechanisms, sites, factors that affect their absorption and their significant functions, deficiency diseases, and hypervitaminosis.

Table 3.4 Mineral absorption mechanisms, sites, factors that affect their absorption, and their significant functions and deficiency diseases.

Mineral	Method of absorption	Absorption site and approximate % absorbed	Factors that affect absorption	Significant functions and deficiency diseases
Calcium	Absorption of Ca is *via* both passive paracellular absorption (higher concentration) and active transcellular absorption mechanisms [the Ca^{2+} ion concentration in the chyme is low (<5 mM)]. Ca is carried by a specific Ca-binding protein (CaBP), calbindin, in the cytoplasm and then the Ca^{2+} ions are pumped out of the cell across the basolateral border by primary active transport which involves Ca^{2+}-ATPase	Ca is actively absorbed in the duodenum (major site) and jejunum. Further, the absorption of Ca is dependent on and regulated by the parathyroid hormone and vitamin D (calcitriol). Absorbed up to 60% in infants and children and ≤15–20% in adults	The absorption efficiency of Ca is negatively affected by vitamin D deficiency, the aging process, chelating effect (due to the excessive intake of oxalate-, phytic acid-, and dietary fiber-containing foods), and competitive effect (due to excessive intake of other divalent cations, *e.g.* Mg^{2+}). However, Ca absorption is enhanced when intake is low or needs are high, at lower pH, and with higher dietary intake of lactose	Essentially, Ca is utilized by the human body for building bones, formation of teeth, and in maintaining bone strength and density. Further, it also aids in various physiological functions such as hormone secretion, blood coagulation, and muscle contraction, and maintains cell membrane integrity and nervous system function. Ca deficiency leads to numerous skeletal abnormalities, osteomalacia, osteoporosis and rickets
Copper	Cu has two absorption processes, namely an active transport and passive diffusion, when the Cu intake is low and high, respectively. In addition, the Cu is cleaved from the food before absorption by HCl and pepsin	Cu is predominantly absorbed in the duodenum of the small intestine. The absorption rate varies from 25 to 60%	The absorption decreases as the Cu intake increases, and also with intake of fiber, phytic acid, and other minerals such as zinc, iron, calcium gluconate, and molybdenum	Cu aids in red blood cell formation, utilization of iron energy production, bone formation, connective tissue synthesis, and antioxidation. It also serves as a cofactor for numerous proteins and enzymes. It plays an important role in

Table 3.4 (*Continued*)

Mineral	Method of absorption	Absorption site and approximate % absorbed	Factors that affect absorption	Significant functions and deficiency diseases
				the cellular functions of skeletal muscle, heart, bone, connective tissue, neural transmission in the brain, and the immune system
Iron	The non-heme ferric form of iron (Fe^{3+}) is cleaved by HCl from the food and then converted into its ferrous form (Fe^{2+}) by ascorbate and the iron reductases (present in the brush border of the duodenum). Further, the iron (Fe^{2+}) is actively absorbed into epithelial cells by the divalent metal cation transporter (DMT1), which transports Fe^{2+} along with cotransport of H^+. However, heme iron is absorbed intact into the enterocytes and converted to its ferrous form intracellularly. Intracellularly, the enzyme ferroxidase converts Fe^{2+} to Fe^{3+}, which is then bound to iron-binding protein complexes that aid passage through the cytosol	Fe is absorbed in the duodenum in the presence of bile. The absorption percentage of heme Fe is 10–30% and that of non-heme iron is 2–10%. Further, the transporter protein IREG1 aids the basolateral transport (export) of Fe^{3+} from the cell and enters into local capillaries for the circulation chaperoned by its plasma carrier protein, transferrin	Various dietary components, such as phytic acid, oxalic acid, Ca, Zn, and Mn, form insoluble complexes with Fe and reduce absorption. In comparison with Fe^{2+}, Fe^{3+} has a greater tendency to form insoluble complexes with other ions in the lumen, which hinders Fe absorption. Inadequate Fe stores or increased needs (growth, pregnancy) increase absorption. Several compounds that increase acidity, such as vitamin C and lactic, aspartic, and glutamic acids, increase absorption	Fe aids in the formation of hemoglobin and myoglobin. In addition, it supports red blood cell formation and the prevention of Fe deficiency anemia. It also acts as a cofactor in various enzyme-catalyzed reactions. Fe deficiency leads to anemia

Magnesium	Mg is absorbed by two mechanisms, namely active transport when there are low luminal Mg concentrations and passive transport when there are higher concentrations. Further, when the intake is reduced, the absorption percentage increases	Mg is absorbed throughout the small intestine and the distal small intestine is the predominant absorption site. Generally, around 30–40% of intake is absorbed	Mg absorption is inversely associated with its intake. Further, a high intake of dietary Ca, P, and proteins reduces magnesium absorption. Other dietary factors such as phytate, saturated animal fats, and fiber also reduce the absorption of Mg	Mg aids in the regulation of blood pressure and sugar and the formation of bones, strengthens the immune and nervous system function, and modulates cardiovascular excitability (heart rhythm). It also plays a crucial role in the activation of enzymes for glucose, DNA, and protein metabolism, and in fatty acid synthesis. Mg deficiency can occur due to poor intake of Mg-rich foods, older age, and long-term use of diuretic medication. Acute hypomagnesemia can occur as a result of electrolyte shifts such as those seen in refeeding syndrome
Zinc	Dietary Zn is usually bound to proteins, amino acids, and nucleic acids, which are hydrolyzed by HCl and enteric enzymes prior to absorption. Zn is absorbed *via* active transport at low luminal Zn concentrations and passive transport at high concentrations. Zn homeostasis is largely	Free ionic Zn is mostly absorbed in the duodenum at low pH. Low intakes of Zn are absorbed more efficiently than higher intakes. Usually, around 20–40% of Zn is absorbed	Various endogenous factors, such as citric acid, picolinic acid (a tryptophan metabolite), histidine, and cysteine, serve as ligands and enhance Zn absorption. Certain animal proteins in the diet enhance zinc absorption. Riboflavin enhances the intestinal uptake of Zn by	Zn aids in the synthesis of proteins, growth of genital organs, immunity, growth and development, nervous system function, reproduction, taste and smell. Intestinal disorders such as Crohn's disease, ulcerative colitis, celiac sprue, short-bowel syndrome, jejunal

Table 3.4 (*Continued*)

Mineral	Method of absorption	Absorption site and approximate % absorbed	Factors that affect absorption	Significant functions and deficiency diseases
	regulated by its uptake and loss through the small intestine. Zn is absorbed into the enterocytes by a carrier-mediated process and metallothioneine (a sulfur-rich metal-binding protein present in the intestinal mucosa) is thought to be involved in the Zn regulatory mechanism. From the enterocytes, Zn is transferred to the portal circulation mostly bound to albumin, but also to globulin and transferrin		acting as a carrier and transferring Zn from outer surface layers into inner membrane layers. Further, absorption of Zn is hindered by Fe, Ca, phytic acid, polyphenols, and an excessive intake of dietary fiber. In addition, inflammation increases metallothionein synthesis, which may reduce Zn absorption	bypass, cystic fibrosis, and pancreatic insufficiency lead to malabsorption of Zn. Zn deficiency risk factors include trauma, burns, surgery, chronic diarrhea, chronic alcoholism, poor intake of animal and dairy products, renal disease, and excess Fe or Cu supplementation
Phosphorus	Initially, the P present in foods has to undergo a hydrolysis process with pancreatic or intestinal phosphatases prior to absorption. Further, the P is mostly absorbed as inorganic phosphate through two mechanisms, namely cotransportation with sodium, which is dependent on and	Phosphate is absorbed efficiently in the duodenum of the small intestine followed by the jejunum and ileum. However, owing to the greater length of the jejunum, the majority of phosphate is absorbed there.	In the case of animal-based P, 40–60% is absorbed, which varies based on the degree of gastrointestinal vitamin D receptor activation. Inorganic P present in processed foods such as processed cheese and some soda (cola) drinks is completely absorbed (~100%).	Phosphorus, usually in the form of creatinine, phosphate, and ATP, plays a crucial role in bone formation, cellular metabolism, pH maintenance, hormone secretion, energy metabolism, glycolysis, and muscle and nerve function.

	enhanced by vitamin D, and a passive diffusion process through the paracellular pathway due to electrochemical gradients that occur during high intake of phosphate. Finally, it is likely that the absorbed phosphate exits the basolateral membrane by facilitated diffusion *via* unknown phosphate transporters	In general, the overall phosphorus absorption varies from 60 to 70%	The absorption of plant P by the human GI tract is less in spite of plant-based foods being rich in P owing to their association with phytates	Deficiency and poor consumption of P can result in anorexia, impaired growth, calcification of bones, osteomalacia (adults), rickets (children), skeletal demineralization, weakness, cardiac arrhythmias, respiratory insufficiency, increased erythrocytes, lymphocyte dysfunction, and nervous system disorders
Sodium	There are three main mechanisms involved in Na^+ absorption and transportation: (i) Nutrient-coupled Na^+ absorption mediated by Na^+-dependent nutrient cotransporters such as sugar or amino acids that are either transported actively or powered by Na^+/K^+-ATPase transport protein (ii) Electroneutral NaCl absorption mediated principally *via* the Na^+/H^+ exchange mechanism (iii) Colon-predominant electrogenic Na^+ absorption that occurs *via* the epithelial Na^+ channels (ENaC)	In the small intestine Na^+ is absorbed in the duodenum and jejunum by a variety of Na^+–nutrient cotransporters by an active mechanism and is not physiologically regulated. In the ileum Na^+ absorption occurs against very steep electrochemical gradients (110 mequiv L^{-1}, 5–15 mV), *via* efficiently active Na transport across a membrane that is relatively impermeable to Na	Various factors such as the effects of water flow, Na concentration, presence of glucose, galactose, and bicarbonate, and variations in the amino acid composition of luminal fluid affect Na absorption in the small intestine	Na aids in the regulation of water balance by maintaining osmotic pressure, maintains acid–base balance, regulates blood pressure, membrane potential muscle, and nerve function, maintains heartbeat, and actively transports molecules across cell membranes. Lack of Na leads to nausea, vomiting, fatigue, headache, loss of appetite, short-term memory loss, cramps, seizures, muscle weakness, and coma

Table 3.4 (*Continued*)

Mineral	Method of absorption	Absorption site and approximate % absorbed	Factors that affect absorption	Significant functions and deficiency diseases
Potassium	Dietary K is absorbed by passive diffusion as K is intrinsically soluble and quickly dispersed in the luminal water of the upper digestive tract	The small intestine is the primary site of K absorption. Approximately 90% of the ingested K is absorbed	The factors that significantly affect K absorption are still unknown but research has suggested that potentially there could be many as yet unidentified inhibitors of or food matrix effects on K absorption	K aids in various physiological functions such as synthesis of glycogen, proteins, and electrolytes, is involved in nerve impulse transmission and muscle formation, maintains the balance of fluids and minerals in and out of the body, and maintains normal blood pressure. K deficiency leads to gastrointestinal disorders, tingling or numbness, muscle weakness, palpitations, hyperaldosteronism, depression, confusion, psychosis, and hallucinations. Severe hypokalemia is an extreme K deficiency condition that may result in muscular paralysis or abnormal heart rhythms (cardiac arrhythmias), which can be fatal

Chloride	The absorption of Cl^- from the intestinal lumen proceeds through three distinct mechanisms: (i) Passive paracellular pathway (ii) Electroneutral pathways that involve coupled Na^+/H^+ and Cl^-/HCO_3^- exchange (iii) HCO_3^--dependent Cl^- absorption	The passive pathway is predominant in the small intestinal epithelium especially in the duodenum and jejunum. In addition, the jejunum is the predominant site of Cl^- absorption. Further, the main route of Cl^- absorption in the ileum and colon is the electroneutral exchange pathway	Factors such as water and Na transport affect chloride absorption in the jejunum	Chloride aids in the regulation of water distribution, in the maintenance of acid–base balance, and in the transmission of nerve impulses
Molybdenum	Mo is absorbed *via* active transportation when the serum Mo concentration is low and *via* passive diffusion when its serum concentration is high	The exact region where Mo absorption occurs in the intestinal tract is not known. In general, the overall Mo absorption varies from 88 to 93%	The absorption of Mo is affected by the nature and complexity of food. For example, Mo ingested in liquid form was rapidly and totally absorbed whereas the rates and extent of absorption were lower for composite meals. Further, increasing the levels of administration and addition of black tea reduced Mo absorption	Mo is a structural constituent of molybdopterin, a cofactor that is required for the function of the enzymes sulfite oxidase, which aids in the metabolism of sulfur-containing amino acids and heterocyclic compounds, and xanthine oxidase, aldehyde oxidase, and mitochondrial amidoxime reducing component, which are involved in the metabolism of drugs and toxins

Table 3.4 (*Continued*)

Mineral	Method of absorption	Absorption site and approximate % absorbed	Factors that affect absorption	Significant functions and deficiency diseases
Selenium	Dietary Se is predominantly absorbed *via* active transport through a sodium pump; however, the intestinal absorption mechanism of Se differs depending on the chemical form of the element. In addition, Se in the form of selenoproteins (selomethionine and selenocysteine) is readily absorbed	The absorption of Se in humans occurs mainly in the duodenum of the small intestine and the cecum of the large intestine. The absorption ranges from 50 to 100%	The absorption of Se increases with its inadequacy or deficiency	Se acts as a cofactor for many enzymes, such as glutathione peroxidase, thioredoxin reductase, and deiodinases. It also binds with toxic substances such as As, Cd, and Hg to make them less harmful. Further, it plays a crucial role in many body functions such as the formation of thyroid hormones, antioxidant defense, reproduction, muscle formation, cancer prevention, and stimulation of the immune system. Lack of Se can lead to degeneration of skeletal muscles and reproductive disorders

Manganese	The exact absorption process and mechanism that control Mn absorption are not completely understood but appear to be similar to those for other minerals	Only ~1–5% of dietary Mn is absorbed in humans	The efficiency of Mn absorption increases with low Mn intakes and decreases with higher intakes to prevent toxicity. Further, non-heme Fe intake may compete for binding sites and reduce Mn absorption due to competition	Mn acts as a cofactor for enzymes such as manganese superoxide dismutase, arginase, and pyruvate carboxylase. Thus, Mn is involved in the metabolism of amino acids, cholesterol, glucose, and carbohydrates, scavenges reactive oxygen species, and aids in bone formation, reproduction and immune response. Deficiency of Mn can cause bone demineralization and poor growth in children, skin rashes, hair depigmentation, decreased serum cholesterol, alterations in lipid and carbohydrate metabolism, and abnormal glucose tolerance

Table 3.5 Vitamin absorption mechanisms, sites, factors that affect their absorption, and their significant functions and deficiency diseases.

Vitamin	Method of absorption	Absorption site and approximate % absorbed	Factors that affect absorption	Significant functions, deficiency diseases and hypervitaminosis
Vitamin A (retinol)	Enzymes such as proteases and lipases are necessary to hydrolyze proteins to release vitamin A and hydrolyze retinyl esters, respectively. Fat and bile acids are essential for the incorporation of vitamin A into micelles and micellar solubilization. Further, in lymphatic uptake the retinol is taken up by mucosal cells by a saturable process involving transmembrane protein STRA6, and/or retinoid-binding protein 2 (RBP2). However, the absorption of carotenoids from micelles is thought to involve direct diffusion through the plasma membranes of the enterocytes, and the process may actually be carrier mediated	Around 80–90% of dietary vitamin A is absorbed in humans, mainly in the duodenum and jejunum of the small intestine	Fat malabsorption, β-carotene cleavage, biliary atresia, celiac disease, obstructive jaundice, very low-fat diet, and pancreatic insufficiency (pancreatitis, selenium deficiency, cystic fibrosis, cancer) lead to decreased absorption and/or malabsorption	Vitamin A plays an indispensable role in performing various physiological functions such as vision, gene expression, and reproduction. It also aids in the maintenance of cellular differentiation, antioxidant properties, and cell-mediated immunity. Deficiency of vitamin A causes night blindness, keratomalacia (degeneration of the cornea), follicular hyperkeratosis, xerophthalmia, and keratinization of the skin. Excessive doses and higher storage levels lead to hypertension, skin exfoliation, hepatocellular necrosis, alopecia, ataxia, bone and muscle pain, dermatitis, cheilitis, conjunctivitis, and hyperlipidemia

Vitamin D (calcitriol-D_3)	Fat and bile acids play a crucial role in incorporating them into micelles and vitamin D is absorbed by a non-saturable passive diffusion process in the lumen of the small intestine	Approximately 50% of dietary vitamin D is absorbed in the duodenum and jejunum of the small intestine. The absorption occurs faster in the upper portions of the small intestine (duodenum and ileum) and the greatest amount of vitamin D is probably absorbed in the distal region where food has a longer transit time	Fat malabsorption and pancreatic or biliary insufficiency lead to decreased absorption	Vitamin D plays a significant role in the maintenance of serum Ca and P levels, bone health, and parathyroid function. Also, it plays crucial role in homeostasis by enhancing intestinal absorption, cell maturation in the small intestine, and insulin secretion and promotes differentiation in certain epithelia. Deficiency of vitamin D causes rickets (disordered bone modeling, poor mineralization of bone) in children and osteomalacia in adults, deformity of bone and pathological fractures, and decreased serum concentrations of Ca and phosphate. High doses and storage lead to metastatic calcification, renal damage, and mental disorders
Vitamin E (α-tocopherol)	Vitamin E is absorbed by a non-saturable, micelle-dependent passive diffusion mechanism that is reliant on micellar solubilization, adequate presence of fat in the lumen, absorption of lipids, and the presence of bile salts and pancreatic juice	Generally, around 20–80% absorption efficiencies have been reported for vitamin E. It is mostly absorbed in the duodenal and jejunal areas of the small intestine	Pancreatic or biliary insufficiency leads to malabsorption. The absorption decreases as intake increases and the absorption efficiency can be $<$10% at pharmacological doses	Vitamin E acts as an antioxidant and free radical scavenger in cell membranes. It acts in conjunction with other antioxidants such as Se. Deficiency of vitamin E will cause rare serious neurological dysfunction, red

Table 3.5 (*Continued*)

Vitamin	Method of absorption	Absorption site and approximate % absorbed	Factors that affect absorption	Significant functions, deficiency diseases and hypervitaminosis
				blood cell fragility (hemolytic anemia), peripheral neuropathies, ophthalmoplegia, and destruction of posterior columns of the spinal cord, and it suppresses cell-mediated immunity. Higher doses and excessive storage lead to impaired white blood cell function and increase the incidence of hemorrhagic stroke
Vitamin K (phylloquinone-K_1)	The K vitamers such as phylloquinones, menaquinones, and menadione are absorbed across the brush border *via* non-carrier-mediated passive diffusion, and their diffusion rates are affected by the micellar contents of lipids and bile salts	The mixtures of phylloquinones and menaquinones that diets typically contain appear to be absorbed with a wide range of efficiencies of 5–70%. However, as the free vitamer they are well absorbed (~80%). The absorption occurs in the duodenum, jejunum, and distal part of the small intestine and in the colon	Pancreatic or biliary insufficiency will lead to malabsorption. Further, conditions resulting in impaired luminal micelle formation such as dietary mineral oil, pancreatic exocrine dysfunction, and bile stasis impair the enteric absorption of the vitamin. In addition, vitamin K may be poorly bioavailable from foods such as green leafy vegetables owing to its association with the thylakoid membrane in chloroplasts	Vitamin K plays a crucial role in the synthesis of proteins involved in blood clotting by acting as a coenzyme for enzymes of blood clotting and bone metabolism. In addition, it also acts as a cofactor in the post-translational carboxylation of glutamic acid residues in many proteins. Its deficiency causes impaired blood clotting, hemorrhagic disease of the newborn, and fat malabsorption in adults. High doses and storage lead to dyspnea, flushing, and cardiovascular collapse

Vitamin C (ascorbic acid)	Vitamin C is absorbed by both passive and active processes. At high doses, passive diffusion plays a significant role in absorption. At low doses, saturable, carrier-mediated active transport mechanisms such as Na^+-dependent vitamin C transporters (SVCTs) and glucose transporter 1 (GLUT1) play a crucial role in absorption	The absorption efficiency of vitamin C is high, in the range 80–90%. It is absorbed along the entire length of the small intestine and mainly in the ileum region	The absorption of vitamin C decreases to <50% at doses above 1 g day^{-1}. Further, D-isoascorbic acid lead to vitamin C malabsorption	Vitamin C acts as an antioxidant, promotes iron absorption and the production of noradrenaline, and inhibits the production of nitrosamines in the stomach, Further, it aids in the biosynthesis of collagen, carnitine, bile acids, and norepinephrine. In addition, it also aids in enhancing proper functioning of the hepatic oxygenase system and increases the intestinal absorption of non-heme iron. Deficiency of vitamin C leads to scurvy, impaired wound healing, loss of dental cement, subcutaneous hemorrhaging, fatigue, defects in ossification and bone growth (in infants), bleeding into body cavities, and abnormalities in connective tissues. Higher doses can lead to an increased risk of oxalate kidney stones
Vitamin B_1 (thiamine)	The phosphorylated dietary form of thiamine has to cleaved by phosphatases to release free thiamine prior to absorption. Further, the energy-dependent active transport mechanism is used for absorption at low luminal	The absorption of thiamine usually occurs in the duodenum, but there is no available literature on its absorption percentage/efficiency	Various factors decrease the absorption of thiamine, such as pyrithiamine (antivitamin), excess ethanol/alcohol consumption (interferes with transport), and folate deficiency	Thiamine serves as a coenzyme for enzymes such as ketoglutarate dehydrogenase and transketolase and helps to maintain proper functioning of the heart and nervous and digestive systems.

Table 3.5 (*Continued*)

Vitamin	Method of absorption	Absorption site and approximate % absorbed	Factors that affect absorption	Significant functions, deficiency diseases and hypervitaminosis
	concentrations (<2 μM) and passive diffusion transport is used at higher concentrations (*e.g.* a 2.5 mg dose for humans)		(interferes with the replication of enterocytes)	Thiamine deficiency leads to beriberi (peripheral nerve damage), Wernicke–Korsakoff syndrome (central nerve damage), cardiovascular dysfunction, cerebral dysfunction, impairment of nerve impulse transmission, congestive heart failure, low peripheral vascular resistance, muscle weakness, polyneuritis, inadequate ATP (adenosine triphosphate) synthesis, and abnormal carbohydrate metabolism
Vitamin B_2 (riboflavin)	Riboflavin is liberated in its free form due to the proteolytic action of hydrochloric acid and phosphatases in releasing riboflavin coenzymes from the protein complexes and also the subsequent hydrolytic activities of non-specific brush border pyrophosphatases. The free form of riboflavin is absorbed *via* an energy-dependent carrier-mediated process through highly	Riboflavin is absorbed across the proximal small intestine and colon. However, it is significantly absorbed in the jejunum area of the small intestine. Around 95% of riboflavin is absorbed	The absorption of riboflavin is proportional to intake. The absorption increases in the presence of riboflavin deficiency, bile salts, and factors that stimulate intestinal motility. In addition, the consumption of alcohol, psyllium fiber, and divalent metals (Cu, Zn, Fe, Mn) that form chelates with riboflavin reduce the absorption of riboflavin	Riboflavin plays a significant role in carbohydrate, amino acid, and lipid metabolism, aids in the conversion of vitamin B_6 to its active form, serves as a coenzyme in redox reactions of fatty acids and the tricarboxylic acid (TCA) cycle, and acts as a prosthetic group of flavoproteins. Deficiency of riboflavin causes lesions at the corner of the mouth, lips, and tongue, seborrhoeic dermatitis, hyperemia, edema of

	specific riboflavin transporters located on the enterocyte brush border, basolateral surface, and intracellular vesicles			nasopharyngeal mucosa, cheilosis, angular stomatitis, glossitis, and normochromic and normocytic anemia
Vitamin B_3 (niacin/ nicotinic acid)	At low concentrations, niacin is absorbed *via* Na^+-dependent, carrier-mediated facilitated diffusion, by the organic ion transporter 10 (OAT10). At high doses, it appears to be absorbed *via* a passive diffusion mechanism	Niacin is absorbed in the stomach and small intestine	Niacin can be synthesized from tryptophan, although the process is inadequate to meet needs alone. Further, the synthesis of niacin from tryptophan requires vitamins B_2 and B_6	Niacin is essential for the biosynthesis of the pyridine nucleotides NAD(H) and NADP(H). Hence the vitamin has important roles in virtually all aspects of metabolism. It also serves as a coenzyme for several dehydrogenases. Deficiency of niacin leads to several disorders, such as pellagra (with photosensitive dermatitis, depressive psychosis, and diarrhea), glossitis, stomatitis, vaginitis, vertigo, and burning dysesthesias. Higher doses and storage lead to vasomotor phenomenon (flushing), hyperglycemia, parenchymal liver damage, and hyperuricemia
Vitamin B_5 (pantothenic acid)	Dietary coenzyme A (CoA) and acyl carrier protein (ACP) are hydrolyzed in the intestinal lumen by hydrolases (pyrophosphatase, phosphatase) to release the pantothenic acid prior to absorption.	Limited data suggest that the absorption efficiency of pantothenic acid from foods varies from 40 to 61%. It is absorbed in the small intestine and gut	Alcohol intake and inflammatory bowel disease may decrease the absorption vitamin B_5	Pantothenic acid acts a functional part of coenzyme A and acyl carrier protein and aids in the synthesis and β-oxidation of fatty acids. In addition, it is involved in the synthesis of cholesterol, steroid hormones, vitamins A and D, several amino acids,

Table 3.5 (*Continued*)

Vitamin	Method of absorption	Absorption site and approximate % absorbed	Factors that affect absorption	Significant functions, deficiency diseases and hypervitaminosis
	Free pantothenic acid is absorbed by two mechanisms, active transport (saturable mechanism facilitated by Na^+-dependent multivitamin transporter) and passive diffusion, at low and high luminal concentrations, respectively			and δ-aminolevulinic acid. It acts as a precursor for the corrin ring of vitamin B_{12}, the porphyrin ring of heme, and cytochromes. Deficiency of pantothenic acid causes neuromotor disorders, mental depression, GI complaints, increased insulin sensitivity, fatigue, abdominal pain, vomiting, insomnia, and diarrhea
Vitamin B_6 (pyridoxine/ pyridoxal phosphate)	The absorption of vitamin B_6 depends on the conversion of ingested forms into pyridoxine, pyridoxal, and pyridoxamine by digesting the binding proteins through dephosphorylation by phosphatase enzyme prior to absorption *via* passive diffusion	The absorption efficiency of vitamin B_6 is around 75%	The absorption of vitamin B_6 is pH dependent so any deviation of the pH conditions will lead to malabsorption	Vitamin B_6 acts as a coenzyme in amino acid metabolism, inhibits receptor affinity to DNA and thereby modulates steroid activity, aids in the synthesis of niacin from tryptophan, and also aids in the synthesis of several neurotransmitters and δ-aminolevulinic acid. Its deficiency leads to convulsions, stomatitis, angular cheilosis, glossitis, irritability, depression, confusion, normochromic and normocytic anemia, abnormal electroencephalograms, and disorders in amino acid metabolism

Vitamin B_7 (biotin)	Prior to absorption *via* carrier-mediated diffusion, the protein-bound biotin has to be cleaved and reduced to its free form by the action of enzymes such as protease and biotinidase	The absorption of biotin takes place in the small intestine (mostly in the distal region – jejunum and ileum) and gut	Several disease conditions such as inflammatory bowel disease and achlorhydria (a condition in which the stomach does not produce hydrochloric acid) lead to malabsorption. Further, avidin (biotin-binding protein) found in raw egg whites may impair absorption	Biotin aids in the metabolism of carbohydrates, amino acids, and lipids, and acts as a coenzyme for carboxylation reactions in gluconeogenesis and fatty acid synthesis. Biotin deficiency leads to impaired fat and carbohydrate metabolism, dermatitis, alterations in mental status, myalgias, hyperesthesias, anorexia, seborrheic dermatitis, alopecia, lactic acidosis, and organic aciduria
Vitamin B_9 (folic acid)	The folates present in foods are usually in their polyglutamate form and they must be cleaved into their monoglutamate form by zinc-dependent brush border enzymes (folyl conjugases) before absorption *via* active transport. Passive transport may take place at high doses. Further, some forms of folate that are excreted in bile may be reabsorbed *via* enterohepatic circulation	The folates are predominantly absorbed in the jejunum area of the small intestine. The absorption efficiency of dietary folate can range between 50 and 100%	On average, around 50% of folate naturally occurring in foods is absorbed. However, the absorption efficiency of supplemented folate is around 80–100%. Further, several factors, such as Zn deficiency, loss of conjugase activity, alcohol intake, and changes in enteric pH, lead to reduced absorption of folates	Folic acid aids in the metabolism of nucleic and amino acids and the synthesis of DNA and red blood cells, and acts as a coenzyme in single-carbon metabolism. In addition, it plays an essential role in the *de novo* synthesis of nucleotides and acts as an integral component for the regeneration of the universal methyl donor. Deficiency of folic acid leads to megaloblastic anemia, neural tube defects in babies, glossitis, and diarrhea

Table 3.5 (*Continued*)

Vitamin	Method of absorption	Absorption site and approximate % absorbed	Factors that affect absorption	Significant functions, deficiency diseases and hypervitaminosis
Vitamin B_{12} (cobalamin)	The naturally occurring vitamin B_{12} present in foods is usually bound in coenzyme form to proteins, and the action of pepsin and HCl in the stomach releases vitamin B_{12} from the food. Further, it forms a complex with the intrinsic factor (secreted by gastric parietal cells) in the duodenum and then it is largely absorbed *via* a carrier-mediated active transport mechanism. In addition, vitamin B_{12} excreted in bile may be reabsorbed *via* enterohepatic circulation	The absorption efficiency of vitamin B_{12} is around 50%. It is mainly absorbed in the ileum region of the small intestine	The absorption of cobalamin is affected and reduced by conditions that affect the structure and/or function (such as impaired gastric parietal cell function or with chronic use of proton pump inhibitors) of the stomach or ileum. The absorption efficiency of supplemented vitamin B_{12} decreases significantly with doses higher than 1–2 μg, as the capacity of the intrinsic factor is exceeded	Vitamin B_{12} plays a crucial role in the conversion of folate to the active form and in the metabolism of nucleic and amino acids and aids in the maturation of red blood cells. In addition, it also acts as a coenzyme in the metabolism of propionate, amino acids, and single-carbon units. Deficiency of cobalamin leads to pernicious anemia, loss of intestinal absorption, pancreatic insufficiency, and atrophic gastritis. In addition, it causes several mental disorders, depression, psychoses, and hematological and neurological complications

3.9 Conclusion

In this chapter we have summarized the anatomy and histology of the small intestine, and various physiological roles played by the small intestine such as mucosal immunity and barrier function, intestinal motility, and various small intestinal secretions (water, electrolytes, mucus, brush border enzymes). Further, digestion and absorption of macronutrients, absorption of micronutrients, water, and electrolytes, and various mechanisms, routes, and specific transport proteins required in particular for the absorption of specific nutrients have been discussed in detail. The small intestine has an indispensable and irreplaceable role in food digestion and nutrient absorption in humans. Along with the small intestine, various accessory digestive organs such as the liver, gallbladder, and pancreas play a vital part in the digestion and absorption process. However, digestion in the stomach is only preparatory and dispensable. Nowadays, various techniques such as fluoroscopy, wireless pH capsule, and scintigraphy are widely used to study the motility of the small intestine. It is well acknowledged that the small intestine manometry technique provides the most insightful physiological data about intestinal motility. Additionally, intraluminal manometry is used to identify the pressure waves produced by the circular muscle layer. The small intestine plays a critical role in inhibiting detrimental microorganisms from infiltrating the body with the aid of several characteristic features, including peristaltic waves, the ileocecal valve, the epithelial barrier, and small intestinal immune cells. More research must be carried out in the future to identify and describe the precise transport system or processes involved in the digestion and absorption of different nutrients.

References

1. Z. A. Alicikus and B. Aydin, Toxicity Management for Upper Abdomen Tumors in Radiation Oncology, in *Prevention and Management of Acute and Late Toxicities in Radiation Oncology: Management of Toxicities in Radiation Oncology*, 2020, p. 171.
2. W. Liu, A. Ye, F. Han and J. Han, Advances and challenges in liposome digestion: Surface interaction, biological fate, and GIT modeling, *Adv. Colloid Interface Sci.*, 2019, **263**, 52–67.
3. Z. F. Ma and Y. Y. Lee, Small intestine anatomy and physiology, in *Clinical and Basic Neurogastroenterology and Motility*, Elsevier, 2020, pp. 101–111.
4. J. Campbell, J. Berry and Y. Liang, Anatomy and physiology of the small intestine, in *Shackelford's Surgery of the Alimentary Tract, 2 Volume Set*, Elsevier, 2019, pp. 817–841.
5. B. Osuntokun and S. A. Kocoshis, Anatomy and physiology of the small and large intestine, in *Pediatric Gastrointestinal and Liver Disease*, Elsevier, 2006, pp. 459–474.

6. M. E. Smith and D. G. Morton, *The Digestive System: Systems of the Body Series*, Elsevier Health Sciences, 2011.
7. N. Volk and B. Lacy, Anatomy and physiology of the small bowel, *Gastrointest. Endoscopy Clin.*, 2017, **27**(1), 1–13.
8. J. J. Feher, *Quantitative Human Physiology: An Introduction*, Academic Press, 2017.
9. A. J. Kastl Jr, N. A. Terry, G. D. Wu and L. G. Albenberg, The structure and function of the human small intestinal microbiota: current understanding and future directions, *Cell. Mol. Gastroenterol. Hepatol.*, 2020, **9**(1), 33–45.
10. G. B. Huffnagle and M. Noverr, *GI Microbiota and Regulation of the Immune System*, Springer Science & Business Media, 2008, vol. 635.
11. J. Feher, 8.3—Intestinal and colonic chemoreception and motility, in *Quantitative Human Physiology*, Academic Press, Boston, 2nd edn, 2017.
12. B. Greenwood-Van Meerveld, A. C. Johnson and D. Grundy, Gastrointestinal physiology and function, in *Gastrointestinal Pharmacology*, 2017, pp. 1–16.
13. O. Gouseti, G. M. Bornhorst, S. Bakalis and A. Mackie, *Interdisciplinary Approaches to Food Digestion*, Springer, 2019.
14. G. M. Bornhorst, Gastric mixing during food digestion: Mechanisms and applications, *Annu. Rev. Food Sci. Technol.*, 2017, **8**, 523–542.
15. D. Hooton, R. Lentle, J. Monro, M. Wickham and R. Simpson, The secretion and action of brush border enzymes in the mammalian small intestine, in *Reviews of Physiology, Biochemistry and Pharmacology*, 2015, pp. 59–118.
16. S. Redhai and M. Boutros, The Role of Organelles in Intestinal Function, Physiology, and Disease, *Trends Cell Biol.*, 2021, **31**(6), 485–499.
17. N. G. MacFarlane, Digestion and absorption, *Anaesth. Intensive Care Med.*, 2018, **19**(3), 125–127.
18. J. Feher, 8.5 Digestion and Absorption of the Macronutrients, *Quantitative Human Physiology*, Academic Press, Boston, 2017.
19. J. Bernier-Latmani and T. V. Petrova, Intestinal lymphatic vasculature: structure, mechanisms and functions, *Nat. Rev. Gastroenterol. Hepatol.*, 2017, **14**(9), 510–526.
20. M. H. Montrose, Small Intestine, Absorption and Secretion, *Encyclopedia of Gastroenterology*, 2004, pp. 399–404.
21. A. M. Mowat and W. W. Agace, Regional specialization within the intestinal immune system, *Nat. Rev. Immunol.*, 2014, **14**(10), 667–685.
22. A. A. Argikar and U. A. Argikar, The mesentery: an ADME perspective on a 'new' organ, *Drug Metab. Rev.*, 2018, **50**(3), 398–405.
23. J. C. Coffey and D. P. O'Leary, The mesentery: structure, function, and role in disease, *Lancet Gastroenterol. Hepatol.*, 2016, **1**(3), 238–247.
24. I. Campbell, The mouth, stomach and intestines, *Anaesth. Intensive Care Med.*, 2015, **16**(1), 37–39.

25. B. Shames, Anatomy and physiology of the duodenum, in *Shackelford's Surgery of the Alimentary Tract, 2 Volume Set*, Elsevier, 2019, pp. 786–803.
26. L. M. Bass and B. K. Wershil, Anatomy, histology, embryology and developmental anomalies of the small and large intestine, in *Sleisenger and Fordtran's Gastrointestinal and Liver Disease*, Saunders, Elsevier Inc., Philadelphia, PA, 10th edn, 2016, p. 1649.
27. F. Campbell and C. S. Verbeke, Embryology, Anatomy, and Histology, in *Pathology of the Pancreas*, Springer, 2021, pp. 3–23.
28. F. Campbell and C. S. Verbeke, *Pathology of the Pancreas: A Practical Approach*, Springer Nature, 2020.
29. S. McMahon, The Pancreas, in *Clinical Embryology*, Springer, 2019, pp. 365–370.
30. S. A. Hewes, R. L. Wilson, M. K. Estes, N. F. Shroyer, S. E. Blutt and K. J. Grande-Allen, *In vitro* models of the small intestine: Engineering challenges and engineering solutions, *Tissue Eng., Part B*, 2020, **26**(4), 313–326.
31. J. M. Allaire, S. M. Crowley, H. T. Law, S.-Y. Chang, H.-J. Ko and B. A. Vallance, The intestinal epithelium: central coordinator of mucosal immunity, *Trends Immunol.*, 2018, **39**(9), 677–696.
32. N. F. Shroyer and S. A. Kocoshis, Anatomy and physiology of the small and large intestines, in *Pediatric Gastrointestinal and Liver Disease*, Elsevier, 2011, pp. 324–336.
33. R. Lee and E. A. Mezoff, Anatomy and Physiology of the Small and Large Intestines, in *Pediatric Gastrointestinal and Liver Disease*, 2021, pp. 308–320.e3.
34. V. S. Chandan, Normal Histology of Gastrointestinal Tract, in *Surgical Pathology of Non-neoplastic Gastrointestinal Diseases*, Springer, 2019, pp. 3–18.
35. G. Smith and R. Watson, *Gastrointestinal Nursing*, John Wiley & Sons, 2008.
36. S. Furuya and K. Furuya, Subepithelial fibroblasts in intestinal villi: roles in intercellular communication, *Int. Rev. Cytol.*, 2007, **264**, 165–223.
37. S. Keshav, *The Gastrointestinal System at a Glance*, John Wiley & Sons, 2009.
38. R. W. Kapp Jr, 5 Gastrointestinal Tract as, in *Toxicology of the Gastrointestinal Tract*, 2007, p. 107.
39. A. deFoneska and J. D. Kaunitz, Gastroduodenal mucosal defense, *Curr. Opin. Gastroenterol.*, 2010, **26**(6), 604–610.
40. Y. Akiba and J. D. Kaunitz, Luminal chemosensing and upper gastrointestinal mucosal defenses, *Am. J. Clin. Nutr.*, 2009, **90**(3), 826S–831S.
41. V. Calderaro and F. Rossi, Intestinal chloride secretion: cyclic AMP and Ca^{2+} interactions, in *Regulatory Mechanisms—in—Gastrointestinal Function*, CRC Press, 2017, pp. 27–98.

42. K. E. Barrett and S. J. Keely, Chloride secretion by the intestinal epithelium: molecular basis and regulatory aspects, *Annu. Rev. Physiol.*, 2000, **62**(1), 535–572.
43. R. A. Wapnir and S. Teichberg, Regulation mechanisms of intestinal secretion: implications in nutrient absorption, *J. Nutr. Biochem.*, 2002, **13**(4), 190–199.
44. A. M. Collaco, R. L. Jakab, N. E. Hoekstra, K. A. Mitchell, A. Brooks and N. A. Ameen, Regulated traffic of anion transporters in mammalian Brunner's glands: a role for water and fluid transport, *Am. J. Physiol.: Gastrointest. Liver Physiol.*, 2013, **305**(3), G258–G275.
45. A. S. Verkman, More than just water channels: unexpected cellular roles of aquaporins, *J. Cell Sci.*, 2005, **118**(15), 3225–3232.
46. J. R. Thiagarajah and A. S. Verkman, CFTR pharmacology and its role in intestinal fluid secretion, *Curr. Opin. Pharmacol.*, 2003, **3**(6), 594–599.
47. R. Santaolalla and M. T. Abreu, Innate immunity in the small intestine, *Curr. Opin. Gastroenterol.*, 2012, **28**(2), 124.
48. G. L. Simon and S. L. Gorbach, Intestinal flora in health and disease, *Gastroenterology*, 1984, **86**(1), 174–193.
49. J. Holmgren and C. Czerkinsky, Mucosal immunity and vaccines, *Nat. Med.*, 2005, **11**(4), S45–S53.
50. M. Million, J. Tomas, C. Wagner, H. Lelouard, D. Raoult and J.-P. Gorvel, New insights in gut microbiota and mucosal immunity of the small intestine, *Hum. Microbiome J.*, 2018, **7**, 23–32.
51. B. Johansson-Lindbom and W. W. Agace, Generation of gut-homing T cells and their localization to the small intestinal mucosa, *Immunol. Rev.*, 2007, **215**(1), 226–242.
52. S. Fagarasan, Evolution, development, mechanism and function of IgA in the gut, *Curr. Opin. Immunol.*, 2008, **20**(2), 170–177.
53. H. P. Parkman, P. Orthey and A. H. Maurer, Gastric Motility, in *Encyclopedia of Gastroenterology*, ed. E. J. Kuipers, Academic Press, Oxford, 2nd edn, 2020, pp. 598–612. Available from: http://www.sciencedirect.com/science/article/pii/B978012801238311092X.
54. G. W. Hennig and N. J. Spencer, Physiology of Gastric Motility Patterns, in *Physiology of the Gastrointestinal Tract*, ed. H. M. Said, Academic Press, 6th edn, 2018, ch. 21, pp. 469–484. Available from: http://www.sciencedirect.com/science/article/pii/B9780128099544000219.
55. E. B. Chang and P. S. Leung, Gastrointestinal motility, in *The Gastrointestinal System*, Springer, 2014, pp. 35–62.
56. H. Gregersen, Geometry, Structure and Motor Function of the Gastrointestinal Tract, in *Biomechanics of the Gastrointestinal Tract*, Springer, 2003, pp. 11–43.
57. R. K. Avvari, Biomechanics of the small intestinal contractions, in *Digestive System-Recent Advances*, IntechOpen, 2019.
58. C. Olsson and S. Holmgren, The control of gut motility, *Comp. Biochem. Physiol., Part A: Mol. Integr. Physiol.*, 2001, **128**(3), 479–501.

59. H. Isackson and C. C. Ashley, Secretory functions of the gastrointestinal tract, *Surgery*, 2020, **38**(6), 289–296.
60. M. B. Hansen, Neurohumoral control of gastrointestinal motility, *Physiol. Res.*, 2003, **52**(1), 1–30.
61. S. D. Aros and M. Camilleri, Small-bowel motility, *Curr. Opin. Gastroenterol.*, 2001, **17**(2), 140–146.
62. M. P. Jones and S. Wessinger, Small intestinal motility, *Curr. Opin. Gastroenterol.*, 2006, **22**(2), 111–116.
63. J. C. Ozougwu, Physiology of the liver, *Int. J. Res. Pharm. Biosci.*, 2017, **4**(8),13–24.
64. L. L. Yang, Anatomy and Physiology of the Liver, in *Anesthesia for Hepatico-Pancreatic-Biliary Surgery and Transplantation*, Springer, 2021, pp. 15–40.
65. L. McErlean, The digestive system, in *Fundamentals of Anatomy and Physiology: for Nursing and Healthcare Students*, 2016.
66. A. S. Bodzin and T. B. Baker, Anatomy and Physiology of the Liver, in *Shackelford's Surgery of the Alimentary Tract, 2 Volume Set*, Elsevier, 2019, pp. 1386–1397.
67. P. F. Saldinger and O. E. Bellorin-Marin, Anatomy, embryology, anomalies, and Physiology of the biliary tract, in *Shackelford's Surgery of the Alimentary Tract, 2 Volume Set*, Elsevier, 2019, pp. 1249–1266.
68. C. Housset, Y. Chrétien, D. Debray and N. Chignard, Functions of the Gallbladder, *Compr. Physiol.*, 2011, **6**(3),1549–1577.
69. M. Camilleri, Integrated upper gastrointestinal response to food intake, *Gastroenterology*, 2006, **131**(2), 640–658.
70. P. A. Dawson, Bile formation and the enterohepatic circulation, *Physiology of the Gastrointestinal Tract*, 2018, pp. 931–956.
71. J. Maldonado-Valderrama, P. Wilde, A. Macierzanka and A. Mackie, The role of bile salts in digestion, *Adv. Colloid Interface Sci.*, 2011, **165**(1), 36–46.
72. M. L. Chen, K. Takeda and M. S. Sundrud, Emerging roles of bile acids in mucosal immunity and inflammation, *Mucosal Immunol.*, 2019, **12**(4), 851–861.
73. M. Mañas, E. M. de Victoria, A. Gil, M. D. Yago and J. C. Mathers, The Gastrointestinal Tract, in *Nutrition and Metabolism*, 2010, vol. 2, p. 205.
74. J. Fusco, Y. El-Gohary and G. K. Gittes, Anatomy, Physiology, and Embryology of the Pancreas, in *Shackelford's Surgery of the Alimentary Tract, 2 Volume Set*, Elsevier, 2019, pp. 1062–1075.
75. J. Feher, Pancreatic and Biliary Secretion, in *Quantitative Human Physiology*, 2017, pp. 810–820.
76. D. C. Whitcomb and M. E. Lowe, Human pancreatic digestive enzymes, *Dig. Dis. Sci.*, 2007, **52**(1), 1–17.
77. M. H. Stipanuk and M. A. Caudill, *Biochemical, Physiological, and Molecular Aspects of Human Nutrition-E-book*, Elsevier Health Sciences, 2018.

78. R. E. McConnell, J. N. Higginbotham, D. A. Shifrin Jr, D. L. Tabb, R. J. Coffey and M. J. Tyska, The enterocyte microvillus is a vesicle-generating organelle, *J. Cell Biol.*, 2009, **185**(7), 1285–1298.
79. J. R. Pappenheimer, M. L. Karnovsky and J. E. Maggio, Absorption and excretion of undegradable peptides: role of lipid solubility and net charge, *J. Pharmacol. Exp. Ther.*, 1997, **280**(1), 292–300.
80. H. Daniel, Molecular and integrative physiology of intestinal peptide transport, *Annu. Rev. Physiol.*, 2004, **66**, 361–384.
81. B. Caballero, *Encyclopedia of Human Nutrition*, Elsevier, 2005.
82. A. Ferreira-Lazarte, P. Gallego-Lobillo, F. J. Moreno, M. Villamiel and O. Hernandez-Hernandez, *In vitro* digestibility of galactooligosaccharides: effect of the structural features on their intestinal degradation, *J. Agric. Food Chem.*, 2019, **67**(16),4662–4670.
83. A. Peredo-Lovillo, H. E. Romero-Luna and M. Jiménez-Fernández, Health promoting microbial metabolites produced by gut microbiota after prebiotics metabolism, *Food Res. Int.*, 2020, **136**, 109473.
84. B. E. Goodman, Insights into digestion and absorption of major nutrients in humans, *Adv. Physiol. Educ.*, 2010, **34**(2), 44–53.
85. A. D. Jackson and J. McLaughlin, Digestion and absorption, *Surgery*, 2009, **27**(6), 231–236.
86. P. Sethupathy, J. A. Moses and C. Anandharamakrishnan, Food Oral Processing and Tribology: Instrumental Approaches and Emerging Applications, *Food Rev. Int.*, 2020, 1–34.
87. R. P. Ferraris, Dietary and developmental regulation of intestinal sugar transport, *Biochem. J.*, 2001, **360**(2), 265–276.
88. T. M. Devlin, *Textbook of Biochemistry: With Clinical Correlations*, 2011.
89. M. D. Sitrin, Digestion and absorption of carbohydrates and proteins, in *The Gastrointestinal System*, Springer, 2014, pp. 137–158.
90. Z. A. Bhutta, K. Sadiq and T. Aga, Protein digestion and bioavailability, in *Encyclopedia of Human Nutrition*, 2013, vol. 4, pp. 66–73.
91. B. M. Carlson, The Digestive System, in B. M. Carlson, ed. *The Human Body*, Academic Press, 2019, ch. 12, pp. 321–355. Available from: http://www.sciencedirect.com/science/article/pii/B9780128042540000120.
92. Y. D. Bhutia and V. Ganapathy, Protein digestion and absorption, in *Physiology of the Gastrointestinal Tract*, Elsevier, 2018, pp. 1063–1086.
93. W. F. Boron and E. L. Boulpaep, *Medical Physiology E-book*, Elsevier Health Sciences, 2016.
94. M. Boland, M. Golding and H. Singh, *Food Structures, Digestion and Health*, Academic Press, 2014.
95. A. Acevedo-Fani and H. Singh, Biophysical insights into modulating lipid digestion in food emulsions, *Prog. Lipid Res.*, 2021, 101129.
96. J. Keller, Gastrointestinal Digestion and Absorption, in *Encyclopedia of Biological Chemistry*, ed. W. J. Lennarz and M. D. Lane, Academic Press, Waltham, 2nd edn, 2013, pp. 354–359. Available from: http://www.sciencedirect.com/science/article/pii/B9780123786302001067.

97. J. Chen, Food oral processing: Mechanisms and implications of food oral destruction, *Trends Food Sci. Technol.*, 2015, **45**(2), 222–228.
98. M. Golding and T. J. Wooster, The influence of emulsion structure and stability on lipid digestion, *Curr. Opin. Colloid Interface Sci.*, 2010, **15**(1–2), 90–101.
99. A. Rashidinejad, S. Marze and H. Singh, Lipid Digestion and Bioaccessibility of Lipid-Soluble Compounds, in *Bioaccessibility and Digestibility of Lipids from Food*, Springer, 2021, pp. 171–203.
100. C. T. Phan and P. Tso, Intestinal lipid absorption and transport, *Front. Biosci.*, 2001, **6**(5), D299–D319.
101. K. Hokkanen, A. Tirronen and S. Ylä-Herttuala, Intestinal lymphatic vessels and their role in chylomicron absorption and lipid homeostasis, *Curr. Opin. Lipidol.*, 2019, **30**(5), 370–376.
102. H. Mu and C.-E. Høy, The digestion of dietary triacylglycerols, *Prog. Lipid Res.*, 2004, **43**(2), 105–133.
103. R. Costa-Pinto and D. Gantner, Macronutrients, minerals, vitamins and energy, *Anaesth. Intensive Care Med.*, 2020, **21**(3), 157–161.
104. M. Guo, *Functional Foods: Principles and Technology*, Elsevier, 2013.
105. J. Doley, Vitamins and minerals in older adults: causes, diagnosis, and treatment of deficiency, in *Nutrition and Functional Foods for Healthy Aging*, Elsevier, 2017, pp. 125–137.
106. T. K. Basu and D. Donaldson, Intestinal absorption in health and disease: micronutrients, *Best Pract. Res., Clin. Gastroenterol.*, 2003, **17**(6), 957–979.
107. P. R. Kiela and F. K. Ghishan, Physiology of intestinal absorption and secretion, *Best Pract. Res., Clin. Gastroenterol.*, 2016, **30**(2), 145–159.
108. K. M. Hambidge, Micronutrient bioavailability: Dietary Reference Intakes and a future perspective, *Am. J. Clin. Nutr.*, 2010, **91**(5), 1430S–1432S.
109. S. K. Chang, How Food Structure and Processing Affect the Bioavailability of Nutrients and Antioxidants, in *Encyclopedia of Food Chemistry*, 2019.
110. S. J. Fathima, I. Nallamuthu and F. Khanum, Vitamins and minerals fortification using nanotechnology: bioavailability and Recommended Daily Allowances, in *Nutrient Delivery*, Elsevier, 2017, pp. 457–496.
111. H. M. Said, Intestinal absorption of water-soluble vitamins in health and disease, *Biochem. J.*, 2011, **437**(3), 357–372.

CHAPTER 4

Large Intestine and Gut–Brain–Microbiota Interactions

4.1 Introduction

The large intestine is present in the lower gastrointestinal (GI) tract, and it begins at the ileocecal junction of the GI tract (*i.e.* after the ileum of the small intestine) and ends at the anus. The large intestine can be subdivided into several segments, namely the cecum and vermiform appendix, colon, rectum, anal canal, and anus.[1] The large intestine plays a crucial role in carrying out several major functions of the digestive system, such as digesting previously undigested materials through colonic fermentation, biosynthesis of essential vitamins and other bioactive metabolites, and managing the intestinal fluid and electrolyte balance along with the small intestine, and aids in the formation, storage, transportation, and removal of digestive waste. The colonic motility (motor function) aids in mixing the luminal content and propels it towards the anus for defecation.[2] Further, the large intestine acts as a suitable place for microbial colonization and colonic fermentation due to the long transit time, conducive pH, low cell turnover, and its redox potential.[3]

Additionally, the gut microbiome and its metabolites significantly impact other human systems, namely the nervous system, immune system, endocrine system, and indirectly other organ systems, and function through its bidirectional connection between the brain and gut.[4,5] Alterations in the composition of the gut microbiota in terms of both number and diversity and bacterial metabolites have profound impacts on the host health and metabolism. In recent years, several studies have shown a significant association between gut dysbiosis, health, and diseases. Several diseases, such as neurodegenerative disorders [Parkinson's disease, Alzheimer's disease,

Food Chemistry, Function and Analysis No. 42
Food Digestion and Absorption: Its Role in Food Product Development
By C. Anandharamakrishnan, Jeyan Arthur Moses and S. Priyanka

Published by the Royal Society of Chemistry, www.rsc.org

multiple sclerosis (MS), attention deficit hyperactivity disorder (ADHD), Huntington's disease, amyotrophic lateral sclerosis (ALS), dementia, stress, and depression],[6] mental health (mood, anxiety, social behavior and autism, and age-related cognitive decline),[7,8] inflammatory bowel disease (IBD),[9] celiac disease,[10] type I diabetes,[11] food allergies,[12] type II diabetes,[13] obesity,[14] cardiovascular diseases,[15] non-alcoholic fatty liver disease,[16] chronic constipation,[17] and colorectal cancer,[18] have been widely researched with respect to gut microbiota dysbiosis and the gut–brain–microbiota axis. However, the characteristics of the disease conditions, changes in gut microbiota composition, and relevant mechanisms are outside the scope of this chapter. In this chapter, the impact of specific dietary patterns, food components, and therapeutic interventions such as prebiotics, probiotics, and synbiotics on overall health and specific disease conditions are discussed.

4.2 Gross Anatomy of the Large Intestine

The large intestine is around 150 cm long in adults and around 60 cm long in newborn children.[19] The diameter of the large intestine is around 6 cm but it is widest at the cecum, then it reduces gradually along its length until the rectum.[20] However, the thickness of the colonic wall remains the same along the whole length of the large intestine.[21] The large intestine is suspended in the posterior abdominal wall with the aid of the mesentery. The mesentery restricts the mobility of the cecum and ascending and descending colon as it is relatively narrow, but it is comparatively broad around the transverse and sigmoid colon, allowing considerable movement.[21,22] Figure 4.1 depicts the gross anatomical features of the large intestine. Further, the large intestine has several distinctive anatomical and histological features compared with the small intestine, which is illustrated in Figure 4.2.

4.2.1 Cecum and Appendix

The cecum is the blind-ended pouch that lies below the ileocecal valve and extends downwards towards the vermiform appendix.[21] The remnant and unabsorbed portion of the chyme that leaves the small intestine arrives at the large intestine's cecum *via* the ileocecal valve.[23] The cecum portion is roughly 6 cm in length and 7.5 cm in width.[24] The appendix is believed to be the trace of a redundant organ and it looks like a worm with its narrow and elongated structure.[3] The length of the appendix can range anywhere between 2 and 20 cm, but it is the longest in childhood and shrinks throughout adult life.[25] The appendix was recently found to have a vital role in the mucosal immune system (such as extrathymically derived T lymphocyte- and B lymphocyte-mediated immune responses) and acts as a suitable location for the buildup and reproduction of intestinal bacteria owing to its narrow and twisted shape.[26]

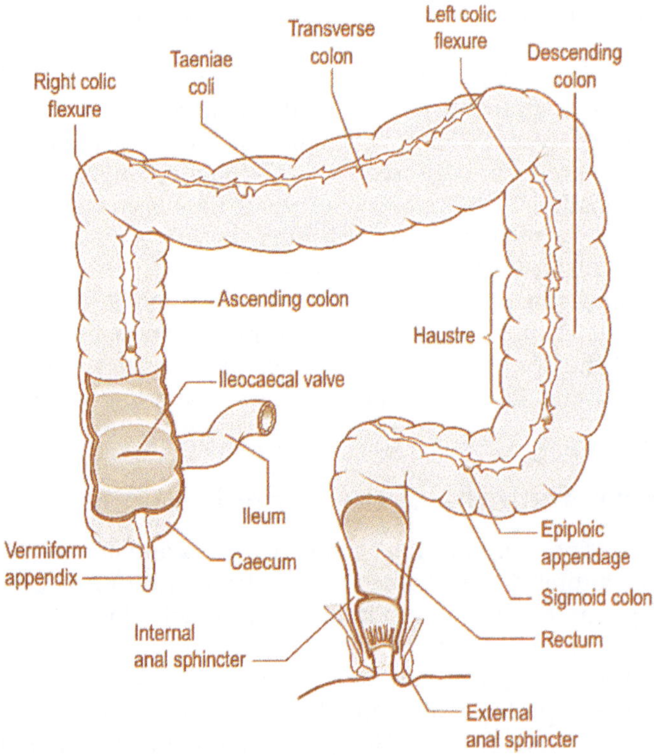

Figure 4.1 Gross anatomical features of the large intestine. Reproduced from ref. 28 with permission from Elsevier, Copyright 2010.

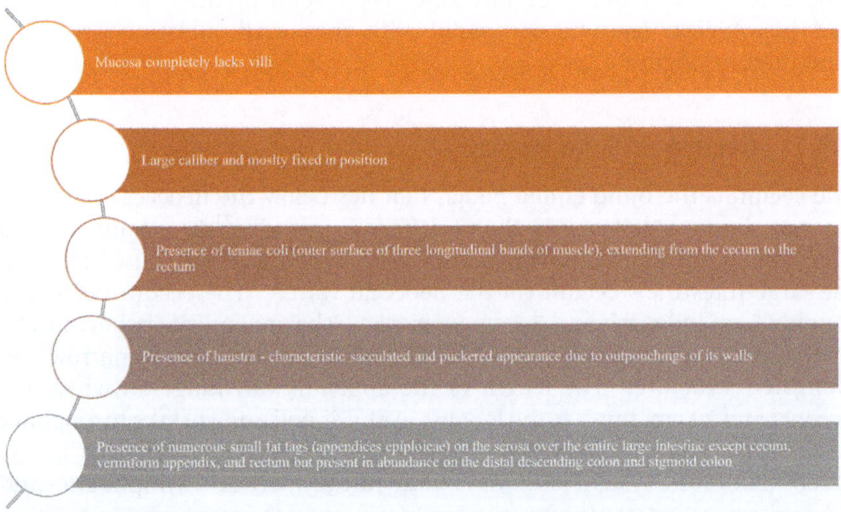

Figure 4.2 Distinctive anatomical and histological features of the large and small intestine.

4.2.2 Colon

The colon segment is seamlessly joined with the cecum, and it is the large intestine's longest segment.[3] The colon is divided into four parts, the ascending colon, transverse colon, descending colon, and sigmoid colon, which measure around 20, 40–50, 25–45, and 40 cm in length, respectively, in adults.[25,27] The food residues from the small intestine after reaching the cecum travel up the ascending colon, which is located on the right side of the abdomen.[28] The ascending colon bends over near the liver at the hepatic flexure to transform into the transverse colon, which dips down towards the pelvis and crosses the abdomen medially to reach the left abdominal quadrant.[29] The transverse colon becomes the descending colon just above the spleen at the left splenic flexure and runs downwards in the abdomen's left side.[22] At the pelvic brim, the descending colon is converted into the sigmoid colon and it is the narrowest segment of the entire large intestine with a diameter of about 2.5 cm.[24]

4.2.3 Rectum, Anal Canal, and Anus

The rectum is present next to the sigmoid colon, and it is *ca.* 15–20 cm in length. Further, the rectum ends at the rectal ampulla, and it connects to the anal canal and anus.[30] The major function of the rectum is to store the feces just before defecation, so in general the rectum is usually empty.[31] The anal canal is around 4–5 cm in length and opens to the exterior of the body at the anus.[32]

4.3 General Physiology of the Large Intestine

The major functions of the large intestine are water and electrolyte absorption, mucus secretion, chemical digestion and fermentation by gut microbiota, absorption of nutrients and other products of bacterial action, gut motility, and formation, storage, transportation, and elimination of feces.

4.3.1 Formation, Storage, Transportation, and Elimination of Feces

In general, in one day the large intestine receives around 1.5–2 L of chyme from the small intestine within 8–9 h of ingestion.[33] The small intestine usually absorbs around 90% of the ingested water. The remaining water in the chyme is absorbed in the large intestine and the liquid chyme is converted to semi-solid feces and gradually becomes solid as it moves forward in the colon.[34] The feces typically consist of undigested food materials (including fiber, unabsorbed fats, and proteins), microorganisms, mostly bacteria, inorganic waste, intestinal mucosa (old epithelial cells), bile pigments, small amounts of electrolytes, and water.[3] The mucus secreted by the goblet cells of the large intestine aids in the lubrication of feces and the main stimulation factor for its secretion is distension and mechanical irritation of the intestinal walls.[28] The presence of an adequate amount of water, dietary fiber, and thick mucus secretion aids in the smooth movement of feces through the GI tract. Also, feces have a characteristic brown color due to the presence of stercobilin and urobilin, which are the hemoglobin breakdown products from old red blood cells.[35] The feces are stored in the colon before defecation because their movement in the colon is a sluggish process and it may take anywhere from 3 days (to clear around 70%) to 7 days (a week) for the complete expulsion of all the remnants of a meal, and also the colonic transit is slower in women than men.[36] In addition, the frequency of defecation can vary anywhere between three times per day to three times per week.[37]

4.3.2 Intestinal Fluid and Electrolyte Balance

The epithelial lining of the entire intestine from the duodenum to the distal colon can perform both absorption and secretion of electrolytes and water, resulting in the net absorption of fluid and electrolytes and their balance.[21] Both the small and large intestine play a very crucial and significant role in the absorption of water and electrolytes such as Na^+, H^+, K^+, Cl^-, and HCO_3^-. On average, around 9–10 L of water flows through the GI tract in a day.[21] This huge quantity of water mainly comes from daily dietary consumption ($\sim$2 L) and the remainder is from other gastrointestinal secretions such as saliva and gastric, pancreatic, biliary, and intestinal secretions. The large intestine (colon) is tight and water permeability is very low owing to the smaller surface area and higher potential difference compared with the small intestine. Nevertheless, the stomach is almost water impermeable.[19]

The small intestine is responsible for almost 60–80% of water reabsorption, which represents around 7–8 L of water, and the remaining water is reabsorbed by the large intestine, excluding the 0.1–0.2 L of water and also 1 mequiv. of sodium and chloride that are excreted in the feces.[38] Furthermore, the villi size and the absorption surface of the brush border are larger in the jejunum of the small intestine than the ileum, except where localized specific transport mechanisms exist. This results in variations in the rate of water and nutrient transfer along the length of the small intestine.[19] In the case of the large intestine, water and electrolyte absorption occur mainly in the proximal region of the colon.[28]

4.3.2.1 Mechanisms Involved in the Absorption of Water and Electrolytes

The major driving force for water and electrolyte absorption in the intestine is believed to be due to the function of electrolytes, osmotic gradients, and aquaporins.[39,40] Among these, the active transport of Na^+ ions across the basolateral enterocyte membrane is the key contributing factor for the development of an osmotic gradient in the small intestine and colon, and it is powered by Na^+/K^+-ATPase transport protein.[41] Further, Na^+ is predominantly absorbed through active cotransport along with other solutes, namely sugars (glucose and galactose), short-chain peptides, amino acids, and bile acids, in the jejunum.[42] In addition, the K^+ is absorbed in the jejunum and ileum by passive diffusion across a concentration gradient, which developed as a result of water absorption. Thus, Na^+ enters the cell through cotransport with the solutes and then exits at the basolateral membrane into the interstitial fluid and K^+ is pumped back into the cell in exchange *via* the Na^+/K^+-ATPase pump.[28]

Further, the electrochemical gradient because of Na^+ absorption favors the absorption of Cl^- *via* a passive free diffusion mechanism mainly in the duodenum and jejunum. In the lower ileum and the colon, Na^+ can also be actively absorbed into the cell in exchange for H^+ that is secreted into the lumen, *i.e. via* an N^+–H^+ exchange mechanism.[43] A major quantity of bicarbonate present in the intestinal chyme is derived from pancreatic secretion and bile. The H^+ secreted into the lumen in exchange for Na^+ entering the cell combines with bicarbonate ions (HCO_3^-) to form carbonic acid (H_2CO_3).[44] The carbonic acid later dissociates to form water, which remains as part of the chyme in the intestinal lumen, and carbon dioxide (CO_2), which is taken up into the blood circulation and then exhaled. This process is popularly known as the 'active reabsorption of bicarbonate ions'. Further, various factors such as luminal factors, neural inputs, and endocrine and paracrine hormones are involved in the regulation and control of water and electrolyte absorption.[45]

4.3.3 Gut Motility

4.3.3.1 Segmental Motility/Haustral Contractions

Segmental motility (also known as haustration or haustral contractions), caused by contractions and reverse movements, results in the development and re-formation of sac-like segmented structures called the haustral.[28,46] Further, the colonic reverse movements are mainly seen in the upper (proximal) colon. In the case of segmental motility, the colonic material can be propelled in both directions, and the material is transferred over numerous segments, around every 30 min.[3,28] The mixing of colonic contents is mainly governed by contractions and reverse movements of the circular muscle present in the walls of the large intestine.[2,46]

4.3.3.2 Peristalsis and Peristaltic Mass Movement

Apart from haustral contractions, peristalsis and peristaltic mass movement aid in the propulsion of the colonic contents in the large intestine.[47] Peristalsis comprises waves of contractions that travel towards the anus and progressively push the colonic contents.[48] The frequency of the slow wave contractions, which extend from the ascending colon to the descending colon, is 11 cycles min^{-1}, falling to 6 cycles min^{-1} in the sigmoid colon and the rectum. However, generally, the colon experiences local contractions at a rate of 4 cm s^{-1}.[20] In addition, the peristalsis waves increase their speed after a meal from *ca.* 5 cm h^{-1} (at rest) to 10 cm h^{-1}.[28] Additionally, the peristaltic mass movement entails the concurrent contraction of a significant portion of the large intestine (typically the ascending and transverse colon), which propels the contents to a longer distance in a short time (*i.e.* from *ca.* 33 to 75% of the length of the colon in a few seconds).[28,49]

4.3.3.3 Factors that Affect Colonic Motility

Several factors affect the colonic motility, such as food, laxatives, emotion, and sleep–wake cycle.[50,51] If the ingested food contains more indigestible materials such as fibers (pectin, cellulose, gums, mucilage, hemicelluloses, and lignin), they aid the rapid movement of the colonic contents by stimulating the mechanoreceptors in the walls of the large intestine through increasing the volume of the colonic contents.[52] Also, fatty meals have a greater impact than carbohydrate meals on colonic movements.[53] Further, laxatives aid in stimulating colonic motility *via* the activation of chemoreceptors.[54] It is also interesting that colonic motility has a circadian rhythm and sleep inhibits colonic motility whereas waking is a prime stimulant for various colonic functions such as absorption, mixing, and propulsion of colonic contents, together with the formation and storage of feces.[54,55] Meals and waking induce two- and threefold increases in colonic motor activity, respectively, and this response is popularly known as a gastro-colonic response with gastric and cerebral involvement.[56] Various mechanisms, systems, and factors such as neurohormonal mechanisms, myogenic mechanisms, the central nervous system (CNS), the enteric nervous system (ENS), and external neural and hormonal factors work together to ensure well-coordinated colonic motility.[24] Further, various assessment techniques such as radiopaque marker methods, scintigraphic techniques, pH–pressure capsules, and other recording techniques such as manometry and barostat are employed in the investigation of colonic motility/motor function.[38]

4.4 Microbiome in the Entire GI Tract

The group of microorganisms including bacteria, viruses, fungi, and protozoa inhabiting a particular environment is known as a microbiome. A human body contains more microbial cells or genomes than there are humans on the planet.[57] Further, the GI tract is the place in the human system where huge colonization of microorganisms occurs, with *ca.* 10^{14} microbes belonging to over 2000 species.[58] Further, the microbial population in the GI tract varies considerably from the oral cavity to the large intestine.[5] Various factors affect the survival and colonization of the microbes in the GI tract, such as harsh environmental conditions such as pH, the influx of acid, digestive enzymes, and bile, a short transit time (oral cavity, stomach, and small intestine compared with the gut), motility of the GI tract, intermittent food substrate delivery, and the immune system.[59] The microbial populations in the proximal GI tract such as the stomach (10^2–10^3 cells mL^{-1}), duodenum (10^3–10^4 cells mL^{-1}), and jejunum (10^4–10^5 cells mL^{-1}) are considerably lower than those in the ileum (10^8 cells mL^{-1}) and large intestine (10^{11} cells mL^{-1}).[5,60]

4.4.1 Gut Microbiome

The gut microbiome consists of a wide diversity of microorganisms that together are known as 'super-organisms' as they produce a variety of microbial metabolites that have unique, significant, and diverse impacts on the physiological functions of humans.[61] The combined weight of the gut microbiota is around 2 kg, which is greater than the weight of the largest internal organ in the human body (liver).[62] In addition, several characteristics of the large intestine favor microbial proliferation and colonization, such as low acidic to neutral pH, larger volume, lower concentrations of bile salts and enzymes, no digestive secretions, a longer transit time due to slower motility, and lack of a direct immune response (absence of Peyer's patches).[5] It is also intriguing that various population studies have shown that the gut microbiome varies significantly among individuals, each person having a unique profile of microbial species.[63] Every human being has around 500–1000 bacterial species residing in and colonizing their gut. So far there are around 35 000 species from the collective human gut bacterial microbiota, with new species constantly being discovered.[64]

Common types of bacteria found in the gut microbiota include the Firmicutes (such as *Lactobacillus* and *Streptococcus* species) and the Bacteroidetes (such as *Bacteroides vulgatus*, *Bacteroides uniformis*, and *Bacteroides intestinalis*), comprising around 65 and 30%, respectively. The remaining 5% include a wide range of primitive bacterial groups containing Proteobacteria (*e.g. Escherichia coli*) and Actinobacteria such as *Bifidobacteria* species.[65,66] There are only limited data on other microbes such as viruses and fungi. In the case of fungi, certain yeast species, such as *Saccharomyces*, *Malassezia*, and *Candida*, are dominant and together they make up only a small proportion of gut microbiota.[67] Table 4.1 summarizes microbial species present in the entire GI tract. The development of the gut microbiome from the infant to the adult stage and the dominant phylum or genus of bacteria are impacted by various phenomena such as mode of birth delivery (vaginal or cesarean), feeding (breast milk or formulated products), weaning, diet/food habits, diseases, infection, antibiotic treatments, and the environment.[68,69]

4.4.2 Gut–Brain–Microbiota Axis

The extensive bidirectional neural connections and communication between the gut and brain through the gut microbiota and its metabolites, neuroendocrine and neuroimmune pathways, the CNS, the ENS, and the

Table 4.1 Summary of microbial species present in the entire GI tract.[5]

Part of the GI tract	Bacterial population present	Autochthonous/allochthonous[a]
Oral cavity	*Fusobacterium*, *Lactobacillus*, *Gemella* (*e.g. G. haemolysans*), *Granulicatella*, *Veillonella*, *Prevotella*, *Porphyromonas*, *Rothia*, *Streptococcus* (*e.g. S. mitis*), *Neisseria*	Autochthonous microbes exceed allochthonous counterparts
Stomach	*Helicobacter pylori*	Allochthonous: *Gemella* (*e.g. G. haemolysans*), *Veillonella*, *Granulicatella*, *Prevotella*, *Streptococcus* (*e.g. S. mitis*), *Rothia*, *Porphyromonas*, *Fusobacterium*, *Neisseria*, *Lactobacillus*
Small intestine	*Escherichia coli*, *Bacteroides*, *Klebsiella*, *Dorea*, *Enterococcus*, *Weissella*, *Ruminococcus*, *Clostridium*, *Lactobacillus* (some species), *Coprococcus*	Allochthonous: *Lactobacillus*, *Streptococcus* (*e.g. S. mitis*), *Veillonella*, *Granulicatella*
Large intestine	Five major phyla with numerous species: Bacteroidetes, Firmicutes, Verrucomicrobia, Actionobacteria, Proteobacteria	Autochthonous microbes exceed allochthonous counterparts

[a] Autochthonous: endogenously derived and common to a majority of hosts. Allochthonous: derived from outside sources, not found in the majority of hosts, but can be abundant in any one host.

sympathetic and parasympathetic branches of the autonomic nervous system (ANS) are referred to as the gut–brain–microbiota axis.[70,71] Despite their distance apart, both the brain and gut are closely connected and play a very critical role in the maintenance of homeostasis, regulation of mood and behavior, prevention of inflammation and infection, and absorption of nutrients, and influence the sensory, hormonal secretion, and motor functions (gastrointestinal motility) of the GI tract through their crosstalk along the neural, hormonal, and immune pathways.[8,72,73]

The gut microbiota can signal the brain through numerous pathways, such as (1) the vagus nerve (afferent signaling pathways); (2) neuro-immune signaling *via* the manufacture of inflammatory cytokines or the stimulation of the hypothalamic–pituitary–adrenal (HPA) axis to produce corticotropin-releasing hormone (CRH), adrenocorticotropin hormone (ACTH), and cortisol to regulate the immune function; (3) producing short-chain fatty acids (SCFAs) such as acetic, propionic and butyric acids to modulate the hormones (peptide YY and glucagon-like peptide-1) from neuroendocrine cells of the gut; (4) mediating tryptophan metabolism, as it is involved in the downstream production of serotonin [5-hydroxytryptamine (5-HT)] and kynurenine and its metabolites (such as kynurenic acid and quinolinic acid), which have

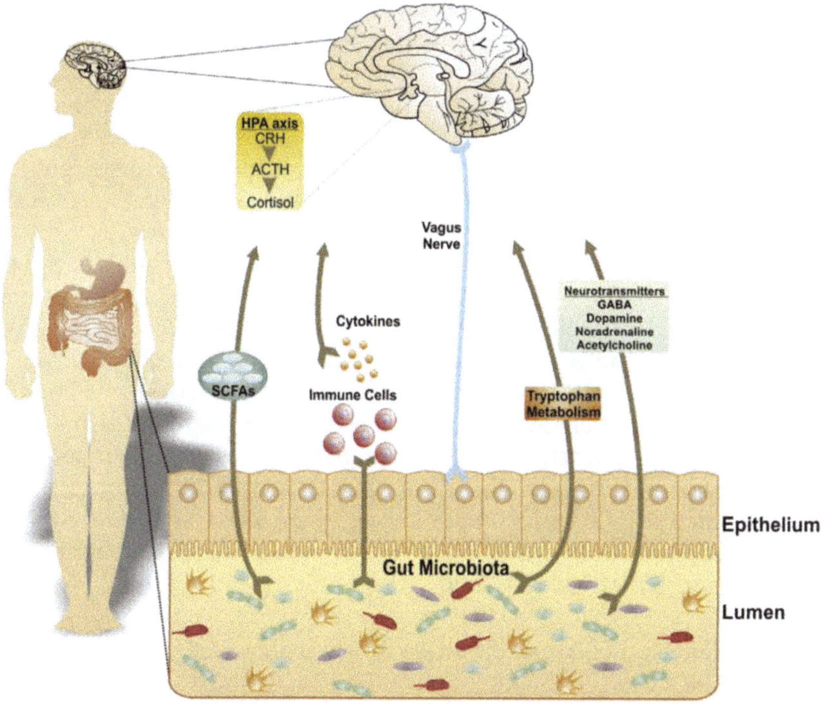

Figure 4.3 The gut–brain–microbiota axis. Reproduced from ref. 8 with permission from Elsevier, Copyright 2016.

been proved to be neuroactive and have significant implications for brain function and behavior; (5) production of neurotransmitters and neuroactive compounds [such as γ-aminobutyric acid (GABA), dopamine, acetylcholine, noradrenaline, and serotonin]; and (6) controlling neuroendocrine function (HPA axis response to stress and depression).[8,73] Figure 4.3 depicts the gut–brain axis.

Various advanced techniques such as visceral sensitivity and central sensitization using intraluminal electrical stimulation or balloon distension, cortical-evoked potentials through distension of different regions of the GI tract (esophagus, rectum, anus) or electrical stimulation, motor-evoked potentials through electrical and magnetic stimulation, functional brain imaging [using functional magnetic resonance imaging (fMRI), positron emission tomography (PET), magnetic resonance spectroscopy (MRS) and magnetoencephalography (MEG)], and vagal tone measurement using an electrocardiogram have been used to investigate and assess the gut–brain axis in humans.[74–76]

4.4.3 Gut Microbiota: Dysbiosis and the Immune System

The symbiotic connection between the gut microbiome and the immune system is well documented.[77] The microorganism composition, *i.e.* the species content within the gut, has a significant role in the maintenance of immune response.[57] There are huge variations in the composition of gut microbiota among healthy individuals owing to a wide range of environmental, physiological, genetic, and psychological factors.[78,79] Gut microbiota imbalance (also known as dysbiosis) could result in the development and aggravation of various health issues, such as inflammation, diseases, autoimmune disorders, and carcinogenesis.[80] Several factors impact the balance of the gut microbiota, such as genetics, mode of birth delivery, dietary habits, lifestyle, exercise, health status, associated diseases, exposure to medicines and antibiotics, age, host immunity, stress, geography, and a variety of other environmental, physiological, and psychological factors.[81–83] Table 4.2 summarizes the various factors that affect the gut microbiome and dominant microbial species at various stages of life.[84,85] The GI tract is protected through specialized immune mechanisms that prevent infection from pathogenic microorganisms while maintaining the colonization of commensal bacteria by avoiding unnecessary defensive responses. The various defense mechanisms and immune systems are the gut epithelial barrier, innate immunity provided by resident phagocytes, and adaptive immunity.[12,57] Table 4.3 summarizes various barrier and defense mechanisms of the GI tract.

Table 4.2 Summary of factors that affect the gut microbiome at various stages of human life.

Stage of life	Factors that affect gut microbiota	Dominant gut microbiota[a]	Other minor factors
Prenatal	Genetics, maternal factors, and gestational age	Enterobacteriaceae Clostridiaceae	Circadian rhythm Association with pets and other animals
Postnatal	Mode of birth delivery	Cesarean delivery: *Bifidobacterium*, *Staphylococcus*, *Corynebacterium*, *Propionibacterium* species ↑ Vaginal delivery: *Lactobacillus*, *Prevotella*, *Sneathia* species, *Clostridium difficile* ↑	Consumption of fermented food Number of siblings Maternal factors Seasons, climate, and weather
Infancy	Modes of infant feeding Weaning foods Medicines and antibiotics	1 month: Enterobacteria ↑ 6 months: Bifidobacteria and *Bacteroides* ↑ 24 months: Firmicutes and Bacteroidetes ↑	
Adulthood	Diet, lifestyle, host genetics, early colonization, age, medication and antibiotics, health and diseases, stress, modulating factors, and geography	Firmicutes ↑ Bacteroidetes ↓	
Elderly		Firmicutes ↓ Bacteroidetes ↑	

[a] High diversity ↑; reduced diversity ↓.

Table 4.3 GI tract barrier and defense systems.

Immune system	Type of cell	Barrier and defense mechanisms
Gut epithelial barrier	Stratified monolayer of epithelial cells	Provides tight junctions through occludin and claudin molecules
	Goblet cells	Mucins of mucus gel, which are high molecular weight glycoproteins, are crucial for maintaining the mucus barrier and also play a crucial role in immune function, including antigen tolerance and sampling. Mucus acts as a physical barrier that prevents microbes or toxins from reaching or harming the epithelial surface and the Peyer's patches
	Paneth cells	In response to inflammatory signals, it strongly enhances innate mucosal immunity by exocytosis of bactericidal granules. Lysozyme suppresses bacterial development by attacking and hydrolyzing glycosidic linkages present in bacterial cell wall peptidoglycans. It secretes antimicrobial proteins such as defensins (α and β), cathelicidins, lysozyme, and phospholipase. The primary job of the α-defensins is to combat intraluminal bacterial and fungal infections
Innate immunity	Microfold cells	Various pathogens (*Salmonella*, *Shigella*, *Vibrio cholerae*, and *Escherichia coli* (O157:H7) exploit the rapid transepithelial transport and the desire for foreign molecules and organisms of M cells in their favor for host invasion to cause several diseases. They destroy the bacteria and viruses through endocytosis
	Intestinal macrophages	They have significant bactericidal activity and are extremely phagocytic. Intestinal pathological inflammation is prevented by mucosal macrophages
	Dendritic cells	They foster an environment that is tolerogenic to commensal bacteria while allowing for strong immune activation against pathogens. They produce retinoic acid, which aids in the maintenance of homeostasis in the small intestine immune environment. Nitric oxide plays a significant role in dendritic cell migration within mucosa-associated lymphoid tissues
Adaptive immunity	Mucosal B cells and T cells – lymphocytes	They are responsible for the secretion of immunoglobulin A (IgA), which aids in maintaining an anti-inflammatory or pro-inflammatory response and also preserves intestinal homeostasis. They also control the surface expression of inflammatory signals by commensal bacteria and inhibit their internalization and increase host tolerance. IgA has been proven to protect against *Vibrio cholera*-induced diarrhea and typhoid infection by *Salmonella typhimurium* by neutralizing cholera toxin and preventing the adherence of microorganisms to the epithelium, respectively.

4.5 Role of the Large Intestine in Food Digestion and Absorption

The large intestine does not secrete any enzymes for the digestion of food/nutrients present in incoming chyme.[3] The chyme entering the large intestine is well digested and nutrients are absorbed (more than 90%), and it mainly consists of complex polysaccharides (dietary fiber), some proteins, and a lesser amount of fat.[86] The gut microbiota has a pivotal role in the chemical digestion of these remaining macronutrients through colonic fermentation. The complex ecosystem of gut microbiota in the large intestine breaks down the substrates (exogenous and endogenous) through various microbial metabolisms to nutrients and metabolites that are used by the host for direct benefits and modulate host immune and organ systems by impacting their physiology.[4] Various factors, such as form and size, cooking and processing, meal composition and various amounts of macronutrients, transit time, and the existence of anti-nutrients, affect the availability of macronutrients to the gut microbiota.[87] Figure 4.4 illustrates the macronutrient metabolism by the human gut microbiome.

4.5.1 Carbohydrates

The anaerobic bacteria present in the colon break down the carbohydrates (dietary fiber and other complex polysaccharides) *via* pyruvate catabolism, chiefly into short-chain fatty acids (SCFAs) (such as formate, acetate, propionate, and butyrate) and comparatively less but substantial quantities of alcohols (including propanol, ethanol, and 2,3-butanediol), and other end

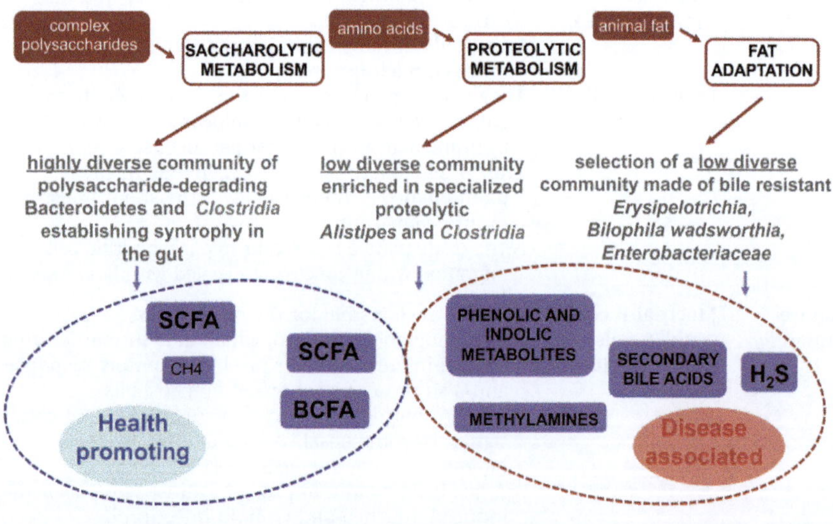

Figure 4.4 Illustration of macronutrient metabolism by the human gut microbiome. Reproduced from ref. 148 with permission from Elsevier, Copyright 2016.

products are lactate, hydrogen, carbon dioxide, and methane.[88,89] SCFAs are the colonic fermentation end products irrespective of the substrate in most cases. Further, each type of SCFA, *e.g.* butyrate, propionate, and acetate, has a specific end route, namely it acts as a predominant energy source for colonocytes, is mostly taken up by the liver and acts as a satiety molecule, and is metabolized in the peripheral tissues (muscle), respectively.[90–92] Of all the SCFAs, butyrate is of the greatest importance and has been widely researched owing to its ability to nourish colonic mucosa and prevent colon cancer through assisting cell differentiation, arresting cell cycles, and inducing apoptosis of transformed colonocytes.[93,94]

4.5.2 Amino Acids

Protein catabolism by a low diverse community of specialized proteolytic gut microbiota mainly depends on the composition of the available amino acids. The amino acid composition reaching the gut is crucially dependent on the source of protein (*i.e.* plant or animal), in addition to various factors that affect the availability of macronutrients such as food processing, macronutrient ratios, and transit time.[95] Protein catabolism products including ammonia, amines, phenols/indoles, and sulfurous compounds produced through amino acid fermentation by the gut microbiota are detrimental to health.[4] Nevertheless, it is worth noting that the most abundant end products of amino acid fermentation in the gut are SCFAs, followed by branched-chain fatty acids (BCFAs).[73] Hence it is either the overall amplified protein fermentation activity or the specific metabolisms that are detrimental to the host, not the protein catabolism itself.

4.5.3 Lipids

Only meager quantities (<5%) of dietary fat end up in the colon and they are metabolized by a low diverse community of bile-resistant microorganisms.[96] The gut microbiota is known to contain lipase enzymes that can digest the triglycerides and phospholipids and free the lipids.[97] However, in the gut's anaerobic environment, the gut microbiota cannot catabolize free lipids.[4] Nevertheless, these free lipids can directly interact with host receptors. For instance, saturated fatty acids activate TLR4 (Toll-like receptor 4), which increases inflammation; in contrast, omega-3 unsaturated fatty acids are TLR4 blockers that prevent inflammation.[98,99] In addition, some free lipids have antimicrobial properties.[100] However, the gut ecosystem also favors the production of numerous metabolic compounds such as secondary bile acids and hydrogen sulfide that are associated with certain diseases.[101]

4.5.4 Vitamin Biosynthesis

Certain essential vitamins that are necessary for the health and survival of humans cannot be synthesized on their own. Their needs can be met through dietary consumption and vitamin biosynthesis by the gut microbiota.[62,73] A well-balanced gut microbiota rich in *Bifidobacteria*, *Escherichia coli*, and *Lactobacilli* can synthesize a variety of water-soluble B vitamins such as vitamin B_{12} (cobalamin), B_9 (folic acid), B_7 (biotin), B_6 (pyridoxal phosphate), B_3 (niacin), B_5 (pantothenic acid), B_2 (riboflavin) and B_1 (thiamine) and fat-soluble vitamins such as vitamin K and the vitamin D group.[102,103]

4.6 Role of Diet in Shaping the Gut Microbiota

Despite various factors that were found to impact the gut microbiota, dietary habits including composition and specific food groups play an indispensable role. Further, diet and specific foods have also been repeatedly proven to be the most critical and modifiable factor that aids in the regulation of gut microbiota throughout the lifespan and for several health conditions.

4.6.1 Dietary Pattern

Diet modifications, even for a short period, have been shown to have a direct influence on the composition of gut microbiota. For instance, switching from either a plant- or animal-based diet to their respective alternative even for 24 h has proven to change the composition of gut microbiota drastically.[104] Further, the composition of the diet also has a significant effect. For example, a fat- or protein-rich diet shows a prominent reduction in the *Bifidobacteria* and butyrate-producing bacteria, and the inclusion of dietary fiber/prebiotics in a high-fat diet has been shown to restore the depleted bacteria.[105,106] There are different types of well-established diet regimens, namely the Mediterranean diet, western diet, rural diet, high-protein diet, vegetarian/vegan diet, high-fat diet, high-fiber diet, and ketogenic diet, and other dietary restrictions such as protein-restricted, intermittent fasting/time-restricted feeding, calorie-restricted, and fasting-mimicking diets that have been widely investigated and shown to have a significant impact on gut–brain–microbiota communication.[73,107–110] Table 4.4 summarizes the impact of various dietary patterns on gut microbiota, its metabolites, and brain and the role of diet in health and diseases.

4.6.2 Effects of Specific Foods and Food Constituents on the Gut–Brain–Microbiota Axis and Health

Recently, the impact of specific foods and food constituents on gut microbiota and health has been extensively researched because of its great significance.[111]

4.6.2.1 Dietary Fibers

Dietary fibers are resistant carbohydrate polymers that cannot be digested or absorbed by the human gastrointestinal system. Dietary fibers play a

Table 4.4 Impact of various dietary patterns on the gut microbiota, its metabolites, and brain and in turn on various health and disease states.

Dietary pattern	Common food groups and characteristic features of the diet	Impact on health and diseases	Gut microbiota composition and metabolites
Mediterranean diet	The majority of food consumed is made up of plant-based foods (*e.g.* fruits, vegetables, grains) and foods rich in monounsaturated or n-3 polyunsaturated fats (*e.g.* seafood and olive oil). Minimal dairy product consumption. Restricted amounts of processed meals, red meat, and poultry. Low to moderate alcohol consumption, especially wine along with foods	Aids in the inhibition of heart disease, diabetes, colorectal cancer, gut dysfunction, and neurodegenerative disease. Inversely associated with the various disease risk factors through favorable alternations of intestinal microbiota and metabolites	Increased Bifidobacteria, *Clostridium,* Bacteroidetes, and *Prevotella* clusters Reduction in Proteobacteria, Firmicutes, and Bacillaceae population. Increases the amount of SCFAs and reduces that of trimethylamine *N*-oxide
Western diet	This diet includes excessive consumption of fried foods, butter, margarine, high-fat dairy products, eggs, red and processed meat, refined cereals, baked goods, and sugary beverages. Little consumption of fruits and vegetables. According to general descriptions, the diet is heavy in fat (saturated fat), high in sugar and salt, and low in fiber	Proven to have numerous undesirable effects on human health. Strongly condemned for its association with many diet-related diseases such as obesity and non-alcoholic fatty liver disease. Connected to strong alterations in behavior profiles (*e.g.* anxiety and depression)	Changes in Firmicutes-to-Bacteroidetes ratio. Increased abundance of Firmicutes and *Collinsella* (a bacterium associated with obesity). Decrease in Bifidobacteria, Bacteroidetes, and *Prevotella.* Reduced levels of SCFAs
Ketogenic diet	In an effort to encourage the generation of ketones, this diet mainly consists of moderate protein, less carbohydrate, and high fat	Widely and successfully used as a dietary intervention approach for neurodegenerative diseases [autism spectrum disorder (ASD), Alzheimer's disease, glucose transporter 1 deficiency syndrome, and autoimmune multiple sclerosis], metabolic diseases, and obesity.	It has a contradictory role with gut microbes as it reduces the gut microbiome diversity and richness. It increases Bacteroidetes and Firmicutes (such as *Lactobacillus* and *Turicinabacter*) and reduces Bifidobacteria.

Table 4.4 (*Continued*)

Dietary pattern	Common food groups and characteristic features of the diet	Impact on health and diseases	Gut microbiota composition and metabolites
		Reduces glycemia and increases insulin sensitivity	It influences the gut mainly through metabolites. Increases the level of GABA (exerts anti-seizure effects). Increases and decreases SCFAs, increases H_2S, reduces lactate (increase in luminal pH)
Vegetarian/vegan diet	This diet contains only plant-based foods consisting of fruits, vegetables, and grains	Shows protective effects against several chronic, metabolic (obesity, diabetes, and cardiovascular risk), and inflammatory diseases (rheumatoid arthritis). The protective health benefits of a vegan diet and gut microbiota are due to reduced levels of inflammation	It uniquely alters the gut microbiota by reducing the pathobionts and increasing the abundance of protective species. Compared with an omnivorous diet, a vegetarian/vegan diet lowers the prevalence of *Bacteroides fragilis*. Compared with an omnivorous diet, it contains a lower ratio (%) of the *Clostridium* cluster XIVa but a larger ratio (%) of *Bacteroides thetaiotaomicron*, *Prevotella*, *Faecalibacterium prausnitzii*, *Bacteroides*, and *Clostridium clostridioforme*
High-fiber diet	Diets rich in plant fibers such as fiber-rich whole grains (barley, brown rice) and other prebiotics [such as GOS (galactooligosaccharides), FOS (fructooligosaccharides), inulin, and pectin]	Improves overall health benefits through modification of gut microbiota	Promotes the expansion of the microbiota (hydrolytic bacteria). Increases *Bifidobacteria* and the Firmicutes-to-Bacteroidetes ratio. Increases production of SCFAs. Decreases *Bacteroidaceae* family, *Bacteroides* spp., and *Clostridium histolitycum* group

significant role in the modulation of the gut microbiota composition as they reach the gut undigested and serve as the source of the substrate of colonic fermentation.[112,113] Various types of dietary fibers that are present in our daily foods are resistant starches, fructooligosaccharides (FOS), inulin, galactooligosaccharides (GOS), resistant maltodextrin, polydextrose, arabinoxylans, and other polysaccharides such as β-glucans and pectins and have been proven to impact the gut–brain–microbiota axis positively.[114,115] The key mechanisms through which the dietary fiber works in symbiosis with the gut microbiota are (i) it encourages the activity and growth of butyrate-producing bacteria (such as *Roseburia*, *Eubacterium rectale*, and *Faecalibacterium prausnitzii*), (ii) it increases the quantity of *Bifidobacteria* and *Lactobacilli*, and (iii) it increases the production of SCFAs and other beneficial metabolites.[111,116] In addition, the consumption of diet and foods that are rich in fiber has been reported to provide various health benefits such as reduction in hyperglycemia, hyperlipidemia, and hypertension, improved immune response, and prevention of diabetes, obesity, cancer, and associated neurological diseases.[117,118]

4.6.2.2 Bioactive Peptides

Bioactive peptides naturally present in foods or derived from food proteins (such as milk, egg, and fish protein-derived peptides and animal and plant protein hydrolyzates) have been demonstrated to maintain gut health and function through modulating gastrointestinal homeostasis *via* the regulation of the barrier function, mucosal immune responses, inflammation, homeostasis of reactive oxygen species (ROS), and gut microbiota.[119] Several studies investigated the effects of various peptides that could act as promising nutraceutical/functional foods in the prevention and delaying of the onset of various neurodegenerative diseases, namely Alzheimer's disease, multiple sclerosis, and Parkinson's disease, by modulating the gut microbiota. The composition of gut microbiota such as Actinobacteria, *Bifidobacterium*, Bacteroidetes, Proteobacteria, *Lactobacillus*, Firmicutes, Verrucomicrobia, *Roseburia*, and *Clostridium* is improved by the peptides, which consequently play a major role in the maintenance of the microbiota–gut–brain axis.[120]

4.6.2.3 Dietary Polyphenols

Dietary polyphenols such as coumarins, flavonoids, lignins, phenolic acids, tannins, and stilbenes (resveratrol) have low bioavailability, so they reach the colon as unabsorbed molecules and are biotransformed into absorbable phenolic metabolites due to the action of colonic microflora.[121] They were shown to exert a prebiotic effect and their intake was proven to positively modify the gut microbiota by suppressing the growth of pathogens, increasing the abundance of beneficial microbes.[122] Also, the consumption of polyphenols reduces the risk of various diseases caused by gut dysbiosis, such as colon cancer, metabolic syndrome (*e.g.* obesity and type 2 diabetes), inflammatory bowel disease, gastroenteritis, and other related infections, and also reduces the production of endotoxin, heightens the proinflammatory response, and improves overall wellbeing.[123] Polyphenols and catechins from various teas such as green, oolong, and black tea have been reported to have antioxidative and anti-inflammatory activities, and to regulate the action of the enzyme protein kinase, respectively.[124] Further, various phenolic acids from cereals and pulses such as *p*-coumaric and ferulic acids are released due to the enzymatic action of colonic bacteria and aid in regulating the immune response and decrease metabolic dysfunctions.[125] The intake of polyphenols modified the gut microbiome by hindering the proliferation of pathogenic bacteria (such as *Escherichia coli*, *Clostridium perfringens*, and *Helicobacter pylori*) and by improving the growth and abundance of probiotic bacterial families (such as Bifidobacteriaceae and Lactobacillaceae).[126]

4.6.2.4 Dietary Flavonoids

Dietary flavonoids are mainly obtained from various fruits and vegetables, especially citrus fruits and juices. The flavonoids are broadly classified into six subgroups, namely flavanones, flavonols, flavones, anthocyanins, flavanonols, and chalcones.[127] The citrus flavonoids such as hesperidin, naringenin, polymethoxyflavones (such as nobiletin, tangeretin, and sinensetin), anthocyanins (such as cyanidin and delphinidin), quercetin, and their metabolic bioactives offer various health benefits (namely reduction in insulin resistance, obesity, colitis-related symptoms, proinflammatory enzyme level, epididymal fat accumulation, hepatic triglyceride concentration, *etc.*), mainly through modulation of barrier permeability, protection of the mucus layer, regulation of the intestinal immune system, combating oxidative stress, improving local/systemic inflammatory status, and positively shaping the microbiome and metabolome.[128] The polymethoxyflavones alter gut dysbiosis by reducing the F/B ratio (Firmicutes-to-Bacteroidetes ratio) through enriching Bacteroidetes and increasing the relative copiousness of *Akkermansia muciniphila* (a gut-protective bacterium) and *Allobaculum.*[129] Further, some flavonoids, such as hesperetin, have also been proven to inhibit the growth of various pathogenic microbes.[130] Further, betacyanin (red pitaya), ellagitannins (pomegranate), and proanthocyanidins (grapes) encouraged the growth of *Akkermansia muciniphila* and regulated metabolic activities, gut health, and associated disorders (such as type 2 diabetes and inflammation).[131]

Table 4.5 summarizes the beneficial effects of specific foods and food constituents on gut microbiota and health.

4.7 Modulation of Gut Microbiota

Modulation of the gut microbiota is very crucial and clinically significant in treating all the diseases and risk factors associated with gut microbiota

Table 4.5 Summary of various foods, food constituents, prebiotics, and their beneficial effects on gut microbiota and health.

Food category	Prebiotics	Gut microbiota modulation	Major health benefits	Ref.
Dietary fiber	*Inulin* Common sources: banana, chicory, artichoke, onion, garlic, barley, rye, yacon tubers, and wheat	Supports the growth of beneficial bacteria, especially *Bifidobacterium* spp., *Anaerostipes* spp., *Actinobacteria*, and *Lactobacillus*. Reduces the abundance of *Bilophila*, *Clostridium* cluster XIVa, *Roseburis*, and *Desulfovibrio*	Improve immunity, increase the absorption of micro- and macro-elements, indirectly impact the levels of plasma lipids, lower blood cholesterol levels, aid in the prevention of cardiovascular disease, increase bone mineral density, reduce the risk of colon cancer, and reduce the risk of osteoporosis	149–152
	Fructooligosaccharides (FOS) Common sources: fruits, vegetables, and cereals, and as reserve material in onion, chicory, asparagus, garlic, tomatoes, and wheat	Increase of the growth of Bifidobacterium spp. and *Lactobacillus* spp. Reduce the counts of undesirable bacteria such as *Escherichia coli* and *Clostridium perfringens*	Fermented into lactate and SCFAs. Reduce total triglycerides. Reduce serum insulin and glucose concentrations, enhance calcium absorption, and improve the immune system	153–155
	Galactooligosaccharides (GOS) Common sources: β-GOS are formed by enzymatic synthesis from lactose present in animal milk and α-GOS are obtained primarily from soybeans	GOS are bifidogenic in nature. Increase the abundance of beneficial gut microbiota *Lactobacillus* spp., *Enterococcus* spp., *Bifidobacterium* spp., and *Clostridium coccoides–Eubacterium rectale.* Reduce the amount of *Clostridium histolyticum*, *Desulfovibrio* spp., Bacteroides, and *Escherichia coli*	Markers of immune function and metabolic syndrome improved. Aid in mineral absorption, stool improvement, weight management, prevention of carcinogenesis, reduction of cholesterol levels, and allergy alleviation	156–159
	Xylooligosaccharides (XOS) Common sources: bamboo shoots, fruits, vegetables, milk, and honey	Aid in the improvement of gut microbiota; reported to be bifidogenic; improve the abundance of some *Bifidobacterium* species (such as *B. adolescentis*), and stimulates the growth of other *Lactobacillus* spp.	Reduce lipid levels, especially low-density lipoprotein (LDL) and total cholesterol. Significantly increase high-density lipoprotein (HDL). Aid in the production of propionate and acids by selective stimulation of bifidogenic bacteria and *Lactobacillus* spp. Promote insulin synthesis through glucagon-like peptide-1 (GLP-1) and inhibit lipolysis	156 and 160–162

Table 4.5 (*Continued*)

Food category	Prebiotics	Gut microbiota modulation	Major health benefits	Ref.
	Pectin Common sources: artichoke, citrus peel, noni, apple pomace, turmeric, sugar beet and chicory pulp, and kiwi fruit	Positively modifies the gut microbiota composition. Increased growth and abundance of *Bacteroides ovatus*, *Faecalibacterium prausnitzii*, *Bacteroides fragilis*, and *Bacteroides finegoldii* and they all provided various beneficial effects to the host *via* metabolism regulation, inflammation modulation, and cancer prevention. Selectively stimulates the growth of Ruminococcaceae, Succinivibrionaceae, Lachnospiraceae, and Bacteroida and the production of their metabolites, which provides various benefits	Aids in the prevention of cholesterol absorption, decreases glucose absorption and secretion of intestinal hormones, generates SCFAs such as acetate and propionate, improves insulin resistance, and helps with metabolism regulation, inflammation modulation, and carcinogen prevention	115 and 163–165
	β-Glucans Common sources: edible mushrooms, yeast, seaweeds, and grains such as oat, barley, and wheat	Support the growth of *Lactobacillus* spp. (such as *L. acidophilus* and *L. casei*) and Bifidobacteria (such as *Bifidobacterium animalis lactis*). β-Glucan-rich durum wheat flour and whole-grain barley pasta lowered the levels of Firmicutes and Fusobacteria. Increased the abundance of *Roseburia hominis*, Clostridiaceae (*Clostridium orbiscindens* and *Clostridium* spp.), and *Ruminococcus* spp.	Provide various clinical benefits such as decreases in serum cholesterol levels and improvements in postprandial glucose metabolism due to their viscous nature. Upon fermentation they yield SCFAs. Promote the production of propionate, which was shown to inhibit cholesterol synthesis due to inhibition of the rate-limiting enzyme HMG CoA reductase. They were also demonstrated to have anticancer, antidiabetic, anti-inflammatory, and immune-modulating effects	158 and 166–168
Dietary polyphenols	*Polyphenols* Common sources: grapes, cranberry, blueberry, sinapine,	Significant reduction in the Firmicutes-to-Bacteroidetes ratio.	Aid in the reduction of diet-induced obesity, insulin resistance, and intestinal inflammation.	123 and 169–171

green tea, orange juice, and sorghum bran	Increased the abundance of *Akkermansia* species (*A. muciniphila*), Lactobacillaceae, and *Blautia*. Enhanced the growth of *Bifidobacterium* spp., Actinobacteria, Proteobacteria, *Lactobacillus*, *Roseburia*, and *Prevotella*	Improve gut health and prevent gut dysbiosis and obesity-mediated metabolic diseases such as insulin resistance, diabetes, and non-alcoholic fatty liver disease	
Flavanols Catechins, epigallocatechin-3-gallate (EGCG) supplement, and epigallocatechin 3-*O*-(3-*O*-methyl)gallate (EGCG3″Me) Sources: green tea, oolong tea, pu-erh tea extract, black tea extract	Encouraged the growth of *Bifidobacterium* spp., *Akkermansia* spp., and *Lactobacillus*/*Enterococcus* groups. Inhibited the growth of *Bacteroides –Prevotella*, *Clostridium histolyticum*, and *Eubacterium–Clostridium* group. Reduced the Firmicutes-to-Bacteroidetes ratio. Increased the abundance of Bacteroidetes and reduced the abundance of Firmicutes	Positively modulate the human gut microbiota and generate SCFAs. Reduce obesity and low-grade inflammation by modulating the abundance of gut microbiota. Reduce serum lipid profile, glucose, and insulin	172–176
Anthocyanidins and proanthocyanidins Sources: bilberry extract, arctic berry extract, cinnamon bark, grape pomace extract, apple, and cranberry extract	Increased the abundance of Bacteroidetes, *Bifidobacterium* spp., *Lactobacillus*, *Akkermansia muciniphila*, and *Roseburia*. Reduced the abundance of Firmicutes and Proteobacteria and the Firmicutes-to-Bacteroidetes ratio. Provided an inhibitory effect on pathogenic bacteria growth	Improve the production of SCFAs by bacterial populations. Aid in the reduction of the intestinal or cecal pH. Reduce serum lipopolysaccharides. Enhance insulin sensitivity and lower fat mass and body weight gain	170, 175 and 177–179
Resveratrol Source: red wine, grapes, and berries	Inhibited the growth of *Enterococcus faecalis*, Enterobacteria, *Sutterella*, Turicibacteraceae, *Moryella*, Lachnospiraceae, and *Akkermansia*. Increased the growth of *Bifidobacterium* spp., Parabacteroides and *Lactobacillus* spp.	Suppresses fat deposition in mice with high-fat-induced obesity. Provides cardioprotective effects, reduces oxidative damage and inflammation, and acts as a neuroprotective, chemopreventive, antiobesity, and antiaging agent	121 and 180–182

dysbiosis. Various methods are used for modulating the gut microbiota, such as prebiotics, probiotics, synbiotics, and fecal matter transplantation (FMT).[69]

4.7.1 Prebiotics

Prebiotics are resistant to digestion, unabsorbed, and selectively fermented by the beneficial colonic microbiota, which has the potential to stimulate selectively the growth and/or activities of one or more species of gut microbes that offer beneficial health effects to the host.[132] Previously, only non-digestible and fermentable carbohydrates, such as FOS, inulin, oligofructose, galactofructose, GOS, xylooligosaccharides, and resistant starches, were regarded as prebiotics.[133] However, other bioactive compounds (resistant to digestion with low bioavailability) present in food, such as polyphenols, have also recently been accepted as a significant prebiotic source.[134] The undigested and unabsorbed phenolics and their postbiotics (*i.e.* their metabolic derivatives in the gut) have been reported to produce prebiotic effects.[135,136] Prebiotics are mostly found in plant-based foods such as fruits, vegetables, whole grains such as cereals and pulses, and nuts. Apart from their beneficial role in gut health, prebiotics provides other nutritional advantages and technological properties (such as improved taste and texture) in the case of food applications.[137] The key characteristic features of prebiotics are as follows:

(i) Selective fermentation by beneficial colonic microbiota, especially probiotics, Bifidobacteria, and Lactobacilli species.
(ii) Resistance of upper GI tract digestion (resistance to gastric acidity and enzyme hydrolysis) and absorption.
(iii) Stability to food processing treatments such as low pH, heat, and Maillard reaction conditions.
(iv) Advantageous to the health of the host.
(v) Selectively stimulate probiotics.

Table 4.5 summarizes various prebiotics and their beneficial effects on gut microbiota and health.

4.7.2 Probiotics

The probiotics are live microorganisms that, when consumed (as oral supplements or as a food product) in suitable amounts, can positively alter the gut microbiota of the host and confer potential health benefits.[138] Probiotic bacteria often come from the genera *Lactobacillus* and *Bifidobacterium*. In addition, some species of other microorganisms such as *Propionibacterium*, *Enterobacteria*, sporulated bacteria, and some yeasts have also been reported to have probiotic effects. Figure 4.5 depicts the common probiotic microorganisms.

Lactic acid bacteria

- *Lactobacillus* spp.: *Lactobacillus acidophilus, L. amylovorus, L. brevis, L. bulgaricus, L. casei, L. cellobiosus, L. crispatus, L. curvatus, L. delbrueckii* subsp. *bulgaris, L. fermentum, L. gallinarum, L. helveticus, L. johnsonii, L. lactis, L. paracasei, L. plantarum, L. reuteri, L. rhamnosus*
- *Streptococcus* spp.: *Streptococcus salivarius* subsp. *thermophiles, S. cremoris, S. intermedius, S. lactis, S. diacetilactis*
- *Lactococcus* spp.: *Lactococcus lactis* subsp. *cremoris*
- *Leuconostoc: Leuconostoc mesenteroides*
- *Pediococcus* spp.

Bifidobacteria

- *Bifidobacterium* spp.: *B. adolescentis, B. animalis, B. bifidum, B. breve, B. essensis, B. infantis, B. laterosporum, B. thermophilum, B. longum*

Propionibacterium

- *Propionibacterium* spp.: *P. acidipropionici, P. freudenreichii, P. jensenii, P. shermani, P. thoenii*

Enterobacteria

- *Enterococcus* spp.: *E. faecalis, E. faecium*

Sporulated bacteria

- *Bacillus* spp.: *B. alcolophilus, B. cereus, B. clausii, B. coagulans, B. subtilis*

Other bacteria

- *Escherichia coli, Sporolactobacillus inulinus*

Yeasts

- *Saccharomyces cerevisae (boulardii), Candida torulopsis*

Molds

- *Aspergillus niger, Aspergillus oryzae*

Figure 4.5 Common probiotic microorganisms.

Further, to be classified as probiotics, the microbes should have certain key characteristics:[139]

(i) Should be a normal inhabitant of the gut microbe of a healthy individual.
(ii) Should survive harsh upper GI tract conditions.
(iii) Should survive and grow in the large intestine (*i.e.* resistant to acids and bile).
(iv) Should be safe for consumption by humans and provide beneficial outcomes to the host.
(v) Should be able to adhere to the intestinal epithelial cell lines.
(vi) Should multiply faster for colonization (permanent or temporary).
(vii) Should produce antimicrobial substances (*i.e.* bacteriocins).
(viii) Should be non-pathogenic.
(ix) Should survive in food and pharmaceutical preparation treatments.

The probiotic microorganisms provide various health benefits through several functional mechanisms, such as through (i) production of various beneficial metabolites (postbiotics) including antimicrobial agents, (ii) enhancing the mucosal barrier integrity, and (iii) modulation of gut microbiota and immune modulation.[140] In addition, probiotic intervention and treatment are used in the treatment and prevention of various illnesses and diseases.[141] Probiotic microorganisms are present in appreciable quantities in various fermented food products, *e.g.* yogurt (*Lactobacillus delbrueckii* subsp. *bulgaricus*, *Streptococcus thermophilus*), cheddar cheese (*Lactococcus lactis* subsp. *lactis*, *Lactococcus lactis* subsp. *cremoris*, *Streptococcus thermophilus*), kefir (*Lactobacillus kefir*, *Lactobacillus kefiranofacies*, *Lactobacillus brevis*), fermented sausages (*Lactobacillus sakei*, *Lactobacillus curvatus*, *Pediococcus acidilactici*, *Pediococcus pentosaceus*), sauerkraut (*Lactobacillus plantarum*, *Lactobacillus brevis*, *Pediococcus acidilactici*, *Pediococcus cerevisiae*, *Leuconostoc mesenteroides*), idli and dosa (*Leuconostoc mesenteroides*, *Enterococcus faecalis*), wine (*Oenococcus oeni*), rice wine (*Lactobacillus sakei*), and kimchi (*Lactobacillus plantarum*, *Leuconostoc mesenteroides*, *Lactococcus brevis*).

4.7.2.1 Postbiotics

Postbiotics are metabolites that are produced by gut microbiota during the colonic fermentation of digestible and non-digestible dietary components and those released by bacterial metabolism. The postbiotics produced can be beneficial or detrimental to the host depending on the link between dietary habits, host health, gut microbiota composition, and the range of metabolites produced.[140,142] Beneficial postbiotics aid in the prevention of diseases, suppress pathogenic microbes, and induce therapeutic responses. The common beneficial postbiotics of probiotic microorganisms are SCFAs, polyamines, tryptophan, vitamins, *p*-cresols, and bacteriocins and they provide various benefits such as regulation of intestinal epithelial cell energy metabolism, promotion of homeostasis, apoptosis, and cell proliferation, strengthening of the intestinal epithelial barrier and disease prevention, regulation of anti-inflammatory and pro-inflammatory responses, provision of antioxidant and antimicrobial properties, and provision of homeostasis among gut microbiota and antimicrobial activity.[135,142]

4.7.3 Synbiotics

Synbiotics are a harmonious combination of probiotics and prebiotics that work synergistically on the host and provide increased health benefits compared with each component separately.[143] Synbiotics are specifically designed to selectively favor the growth and improve survival and colonization of administered probiotics in the gut since the GI tract can be a hostile environment.[144] In the case of some medical conditions, such as

Table 4.6 Summary of prebiotics, probiotics, and synbiotics used in the treatment of various illnesses and diseases.

Condition	Prebiotics, probiotics, and synbiotics administered	Significant outcome	Ref.
Overweight or obesity	Oligofructose-enriched inulin	The treated group showed a significant reduction in body weight *z*-score, percentage body fat and trunk fat, levels of interleukin 6, and serum triglycerides	183
	Bifidobacterium bifidum, *Bifidobacterium lactis*, *Lactobacillus acidophilus*, *Lactobacillus casei*, and *Lactococcus lactis*	There was a significant decrease in waist circumference, waist-to-height ratio, and plasma polyunsaturated fatty acids. Glutathione peroxidase activity increased	184
	Probiotics: *Lactiplantibacillus plantarum* CBT LP3 and *Bifidobacterium breve* CBT BR3 Prebiotic: fructooligosaccharide	Considerable improvements in obesity-related markers were observed	185
Inflammatory bowel disease (IBD)	*Bifidobacterium* BB-12 and *Lactobacillus acidophilus* La-5	*Lactobacillus*, *Bifidobacterium*, and *Bacteroides* populations increased considerably	186
Irritable bowel syndrome (IBS)	*Lactobacillus acidophilus*, *Bifidobacterium breve*, *Lactobacillus rhamnosus*, *Bifidobacterium actis*, *Streptococcus thermophilus*, and *Bifidobacterium longum*	Considerably improved the fecal contents of the majority of probiotic strains. Improved scores for the symptoms of diarrhea in IBS patients	187
Alzheimer's disease (AD)	*Lactobacillus acidophilus*, *Bifidobacterium bifidum*, *Lactobacillus casei*, and *Lactobacillus fermentum*	Enhanced score in the Mini-Mental State Test, but no noticeable effects on oxidative stress, inflammation, fasting plasma glucose, and lipid profiles were found	188
Overweight diabetic patients with coronary heart disease	Prebiotic: Inulin Probiotics: *Lactobacillus acidophilus*, *Bifidobacterium bifidum*, and *Lactobacillus casei*	Significant reductions in plasma glucose and serum insulin concentrations were observed. Enhanced the insulin sensitivity	189
Non-alcoholic fatty liver disease (NAFLD)	*Bifidobacterium bifidum*, *Lactobacillus acidophilus*, *Bifidobacterium lactis*, and *Lactobacillus rhamnosus*	In the therapy group, there were substantial reductions in alanine aminotransferase, mean aspartate aminotransferase, mean cholesterol, low-density lipoprotein-C, and triglycerides, and also waist circumference	190
	Prebiotic: Inulin Probiotics: *Bifidobacterium longum* and *Lactobacillus acidophilus*	Aminotransferase enzymes were enhanced by probiotic and/or prebiotic supplementation. Additionally, probiotic or pro- and prebiotic supplements helped NAFLD patients recover their fatty liver grade	191

Table 4.6 (*Continued*)

Condition	Prebiotics, probiotics, and synbiotics administered	Significant outcome	Ref.
Diabetes	*Lactobacillus acidophilus*, *Bifidobacterium bifidum*, and *Lactobacillus casei*	Remarkably improved quantitative insulin sensitivity and plasma total antioxidant capacity and also a substantial reduction in fasting plasma glucose and serum high-sensitivity C-reactive protein	192
Rheumatoid arthritis	*Lactobacillus acidophilus*, *Bifidobacterium bifidum*, and *Lactobacillus casei*	Substantially improved disease activity score across 28 joints. Significantly reduced high-sensitivity C-reactive protein and serum insulin levels	193
Major depressive disorder	Probiotics: *Lactobacillus helveticus* and *Bifidobacterium longum* Prebiotic: galactooligosaccharide	In the treatment group, the Beck depression inventory (BDI) score improved	194

kidney function and non-alcoholic fatty liver disease, synbiotics did not have any clear benefits, but in the case of other conditions, such as irritable bowel syndrome, high blood pressure, obesity, high fasting blood glucose, selective inflammatory markers, and postoperative infections, synbiotics had positive effects.[145,146]

Table 4.6 summarizes studies that used probiotics, prebiotics, and synbiotics for gut health and in the treatment of various illnesses and diseases.

4.7.4 Fecal Matter Transplantation (FMT)

In FMT, the fecal bacteria from a healthy donor are transplanted to patients with gut dysbiosis or intestinal disease to restore the gut microbiota community and function.[147] FMT has proved efficient in treating several diseases, such as irritable bowel syndrome, recurrent *Clostridium difficile* infection, metabolic diseases, constipation, neuropsychiatric conditions, autoimmune diseases, and colon cancer. Various techniques are used in FMT, such as colonoscopy, enteric-coated capsules, enema, and capsules containing freeze-dried bacteria/feces.[69]

4.8 Conclusion

In recent years, the role of specific foods, individual food constituents, dietary patterns, and therapeutic dietary interventions such as prebiotics, probiotics, and synbiotics has been widely investigated owing to their impact on gut microbiota composition and diversity. It is evident from several studies that the gut microbiota, its metabolites, and interactions between the gut and brain play a critical role in the normal physiological and metabolic function of the host through the regulation of the nervous, immune, and endocrine systems. It is also fascinating to see the extensive connection between the gut, brain, and microbiome and their interrelatedness with health and diseases. Several factors impact the balance of the gut microbiota from birth to old age; however, the effect caused by diet is indispensable and it can be modified in favor of regulating the gut microbiota throughout the lifespan and for numerous health conditions. Therefore, in this chapter, we have given a brief overview of the anatomy and physiology of the large intestine, the essence of gut–brain–microbiota interactions, and the importance of food, dietary patterns, and therapeutic interventions in the modulation of the gut microbiota for health and wellbeing.

References

1. D. Roupar, P. Berni, J. T. Martins, A. C. Caetano, J. A. Teixeira and C. Nobre, Bioengineering approaches to simulate human colon microbiome ecosystem, *Trends Food Sci. Technol.*, 2021, **112**, 808–822.
2. S. A. McQuilken, The mouth, stomach and intestines, *Anaesthesia & Intensive Care Medicine*, 2021, **22**(5), 330–335.
3. Y. Nigam, J. Knight and N. Williams, Gastrointestinal tract 5: the anatomy and functions of the large intestine, *Nurs. Times*, 2019, **115**(10), 50–53.
4. K. Oliphant and E. Allen-Vercoe, Macronutrient metabolism by the human gut microbiome: major fermentation by-products and their impact on host health, *Microbiome*, 2019, **7**(1), 1–15.
5. J. Walter and R. Ley, The human gut microbiome: ecology and recent evolutionary changes, *Annu. Rev. Microbiol.*, 2011, **65**, 411–429.
6. S. R. Sarkar and S. Banerjee, Gut microbiota in neurodegenerative disorders, *J. Neuroimmunol.*, 2019, **328**, 98–104.
7. G. Rogers, D. J. Keating, R. L. Young, M.-L. Wong, J. Licinio and S. Wesselingh, From gut dysbiosis to altered brain function and mental illness: mechanisms and pathways, *Mol. Psychiatry*, 2016, **21**(6), 738–748.
8. P. J. Kennedy, A. B. Murphy, J. F. Cryan, P. R. Ross, T. G. Dinan and C. Stanton, Microbiome in brain function and mental health, *Trends Food Sci. Technol.*, 2016, **57**, 289–301.
9. L. F. Buttó and D. Haller, Dysbiosis in intestinal inflammation: cause or consequence, *Int. J. Med. Microbiol.*, 2016, **306**(5), 302–309.
10. R. Chibbar and L. A. Dieleman, The gut microbiota in celiac disease and probiotics, *Nutrients*, 2019, **11**(10), 2375.
11. A. M. Abdellatif and N. E. Sarvetnick, Current understanding of the role of gut dysbiosis in type 1 diabetes, *J. Diabetes*, 2019, **11**(8), 632–644.
12. Z. Wang, J. Zhong and X. Meng, *et al.*, The gut microbiome-immune axis as a target for nutrition-mediated modulation of food allergy, *Trends Food Sci. Technol.*, 2021, **114**, 116–132.

13. A. Hartstra, M. Nieuwdorp and H. Herrema, Interplay between gut microbiota, its metabolites and human metabolism: Dissecting cause from consequence, *Trends Food Sci. Technol.*, 2016, **57**, 233–243.
14. S. Loman and J.-W. van der Kamp, Insulin resistance as key factor for linking modulation of gut microbiome to health claims and dietary recommendations to tackle obesity, *Trends Food Sci. Technol.*, 2016, **57**, 306–310.
15. E. A. Trautwein, H. P. Peters and D. J. Mela, *et al.*, Is gut microbiota a relevant and competitive dietary target for cardio-metabolic health? Proceedings of an expert workshop, *Trends Food Sci. Technol.*, 2018, **81**, 146–154.
16. Q.-S. Zhang, F.-W. Tian, J.-X. Zhao, H. Zhang, Q.-X. Zhai and W. Chen, The influence of dietary patterns on gut microbiome and its consequences for nonalcoholic fatty liver disease, *Trends Food Sci. Technol.*, 2020, **96**, 135–144.
17. T. Ohkusa, S. Koido, Y. Nishikawa and N. Sato, Gut microbiota and chronic constipation: a review and update, *Front. Med.*, 2019, **6**, 19.
18. X. Fan, Y. Jin, G. Chen, X. Ma and L. Zhang, Gut microbiota dysbiosis drives the development of colorectal cancer, *Digestion*, 2021, **102**(4), 508–515.
19. B. Osuntokun and S. A. Kocoshis, Anatomy and physiology of the small and large intestine, in *Pediatric Gastrointestinal and Liver Disease*, Elsevier, 2006, pp. 459–474.
20. I. Sensoy, A review on the food digestion in the digestive tract and the used *in vitro* models, *Curr. Res. Food Sci.*, 2021, **4**, 308–319.
21. R. Lee and E. A. Mezoff, Anatomy and Physiology of the Small and Large Intestines, in *Pediatric Gastrointestinal and Liver Disease*, 2021, pp. 308–320.e3.
22. V. Mahadevan, Anatomy of the caecum, appendix and colon, *Surgery*, 2020, **38**(1), 1–6.
23. H. T. Debas, Small and large intestine, *Gastrointestinal Surgery: Pathophysiology and Management*, 2004, pp. 239-310.
24. A. Erdogan and Y. Y. Lee, Colon and pelvic floor anatomy and physiology, in *Clinical and Basic Neurogastroenterology and Motility*, 2020, pp. 113–126.
25. *Handbook of Physiology: Section 6: The Gastrointestinal System Volume IV: Intestinal Absorption and Secretion*, ed. S. G. Schultz, M. Field, R. A. Frizzell and B. B. Rauner, Oxford University Press, Oxford, UK, 1991.
26. T. Judge and G. R. Lichtenstein, Is the appendix a vestigial organ? Its role in ulcerative colitis, *Gastroenterology*, 2001, **121**(3), 730–732.
27. H. Eastcott, Gray's Anatomy. Edited by P. L. Williams and R. Warwick. Thirty-sixth edition. 305 × 234 mm. Pp. 1578+xvii. Illustrated. 1980. Edinburgh: Churchill Livingstone. £32·00, *Br. J. Surg.*, 1981, **68**, 218.
28. M. E. Smith and D. G. Morton, *The Digestive System: Systems of the Body Series*, Elsevier Health Sciences, 2011.

29. M. P. Kelley, J. Efron, S. H. Fang, B. Safar and S. Gearhart, Operative Anatomy of the Colon, Rectum, and Anus, in *Shackelford's Surgery of the Alimentary Tract, 2 Volume Set*, 2019, pp. 1662–1675.
30. Cancer AJCO, Colon and rectum, in *AJCC Cancer Staging Manual*, Springer, 2002, pp. 113–123.
31. S. Palit, P. J. Lunniss and S. M. Scott, The physiology of human defecation, *Dig. Dis. Sci.*, 2012, **57**(6), 1445–1464.
32. S. Salimoglu, G. Kilinc and B. Calik, Anatomy of the Colon, Rectum, and Anus, in *Colon Polyps and Colorectal Cancer*, Springer, 2021, pp. 1–22.
33. L. Barducci, J. C. Norton and S. Sarker, *et al.*, Fundamentals of the gut for capsule engineers, *Prog. Biomed. Eng.*, 2020, **2**(4), 042002.
34. P. Cleary, M. Sinnott, B. Hari, S. Bakalis and S. Harrison, Modelling food digestion, in *Modeling Food Processing Operations*, Elsevier, 2015, pp. 255–305.
35. M. S. Stoll, *Formation, metabolism, and properties of pyrrolic compounds appearing in the gut*, CRC Press, Bilirubin, 2018, pp. 103–132.
36. L. Degen and S. Phillips, Variability of gastrointestinal transit in healthy women and men, *Gut*, 1996, **39**(2), 299–305.
37. S. A. Walter, L. Kjellström, H. Nyhlin, N. J. Talley and L. Agréus, Assessment of normal bowel habits in the general adult population: the Popcol study, *Scand. J. Gastroenterol.*, 2010, **45**(5), 556–566.
38. A. E. Bharucha and M. Camilleri, Physiology of the Colon and Its Measurement. *Shackelford's Surgery of the Alimentary Tract, 2 Volume Set*, 2019, pp. 1676–1688.
39. J. Keller, Gastrointestinal digestion and absorption, *Encyclopedia of Biological Chemistry*, 2013.
40. A. Verkman, More than just water channels: unexpected cellular roles of aquaporins, *J. Cell Sci.*, 2005, **118**(15), 3225–3232.
41. J. J. Feher, *Quantitative human physiology: an introduction*, Academic Press, 2017.
42. B. E. Goodman, Insights into digestion and absorption of major nutrients in humans, *Adv. Physiol. Educ.*, 2010, **34**(2), 44–53.
43. J. Campbell, J. Berry and Y. Liang, Anatomy and physiology of the small intestine, in *Shackelford's Surgery of the Alimentary Tract, 2 Volume Set*, Elsevier, 2019, pp. 817–841.
44. A. D. Jackson and J. McLaughlin, Digestion and absorption, *Surgery*, 2009, **27**(6), 231–236.
45. N. F. Shroyer and S. A. Kocoshis, Anatomy and physiology of the small and large intestines, in *Pediatric Gastrointestinal and Liver Disease*, Elsevier, 2011, pp. 324–336.e2.
46. A. E. Bharucha and W. L. Hasler, Motility of the small intestine and colon, in *Yamada's Textbook of Gastroenterology*, Wiley-Blackwell, 2015, pp. 367–385.
47. P. G. Dinning, Recording *in vivo* human colonic motility: what have we learnt over the past 100 Years? in *The Enteric Nervous System*, 2016, pp. 213–222.

48. S. K. Sarna, *Colonic motility: from bench side to bedside*, Morgan & Claypool Life Sciences, 2010, pp. 1–157.
49. J. Kellow, Gastrointestinal motility and defecation, in *Comprehensive Human Physiology*, Springer, 1996, pp. 1289–1308.
50. K. Thiruppathy and A. Emmanuel, Colonic Motility and Physiology, in *Anorectal and Colonic Diseases*, Springer, 2010, pp. 31–41.
51. M. Camilleri and V. Andresen, Current and novel therapeutic options for irritable bowel syndrome management, *Dig. Liver Dis.*, 2009, **41**(12), 854–862.
52. A. Sharma and B. S. Yadav, Ritika. Resistant starch: physiological roles and food applications, *Food Rev. Int.*, 2008, **24**(2), 193–234.
53. S. S. C. Rao, R. Kavelock, J. Beaty, K. Ackerson and P. Stumbo, Effects of fat and carbohydrate meals on colonic motor response, *Gut*, 2000, **46**(2), 205–211.
54. M. Camilleri, Gastrointestinal motility disorders in neurologic disease, *J. Clin. Invest.*, 2021, **131**(4), e143771.
55. H. Duboc, B. Coffin and L. Siproudhis, Disruption of circadian rhythms and gut motility: an overview of underlying mechanisms and associated pathologies, *J. Clin. Gastroenterol.*, 2020, **54**(5), 405.
56. J. Wiley, D. Tatum, R. Keinath and C. Owyang, Participation of gastric mechanoreceptors and intestinal chemoreceptors in the gastrocolonic response, *Gastroenterology*, 1988, **94**(5), 1144–1149.
57. G. C. Manley, Y.-K. Lee and Y. Zhang, Gut microbiota and immunology of the gastrointestinal tract, in *Clinical and Basic Neurogastroenterology and Motility*, Elsevier, 2020, pp. 63–78.
58. V. Lazar, L.-M. Ditu and G. G. Pircalabioru, *et al.*, Aspects of gut microbiota and immune system interactions in infectious diseases, immunopathology, and cancer, *Front. Immunol.*, 2018, **9**, 1830.
59. A. J. Kastl Jr., N. A. Terry, G. D. Wu and L. G. Albenberg, The Structure and Function of the Human Small Intestinal Microbiota: Current Understanding and Future Directions, *Cell. Mol. Gastroenterol. Hepatol.*, 2020, **9**(1), 33–45.
60. C. C. Booijink, E. G. Zoetendal, M. Kleerebezem and W. M. De Vos, Microbial communities in the human small intestine: coupling diversity to metagenomics, *Future Microbiol.*, 2007, **2**(3), 285–295.
61. A. M. Valdes, J. Walter, E. Segal and T. D. Spector, Role of the gut microbiota in nutrition and health, *BMJ*, 2018, **361**, k2179.
62. J. Knight, Z. Bayram Weston and Y. Nigam, Gastrointestinal tract 6: the effects of gut microbiota on human health, *Nurs. Times*, 2019, **115**, 11–46.
63. S. Schloissnig, M. Arumugam and S. Sunagawa, *et al.*, Genomic variation landscape of the human gut microbiome, *Nature*, 2013, **493**(7430), 45–50.
64. C. Barras, Your mystery microbiome, *New Sci.*, 2019, **242**(3225), 28–31.
65. X. Yang, L. Xie, Y. Li and C. Wei, More than 9 000 000 unique genes in human gut bacterial community: estimating gene numbers inside a human body, *PLoS One*, 2009, **4**(6), e6074.

66. B. Das and G. B. Nair, Homeostasis and dysbiosis of the gut microbiome in health and disease, *J. Biosci.*, 2019, **44**(5), 1–8.
67. M. L. Richard and H. Sokol, The gut mycobiota: insights into analysis, environmental interactions and role in gastrointestinal diseases, *Nat. Rev. Gastroenterol. Hepatol.*, 2019, **16**(6), 331–345.
68. T. M. Marques, R. Wall, R. P. Ross, G. F. Fitzgerald, C. A. Ryan and C. Stanton, Programming infant gut microbiota: influence of dietary and environmental factors, *Curr. Opin. Biotechnol.*, 2010, **21**(2), 149–156.
69. N. Hasan and H. Yang, Factors affecting the composition of the gut microbiota, and its modulation, *PeerJ*, 2019, **7**, e7502.
70. Q. Li and J.-M. Zhou, The microbiota–gut–brain axis and its potential therapeutic role in autism spectrum disorder, *Neuroscience*, 2016, **324**, 131–139.
71. A. C. Johnson, T. Louwies, T. Yuan, A. Orock and B. Greenwood-Van Meerveld, Gut and brain interactions, in *Clinical and Basic Neurogastroenterology and Motility*, Elsevier, 2020, pp. 17–30.
72. M. Bistoletti, A. Bosi, D. Banfi, C. Giaroni and A. Baj, The microbiota-gut-brain axis: Focus on the fundamental communication pathways, *Prog. Mol. Biol. Transl. Sci.*, 2020, **176**, 43–110.
73. K. V. Sandhu, E. Sherwin, H. Schellekens, C. Stanton, T. G. Dinan and J. F. Cryan, Feeding the microbiota-gut-brain axis: diet, microbiome, and neuropsychiatry, *Transl. Res.*, 2017, **179**, 223–244.
74. M. A. Secretariat, Functional brain imaging: an evidence-based analysis, *Ontario Health Technol. Assess. Series*, 2006, **6**(22), 1.
75. A. Sasegbon, X. Xiang, D. H. Vasant and S. Hamdy, Investigation of the brain–gut axis, in *Clinical and Basic Neurogastroenterology and Motility*, Elsevier, 2020, pp. 127–143.
76. J. S. Labus, G. Tun, L. A. Kilpatrick, S. S. Rao, E. A. Mayer and K. Tillisch, Neuroimaging and biomarkers in functional gastrointestinal disorders: What the scientists and clinicians need to know about basic neuroimaging, biomarkers, microbiome, gut and brain interactions, in *Clinical and Basic Neurogastroenterology and Motility*, Elsevier, 2020, pp. 31–61.
77. N. Kamada, S.-U. Seo, G. Y. Chen and G. Núñez, Role of the gut microbiota in immunity and inflammatory disease, *Nat. Rev. Immunol.*, 2013, **13**(5), 321–335.
78. O. Jazayeri, S. M. Daghighi and F. Rezaee, Lifestyle alters GUT-bacteria function: linking immune response and host, *Best Pract. Res. Clin. Gastroenterol.*, 2017, **31**(6), 625–635.
79. M. S. R. Rajoka, H. Zhao and N. Li, *et al.*, Origination, change, and modulation of geriatric disease-related gut microbiota during life, *Appl. Microbiol. Biotechnol.*, 2018, **102**(19), 8275–8289.
80. S. Carding, K. Verbeke, D. T. Vipond, B. M. Corfe and L. J. Owen, Dysbiosis of the gut microbiota in disease, *Microb. Ecol. Health Dis.*, 2015, **26**(1), 26191.

81. R. M. Alasmar, K. Varadharajan, M. Shanmugakonar and H. A. Al-Naemi, Gut microbiota and health: understanding the role of diet, *Food Nutr. Sci.*, 2019, **10**(11), 1344.
82. G. A. Cresci and E. Bawden, Gut microbiome: what we do and don't know, *Nutr. Clin. Pract.*, 2015, **30**(6), 734–746.
83. E. Z. Gomaa, Human gut microbiota/microbiome in health and diseases: a review, *Antonie van Leeuwenhoek*, 2020, 1–22.
84. L. Drago, S. Panelli, C. Bandi, G. Zuccotti, M. Perini and E. D'Auria, What pediatricians should know before studying gut microbiota, *J. Clin. Med.*, 2019, **8**(8), 1206.
85. M. Kumar, P. Babaei, B. Ji and J. Nielsen, Human gut microbiota and healthy aging: recent developments and future prospective, *Nutr. Healthy Aging*, 2016, **4**(1), 3–16.
86. A. Mackie, The digestive tract: a complex system, in *Interdisciplinary approaches to food digestion*, Springer, 2019, pp. 11–27.
87. J. M. Wong and D. J. Jenkins, Carbohydrate digestibility and metabolic effects, *J. Nutr.*, 2007, **137**(11), 2539S–2546S.
88. A. Koh, F. De Vadder, P. Kovatcheva-Datchary and F. Bäckhed, From dietary fiber to host physiology: short-chain fatty acids as key bacterial metabolites, *Cell*, 2016, **165**(6), 1332–1345.
89. T. Bhattacharya, T. S. Ghosh and S. S. Mande, Global profiling of carbohydrate active enzymes in human gut microbiome, *PLoS One*, 2015, **10**(11), e0142038.
90. I. Rowland, G. Gibson and A. Heinken, *et al.*, Gut microbiota functions: metabolism of nutrients and other food components, *Eur. J. Nutr.*, 2018, **57**(1), 1–24.
91. C. Shortt, O. Hasselwander and A. Meynier, *et al.*, Systematic review of the effects of the intestinal microbiota on selected nutrients and non-nutrients, *Eur. J. Nutr.*, 2018, **57**(1), 25–49.
92. L. Zheng, Of microbes and men: role of colonic short-chain fatty acids on epithelial barrier, PhD thesis, University of Colorado Denver, Anschutz Medical Campus, 2017.
93. K. Ashwin, A. K. Pattanaik and G. S. Howarth, Polyphenolic bioactives as an emerging group of nutraceuticals for promotion of gut health: A review, *Food Biosci.*, 2021, **44**, 101376.
94. P. Markowiak-Kopeć and K. Śliżewska, The effect of probiotics on the production of short-chain fatty acids by human intestinal microbiome, *Nutrients*, 2020, **12**(4), 1107.
95. C. Yao, J. Muir and P. Gibson, Insights into colonic protein fermentation, its modulation and potential health implications, *Aliment. Pharmacol. Ther.*, 2016, **43**(2), 181–196.
96. S. Ramos and M. Á. Martín, Impact of diet on gut microbiota, *Curr. Opin. Food Sci.*, 2021, **37**, 83–90.
97. H. Mu and C.-E. Høy, The digestion of dietary triacylglycerols, *Prog. Lipid Res.*, 2004, **43**(2), 105–133.

98. S. Huang, J. M. Rutkowsky, R. G. Snodgrass, K. D. Ono-Moore, D. A. Schneider, J. W. Newman, S. H. Adams and D. H. Hwang, Saturated fatty acids activate TLR-mediated proinflammatory signaling pathways, *J. Lipid Res.*, 2012, **53**(9), 2002–2013.
99. P. C. Calder, n-3 fatty acids, inflammation and immunity: new mechanisms to explain old actions, *Proc. Nutr. Soc.*, 2013, **72**(3), 326–336.
100. F. G. Cândido, F. X. Valente, Ł. M. Grześkowiak, A. P. B. Moreira, D. M. U. P. Rocha and R. D. C. G. Alfenas, Impact of dietary fat on gut microbiota and low-grade systemic inflammation: mechanisms and clinical implications on obesity, *Int. J. Food Sci. Nutr.*, 2018, **69**(2), 125–143.
101. T. Requena, M. C. Martínez-Cuesta and C. Peláez, Diet and microbiota linked in health and disease, *Food Funct.*, 2018, **9**(2), 688–704.
102. J. G. LeBlanc, C. Milani, G. S. De Giori, F. Sesma, D. Van Sinderen and M. Ventura, Bacteria as vitamin suppliers to their host: a gut microbiota perspective, *Curr. Opin. Biotechnol.*, 2013, **24**(2), 160–168.
103. K. Yoshii, K. Hosomi, K. Sawane and J. Kunisawa, Metabolism of dietary and microbial vitamin B family in the regulation of host immunity, *Fronti. Nutr.*, 2019, **6**, 48.
104. L. A. David, C. F. Maurice and R. N. Carmody, *et al.*, Diet rapidly and reproducibly alters the human gut microbiome, *Nature*, 2014, **505**(7484), 559–563.
105. J. A. Parnell and R. A. Reimer, Prebiotic fiber modulation of the gut microbiota improves risk factors for obesity and the metabolic syndrome, *Gut Microbes*, 2012, **3**(1), 29–34.
106. K. Brown, D. DeCoffe, E. Molcan and D. L. Gibson, Diet-induced dysbiosis of the intestinal microbiota and the effects on immunity and disease, *Nutrients*, 2012, **4**(8), 1095–1119.
107. C. S. Oriach, R. C. Robertson, C. Stanton, J. F. Cryan and T. G. Dinan, Food for thought: The role of nutrition in the microbiota-gut–brain axis, *Clin. Nutr. Exp.*, 2016, **6**, 25–38.
108. A. Tomova, I. Bukovsky and E. Rembert, *et al.*, The effects of vegetarian and vegan diets on gut microbiota, *Front. Nutr.*, 2019, **6**, 47.
109. C. L. Gentile and T. L. Weir, The gut microbiota at the intersection of diet and human health, *Science*, 2018, **362**(6416), 776–780.
110. A. Paoli, L. Mancin, A. Bianco, E. Thomas, J. F. Mota and F. Piccini, Ketogenic diet and microbiota: friends or enemies?, *Genes*, 2019, **10**(7), 534.
111. T. M. Barber, G. Valsamakis and G. Mastorakos, *et al.*, Dietary Influences on the Microbiota–Gut–Brain Axis, *Int. J. Mol. Sci.*, 2021, **22**(7), 3502.
112. K. Makki, E. C. Deehan, J. Walter and F. Bäckhed, The impact of dietary fiber on gut microbiota in host health and disease, *Cell Host Microbe*, 2018, **23**(6), 705–715.
113. J. L. Carlson, J. M. Erickson, J. M. Hess, T. J. Gould and J. L. Slavin, Prebiotic dietary fiber and gut health: comparing the *in vitro* fermentations of beta-glucan, inulin and xylooligosaccharide, *Nutrients*, 2017, **9**(12), 1361.

114. A. Benítez-Páez, E. M. G. Del Pulgar and L. Kjølbæk, *et al.*, Impact of dietary fiber and fat on gut microbiota re-modeling and metabolic health, *Trends Food Sci. Technol.*, 2016, **57**, 201–212.
115. D. Wu, X. Ye and R. J. Linhardt, *et al.*, Dietary pectic substances enhance gut health by its polycomponent: A review, *Compr. Rev. Food Sci. Food Saf.*, 2021, **20**(2), 2015–2039.
116. M. A. Conlon and A. R. Bird, The impact of diet and lifestyle on gut microbiota and human health, *Nutrients*, 2015, **7**(1), 17–44.
117. J. Buttriss and C. Stokes, Dietary fibre and health: an overview, *Nutr. Bull.*, 2008, **33**(3), 186–200.
118. M. M. Kaczmarczyk, M. J. Miller and G. G. Freund, The health benefits of dietary fiber: beyond the usual suspects of type 2 diabetes mellitus, cardiovascular disease and colon cancer, *Metabolism*, 2012, **61**(8), 1058–1066.
119. X. Bao and J. Wu, Impact of food-derived bioactive peptides on gut function and health, *Food Res. Int.*, 2021, **147**, 110485.
120. S. Wu, A.E.-D. A. Bekhit and Q. Wu, *et al.*, Bioactive peptides and gut microbiota: Candidates for a novel strategy for reduction and control of neurodegenerative diseases, *Trends Food Sci. Technol.*, 2021, **108**, 164–176.
121. D. Serra, L. M. Almeida and T. C. Dinis, Dietary polyphenols: A novel strategy to modulate microbiota-gut-brain axis, *Trends Food Sci. Technol.*, 2018, **78**, 224–233.
122. F. Cardona, C. Andrés-Lacueva, S. Tulipani, F. J. Tinahones and M. I. Queipo-Ortuño, Benefits of polyphenols on gut microbiota and implications in human health, *J. Nutr. Biochem.*, 2013, **24**(8), 1415–1422.
123. V. Gowd, N. Karim, M. R. I. Shishir, L. Xie and W. Chen, Dietary polyphenols to combat the metabolic diseases *via* altering gut microbiota, *Trends Food Sci. Technol.*, 2019, **93**, 81–93.
124. H. Sun, Y. Chen, M. Cheng, X. Zhang, X. Zheng and Z. Zhang, The modulatory effect of polyphenols from green tea, oolong tea and black tea on human intestinal microbiota *in vitro*, *J. Food Sci. Technol.*, 2018, **55**(1), 399–407.
125. S. M. Aravind, S. Wichienchot, R. Tsao, S. Ramakrishnan and S. Chakkaravarthi, Role of dietary polyphenols on gut microbiota, their metabolites and health benefits, *Food Res. Int.*, 2021, **142**, 110189.
126. C. A. Morais, V. V. de Rosso, D. Estadella and L. P. Pisani, Anthocyanins as inflammatory modulators and the role of the gut microbiota, *J. Nutr. Biochem.*, 2016, **33**, 1–7.
127. R. Pei, X. Liu and B. Bolling, Flavonoids and gut health, *Curr. Opin. Biotechnol*, 2020, **61**, 153–159.
128. M. Wang, H. Zhao, X. Wen, C. T. Ho and S. Li, Citrus flavonoids and the intestinal barrier: interactions and effects, *Compr. Rev. Food Sci. Food Saf.*, 2021, **20**(1), 225–251.
129. M. Zhang, J. Zhu and X. Zhang, *et al.*, Aged citrus peel (chenpi) extract causes dynamic alteration of colonic microbiota in high-fat diet induced obese mice, *Food Funct.*, 2020, **11**(3), 2667–2678.

130. P. D. P. M. Barbosa, A. R. Ruviaro, I. M. Martins, J. A. Macedo, G. LaPointe and G. A. Macedo, Effect of enzymatic treatment of citrus by-products on bacterial growth, adhesion and cytokine production by Caco-2 cells, *Food Funct.*, 2020, **11**(10), 8996–9009.
131. F. A. Tomás-Barberán, M. V. Selma and J. C. Espín, Interactions of gut microbiota with dietary polyphenols and consequences to human health, *Curr. Opin. Clin. Nutr. Metab. Care*, 2016, **19**(6), 471–476.
132. A. Ahmad and S. Khalid, Therapeutic aspects of probiotics and prebiotics, in *Diet, Microbiome and Health*, Elsevier, 2018, pp. 53–91.
133. S. H. Al-Sheraji, A. Ismail, M. Y. Manap, S. Mustafa, R. M. Yusof and F. A. Hassan, Prebiotics as functional foods: A review, *J. Funct. Foods*, 2013, **5**(4), 1542–1553.
134. B. B. Cardoso, C. Amorim, S. C. Silvério and L. R. Rodrigues, Novel and emerging prebiotics: Advances and opportunities, in *Advances in Food and Nutrition Research*, Elsevier, 2021, pp. 41–95.
135. C. Klemashevich, C. Wu, D. Howsmon, R. C. Alaniz, K. Lee and A. Jayaraman, Rational identification of diet-derived postbiotics for improving intestinal microbiota function, *Curr. Opin. Biotechnol*, 2014, **26**, 85–90.
136. R. Korczak and J. L. Slavin, Definitions, regulations, and new frontiers for dietary fiber and whole grains, *Nutr. Rev.*, 2020, **78**(Supplement_1), 6–12.
137. B. Zhang, Y. Zhang, H. Li, Z. Deng and R. Tsao, A review on insoluble-bound phenolics in plant-based food matrix and their contribution to human health with future perspectives, *Trends Food Sci. Technol.*, 2020, **105**, 347–362.
138. D. de Paulo Farias, F. F. de Araujo, I. A. Neri-Numa and G. M. Pastore, Prebiotics: Trends in food, health and technological applications, *Trends Food Sci. Technol.*, 2019, **93**, 23–35.
139. P. J. Tomasik and P. Tomasik, Probiotics and prebiotics, *Cereal Chem.*, 2003, **80**(2), 113–117.
140. N. Yeşilyurt, B. Yılmaz, D. Ağagündüz and R. Capasso, Involvement of probiotics and postbiotics in the immune system modulation *Biologics*, 2021, **1**(2), 89–110.
141. T. Mancuskova, A. Medved'ova and M. Ozbolt, The medical functions of probiotics and their role in clinical nutrition, *Curr. Nutr. Food Sci.*, 2018, **14**(1), 3–10.
142. P.eluzio, MdCG, Martinez, J. A., Milagro and F. I., Postbiotics: Metabolites and mechanisms involved in microbiota-host interactions, *Trends Food Sci. Technol.*, 2021, **108**, 11–26.
143. M. De Vrese and J. Schrezenmeir, Probiotics, prebiotics, and synbiotics, *Food Biotechnol.*, 2008, 1–66.
144. K. R. Pandey, S. R. Naik and B. V. Vakil, Probiotics, prebiotics and synbiotics-a review, *J. Food Sci. Technol.*, 2015, **52**(12), 7577–7587.
145. T. Jana, B. W. Acker and B. D. Cash, Probiotics and prebiotics, including fibers and medicinal foods, in *Clinical and Basic Neurogastroenterology and Motility*, Elsevier, 2020, pp. 587–600.

146. B. R. Shah, B. Li, H. Al Sabbah, W. Xu and J. Mráz, Effects of prebiotic dietary fibers and probiotics on human health: With special focus on recent advancement in their encapsulated formulations, *Trends Food Sci. Technol.*, 2020, **102**, 178–192.
147. A. Khoruts and M. J. Sadowsky, Understanding the mechanisms of faecal microbiota transplantation, *Nat. Rev. Gastroenterol. Hepatol.*, 2016, **13**(9), 508–516.
148. S. Rampelli, M. Candela and S. Turroni, *et al.*, Microbiota and lifestyle interactions through the lifespan, *Trends Food Sci. Technol.*, 2016, **57**, 265–272.
149. D. Vandeputte, G. Falony and S. Vieira-Silva, *et al.*, Prebiotic inulin-type fructans induce specific changes in the human gut microbiota, *Gut*, 2017, **66**(11), 1968–1974.
150. G. O. Phillips and P. A. Williams. *Handbook of hydrocolloids*, Elsevier, 2009.
151. P. Glibowski and K. Skrzypczak, Prebiotic and synbiotic foods, in *Microbial production of food ingredients and additives*, Elsevier, 2017, pp. 155–188.
152. S. Hiel, M. A. Gianfrancesco and J. Rodriguez, *et al.*, Link between gut microbiota and health outcomes in inulin-treated obese patients: Lessons from the Food4Gut multicenter randomized placebo-controlled trial, *Clin. Nutr.*, 2020, **39**(12), 3618–3628.
153. P. Kumar and K. K. Dubey, Current perspectives and future strategies for fructooligosaccharides production through membrane bioreactor, in *Applied Microbiology and Bioengineering*, Elsevier, 2019, pp. 185–202.
154. A. Drakoularakou, R. Rastall and G. Gibson, Functional foods for the gut: probiotics, prebiotics and synbiotics, in *Functional Foods*, Elsevier, 2011, pp. 449–470.
155. E. S. Prudencio, E. A. Esmerino, M. C. da Silva and A. G. da Cruz, Probiotic and Prebiotics in Foods: Challenges, in *Innovations and Advances*, Academic Press, 2020.
156. F. Cosme, A. Inês and A. Vilela, Consumer's acceptability and health consciousness of probiotic and prebiotic of non-dairy products, *Food Res. Int.*, 2022, **151**, 110842.
157. J. Vulevic, A. Drakoularakou, P. Yaqoob, G. Tzortzis and G. R. Gibson, Modulation of the fecal microflora profile and immune function by a novel trans-galactooligosaccharide mixture (B-GOS) in healthy elderly volunteers, *Am. J. Clin. Nutr.*, 2008, **88**(5), 1438–1446.
158. S. Mitmesser and M. Combs, Prebiotics: Inulin and other oligosaccharides, in *The microbiota in gastrointestinal pathophysiology*, Elsevier, 2017, pp. 201–208.
159. D. Meyer, Health benefits of prebiotic fibers, *Adv. Food Nutr. Res.*, 2015, **74**, 47–91.
160. Y.-C. Chung, C.-K. Hsu, C.-Y. Ko and Y.-C. Chan, Dietary intake of xylooligosaccharides improves the intestinal microbiota, fecal moisture, and pH value in the elderly, *Nutr. Res.*, 2007, **27**(12), 756–761.

161. A. F. A. Carvalho, P. de Oliva Neto, D. F. Da Silva and G. M. Pastore, Xylo-oligosaccharides from lignocellulosic materials: chemical structure, health benefits and production by chemical and enzymatic hydrolysis, *Food Res. Int.*, 2013, **51**(1), 75–85.
162. A. S. Althubiani, S. B. Al-Ghamdi, F. A. Qais, M. S. Khan, I. Ahmad and H. A. Malak, Plant-Derived Prebiotics and Its Health Benefits, in *New Look to Phytomedicine*, Elsevier, 2019, pp. 63–88.
163. B. Gómez, B. Gullón, R. Yáñez, H. Schols and J. L. Alonso, Prebiotic potential of pectins and pectic oligosaccharides derived from lemon peel wastes and sugar beet pulp: A comparative evaluation, *J. Funct. Foods*, 2016, **20**, 108–121.
164. H. Tan and S. Nie, Deciphering diet-gut microbiota-host interplay: Investigations of pectin, *Trends Food Sci. Technol.*, 2020, **106**, 171–181.
165. A. Koutsos, M. Lima and L. Conterno, *et al.*, Effects of commercial apple varieties on human gut microbiota composition and metabolic output using an *in vitro* colonic model, *Nutrients*, 2017, **9**(6), 533.
166. M. De Angelis, E. Montemurno and L. Vannini, *et al.*, Effect of whole-grain barley on the human fecal microbiota and metabolome, *Appl. Environ. Microbiol.*, 2015, **81**(22), 7945–7956.
167. P. Russo, P. López and V. Capozzi, *et al.*, Beta-glucans improve growth, viability and colonization of probiotic microorganisms, *Int. J. Mol. Sci.*, 2012, **13**(5), 6026–6039.
168. J. A. Nazare, S. Normand, A. Oste Triantafyllou, A. Brac de la Perrière, M. Desage and M. Laville, Modulation of the postprandial phase by β-glucan in overweight subjects: Effects on glucose and insulin kinetics, *Mol. Nutr. Food Res.*, 2009, **53**(3), 361–369.
169. D. E. Roopchand, R. N. Carmody and P. Kuhn, *et al.*, Dietary polyphenols promote growth of the gut bacterium Akkermansia muciniphila and attenuate high-fat diet–induced metabolic syndrome, *Diabetes*, 2015, **64**(8), 2847–2858.
170. F. F. Anhê, D. Roy and G. Pilon, *et al.*, A polyphenol-rich cranberry extract protects from diet-induced obesity, insulin resistance and intestinal inflammation in association with increased Akkermansia spp. population in the gut microbiota of mice, *Gut*, 2015, **64**(6), 872–883.
171. Y. Li, J. Li, Q. Su and Y. Liu, Sinapine reduces non-alcoholic fatty liver disease in mice by modulating the composition of the gut microbiota, *Food Funct.*, 2019, **10**(6), 3637–3649.
172. X. Zhang, X. Zhu and Y. Sun, *et al.*, Fermentation *in vitro* of EGCG, GCG and EGCG3" Me isolated from Oolong tea by human intestinal microbiota, *Food Res. Int.*, 2013, **54**(2), 1589–1595.
173. M. Cheng, X. Zhang, Y. Miao, J. Cao, Z. Wu and P. Weng, The modulatory effect of (-)-epigallocatechin 3-O-(3-O-methyl) gallate (EGCG3 ″Me) on intestinal microbiota of high fat diet-induced obesity mice model, *Food Res. Int.*, 2017, **92**, 9–16.

174. S. M. Henning, J. Yang and M. Hsu, *et al.*, Decaffeinated green and black tea polyphenols decrease weight gain and alter microbiome populations and function in diet-induced obese mice, *Eur. J. Nutr.*, 2018, **57**(8), 2759–2769.
175. A. M. Alves-Santos, C. S. A. Sugizaki, G. C. Lima and M. M. V. Naves, Prebiotic effect of dietary polyphenols: A systematic review, *J. Funct. Foods*, 2020, **74**, 104169.
176. X. Lu, J. Liu and N. Zhang, *et al.*, Ripened Pu-erh tea extract protects mice from obesity by modulating gut microbiota composition, *J. Agric. Food Chem.*, 2019, **67**(25), 6978–6994.
177. T. A. Verediano, H. Stampini Duarte Martino, M. C. Dias Paes and E. Tako, Effects of Anthocyanin on Intestinal Health: A Systematic Review, *Nutrients*, 2021, **13**(4), 1331.
178. B. Tian, J. Zhao and W. An, *et al.*, Lycium ruthenicum diet alters the gut microbiota and partially enhances gut barrier function in male C57BL/6 mice, *J. Funct. Foods*, 2019, **52**, 516–528.
179. M. Van Hul, L. Geurts and H. Plovier, *et al.*, Reduced obesity, diabetes, and steatosis upon cinnamon and grape pomace are associated with changes in gut microbiota and markers of gut barrier, *Am. J. Physiol.: Endocrinol. Metab.*, 2018, **314**(4), E334–E352.
180. M. M. Sung, N. J. Byrne and I. M. Robertson, *et al.*, Resveratrol improves exercise performance and skeletal muscle oxidative capacity in heart failure, *Am. J. Physiol.: Heart Circ. Physiol.*, 2017, **312**(4), H842–H853.
181. L. Shen and H.-F. Ji, Reciprocal interactions between resveratrol and gut microbiota deepen our understanding of molecular mechanisms underlying its health benefits, *Trends Food Sci. Technol.*, 2018, **81**, 232–236.
182. Y. Qiao, J. Sun, S. Xia, X. Tang, Y. Shi and G. Le, Effects of resveratrol on gut microbiota and fat storage in a mouse model with high-fat-induced obesity, *Food Funct.*, 2014, **5**(6), 1241–1249.
183. A. C. Nicolucci, M. P. Hume, I. Martínez, S. Mayengbam, J. Walter and R. A. Reimer, Prebiotics reduce body fat and alter intestinal microbiota in children who are overweight or with obesity, *Gastroenterology*, 2017, **153**(3), 711–722.
184. A. C. Gomes, R. G. M. de Sousa, P. B. Botelho, T. L. N. Gomes, P. O. Prada and J. F. Mota, The additional effects of a probiotic mix on abdominal adiposity and antioxidant Status: a double-blind, randomized trial, *Obesity*, 2017, **25**(1), 30–38.
185. E.-J. Song, K. Han and T.-J. Lim, *et al.*, Effect of probiotics on obesity-related markers per enterotype: a double-blind, placebo-controlled, randomized clinical trial, *EPMA J.*, 2020, **11**(1), 31–51.
186. M. Shadnoush, R. S. Hosseini, A. Khalilnezhad, L. Navai, H. Goudarzi and M. Vaezjalali, Effects of probiotics on gut microbiota in patients with inflammatory bowel disease: a double-blind, placebo-controlled clinical trial, *Korean J. Gastroenterol.*, 2015, **65**(4), 215–221.

187. H. Yoon, Y. S. Park, D. H. Lee, J.-G. Seo, C. M. Shin and N. Kim, Effect of administering a multi-species probiotic mixture on the changes in fecal microbiota and symptoms of irritable bowel syndrome: a randomized, double-blind, placebo-controlled trial, *J. Clin. Biochem. Nutr.*, 2015, **57**(2), 129–134.
188. E. Akbari, Z. Asemi and R. Daneshvar Kakhaki, *et al.*, Effect of probiotic supplementation on cognitive function and metabolic status in Alzheimer's disease: a randomized, double-blind and controlled trial, *Front. Aging Neurosci.*, 2016, **8**, 256.
189. M. Tajabadi-Ebrahimi, N. Sharifi and A. Farrokhian, *et al.*, A randomized controlled clinical trial investigating the effect of synbiotic administration on markers of insulin metabolism and lipid profiles in overweight type 2 diabetic patients with coronary heart disease, *Exp. Clin. Endocrinol. Diabetes*, 2017, **125**(01), 21–27.
190. F. Famouri, Z. Shariat, M. Hashemipour, M. Keikha and R. Kelishadi, Effects of probiotics on nonalcoholic fatty liver disease in obese children and adolescents, *J. Pediatr. Gastroenterol. Nutr.*, 2017, **64**(3), 413–417.
191. L. Javadi, M. Ghavami, M. Khoshbaten, A. Safaiyan, A. Barzegari and B. P. Gargari, The potential role of probiotics or/and prebiotic on serum lipid profile and insulin resistance in alcoholic fatty liver disease: a double blind randomized clinical trial, *Crescent J. Med. Biol. Sci.*, 2017, **4**(3), 131–138.
192. A. Soleimani, M. Z. Mojarrad and F. Bahmani, *et al.*, Probiotic supplementation in diabetic hemodialysis patients has beneficial metabolic effects, *Kidney Int.*, 2017, **91**(2), 435–442.
193. B. Zamani, H. R. Golkar and S. Farshbaf, *et al.*, Clinical and metabolic response to probiotic supplementation in patients with rheumatoid arthritis: a randomized, double-blind, placebo-controlled trial, *Int. J. Rheum. Dis.*, 2016, **19**(9), 869–879.
194. A. Kazemi, A. A. Noorbala, K. Azam, M. H. Eskandari and K. Djafarian, Effect of probiotic and prebiotic *vs.* placebo on psychological outcomes in patients with major depressive disorder: A randomized clinical trial, *Clin. Nutr.*, 2019, **38**(2), 522–528.

CHAPTER 5

Simulating Food Digestion and Absorption

5.1 Introduction

The food science fraternity is very keen to understand the various complex phenomena that occur during the digestion and absorption of foods within the gastrointestinal (GI) tract to gain more knowledge about the fate of foods during digestion. The main driving force for this research is the interconnection between dietary patterns and health and disease, and the thirst to develop innovative and tailored food products for specific population groups. Hence there is a great need to comprehend the digestion and absorption behavior of foods with different length scales and the physical, chemical, and enzymatic changes that occur to food across various compartments of the GI tract.[1] Various approaches, such as *in vitro* mechanical systems (static, semidynamic, and dynamic),[2,3] *in vivo* methods [sensory studies, electromyography (EMG), magnetic resonance imaging (MRI), manometry, and blood and breath tests],[4,5] animal studies,[6] cell culture studies,[7] *ex vivo* techniques,[8] mathematical and computational models,[9] and other advanced tools and techniques (imaging techniques, spectroscopic methods, analytical procedures, and omics approaches),[10] have been utilized alone and in combination in accordance with the research aims taking into account the pros and cons of the methods adopted.

Further, *in vitro* systems that can mimic one or more of the various mechanical, chemical, biological, and physiological processes that occur in the human digestive system are always in high demand. *In vivo* techniques are very precise and effective tools for investigating and understanding the interrelationships between food, digestion and absorption, and health but they are limited owing to the associated challenges and difficulties. Thus, animals and *ex vivo* methods are widely used in food digestion, food

Food Chemistry, Function and Analysis No. 42
Food Digestion and Absorption: Its Role in Food Product Development
By C. Anandharamakrishnan, Jeyan Arthur Moses and S. Priyanka

Published by the Royal Society of Chemistry, www.rsc.org

component and drug absorption, and nutritional studies, as it is not always possible, feasible, or ethical to use humans in the above-mentioned research. Other emerging approaches and advanced techniques, such as microfluidics systems and *in silico* approaches, are expected to serve as beneficial and convenient tools for future food digestion and absorption research, as they bridge the gaps among *in vitro* systems, cell culture models, *ex vivo* models, and animal models.

In this chapter, we summarize various *in vitro*, *ex vivo*, *in vivo*, and *in silico* approaches and advanced tools and techniques that have been utilized in understanding food digestion and absorption at various stages of gastrointestinal digestion, such as oral processing (mouth), gastric digestion (stomach), and intestinal digestion and absorption (small and large intestine).

5.2 *In Vitro* Mechanical Systems

Food digestion research using *in vitro* systems and the development of new *in vitro* systems to imitate various stages of food digestion and absorption have increased exponentially during the last two to three decades owing to ethical concerns, technical restrictions, cost, and the high inter-individual variability of *in vivo* studies.

The various *in vitro* models differ from each other in several respects, such as the following:

(i) Static/dynamic nature. Static systems are typically a bioreactor/beaker setup where only the static chemical and enzymatic environment of each digestion stage is mimicked, whereas in dynamic systems time-related features of the digestion process, namely the mechanical and flow properties, continuous and controlled digestive secretions, mixing, peristalsis, and wall contraction and motility, or dynamic pH changes of each stage of digestion, are mimicked.
(ii) Number (monocompartmental/multicompartmental systems) and type of stages (oral, gastric, small intestine, and large intestine) included in the system.
(iii) Composition of the fluids and secretions (such as electrolytes, enzymes, buffers, and biopolymers) and pH conditions and residence time used in each stage.
(iv) Geometric aspects (shape and volume), application and magnitude of mechanical forces, and flow profiles, and to recreate GI tract features such as grinding, mixing, peristalsis, and motility.
(v) GI tract characteristics specific to certain age groups (infants, children, adults, and elderly people).

5.2.1 Oral Phase

The oral processing of food is a very dynamic and multifactorial step as it includes a wide range of processes such as first bite, mastication, saliva incorporation, bolus formation, and swallowing.[11] Further, it is highly variable based on the type and characteristics of the food, namely physical state (solid/liquid), textural properties, and rheological behavior.[12] Also, the mastication pattern of each individual varies based on their mood, eating habit, dental health, and degree of hunger.[13] In addition, during oral processing, a multitude of phenomena such as structure breakdown, texture perception, particle size reduction, lubrication by saliva, enzymatic interactions, and flavor perception take place simultaneously.[14]

To study the complex and dynamic changes that occur during oral processing, various approaches such as *in vivo*, *ex vivo* (expectorated chewed food samples), and *in vitro* techniques are frequently used.[15] In addition, the oral anatomical features such as the teeth, tongue, temporomandibular joint, and masticatory muscles and physiological factors such as bite force, the flow rate of saliva, and mastication performance play a crucial role in the oral processing of foods.[16,17] In various food digestion studies, the oral stage is often overlooked or oversimplified to just grinding for particle size reduction using kitchen appliances such as commercial meat mincers and commercial/laboratory blenders.[18] Various oral *in vitro* systems (artificial masticators) are applied for specific purposes such as for taste, texture and flavor perception, structure destruction, and bolus formation. They are specifically designed to mimic only certain aspects of oral processing: biting,[19] chewing,[20] mixing,[21] shearing,[22] tongue movements,[23] and compression.[24]

5.2.1.1 *Artificial Masticators*

Artificial masticators are designed and fabricated to simulate various activities that occur during oral processing, namely particle size reduction, incorporation of saliva, bolus formation, tongue movement, forces exerted by the jaw muscles, and feedback mechanism of the oral sensors.[25] They are mostly utilized in fields such as dental studies, food science, and human nutrition. In-depth knowledge regarding the anatomy of the oral cavity, physiology of mastication, sensory control of the visceral motor system by the central nervous system, and feedback mechanisms is crucial to designing and developing artificial masticators. Oral processing of food from the first bite to swallowing and oral clearance is controlled through dynamic and complex feedback control.[18] Table 5.1 details artificial oral mastication systems designed for oral processing studies.

It is necessary to validate the performance of designed artificial masticators with *in vivo* oral processing by analyzing and comparing their bolus properties such as particle size, compound and aroma release, solid loss, biochemical or enzymatic changes, and saliva incorporation.[18,26] Numerous artificial masticators have been developed for research purposes, such as the Waseda Jaw,[27] artificial mouth,[28] AM2 artificial masticator apparatus,[29] Bite Master II,[19] and chewing robot,[30] and they are continuously being improved to mimic the anatomy and physiology of the human mouth. However, the

Table 5.1 Artificial oral mastication systems designed for oral processing studies.

No.	Artificial masticator	Key features	Purpose of the system	Ref.
1	Waseda Jaw	It has a human skull model made of epoxy resin as a skeleton system. It has artificial muscle actuators and various sensors for position, velocity, and force. The whole system is computer controlled	This artificial system was developed to mimic the muscle control systems and jaw motion of mastication to analyze the mechanical effects of mastication on jaw bones in terms of position, force, velocities, and muscle controls	27 and 107
2	Crush chamber	It has a corrugated cylinder and piston setup connected to a lever and it is motor powered. The crushing chamber is temperature controlled and maintained at 37 °C. It is also equipped with a dynamic microphone, force sensors, and gas sensors to obtain acoustic, tactile, and olfactory information during crushing	A "crush chamber" was designed to evaluate various sensory stimuli such as olfactory, acoustic, and tactile, stimuli perceived during the chewing of foods (crisp bread)	108
3	BITE Master II	It is a computer-controlled setup equipped with a biting station with two plates resembling the two jaws (upper-fixed and lower-movable) and they are connected by ball joints. These plates are mounted with dentures/teeth with specific geometry. In addition, they also have force sensors to measure the compression and shearing forces	It was developed to replace the human panels in food texture analysis. It closely mimics the food biting conditions and mastication of humans and measures forces applied by the jaw for foods (cheese) with various textures. It was used to study the interrelationships among food fracture properties such as hardness and mastication behavior	19

Table 5.1 (*Continued*)

No.	Artificial masticator	Key features	Purpose of the system	Ref.
4	Chewing simulator (Bouche Artificielle)	It consists of a sealed and temperature-controlled cylindrical chewing chamber with jaws, teeth, and tongue. The lower jaw and tongue are movable and driven by three motors, whereas the upper jaw is fixed. The jaws are ring shaped and are fitted with artificial molar teeth It is also equipped with a gas sampling arrangement and artificial saliva	It stimulates the flavor release mechanism that occurs during mastication. Flavor perception is measured by assessing the concentration of volatile aromatic compounds released during mastication	20 and 28
5	Chewing robot	It has a four-bar linkage mechanism to adjust to various trajectories, lengths, and jaw movements. It has two jaws with molar teeth; the fixed jaw is mounted on the platform and the moving jaw is attached to the bottom of the linkages. It also has a food manipulation system with actuators to gather and reposition food so that rhythmic chewing is continued uninterrupted. Mastication is accomplished by the combined action of shearing and compression	A robotic human masticatory system was mainly developed to stimulate the kinematics of human oral processing and evaluate the food texture. It reproduces the mastication trajectory and chewing force extracted from various *in vivo* chewing experiments on humans. From the chewing force and jaw positional data obtained during chewing, the food textural characteristics are subsequently estimated	77 and 109
6	AM2 artificial masticator	It consists of a mastication chamber equipped with disc-shaped artificial jaws comprising two projected bevel-shaped teeth positioned diametrically opposite on the circular surface, a temperature controller, and an artificial saliva supply.	It enables the simulation of mastication in diverse oral settings and offers a thorough post-mastication food bolus recovery that is comparable to *in vivo* mastication. The bolus obtained is subjected to various analyses, such as average particle size, to	110

		The mastication action is simulated with a combination of concentric rotation and coaxial translation motions of the discs, which aid in food structure destruction through both shearing and compression	analyze the efficiency of the system and the similarity of the bolus obtained. Additionally, this system allows the control of the various masticatory factors such as the type of constraints and force range applied to the food, masticatory cycle number and duration, temperature of the mastication chamber, and quantity of artificial saliva	
7	Model mouth with artificial tongue	It has a glass chamber where various oral conditions such as temperature, saliva flow, gas flow, ball-shaped tongue, and appropriate oral cavity volumes are recreated. Also, it is equipped with proton transfer reaction mass spectrometry and a headspace volume that is equivalent to the volume of the combined nasal and oral cavities to assess the release of volatile compounds.	It can reproduce tongue pressure patterns and provide online measurement of volatile compounds released during simulated oral processing	23
8	Bolus-oriented artificial oral mastication system	It is equipped with four sections simulating various aspects of oral processing, as follows: Section 1. First bite – it consists of a rotating angled baffle unit and a mesh-bottomed sample holder; Section 2. Dynamic saliva incorporation using peristaltic pumps; Section 3. Chewing – it is made up of revolving chewing discs with sharp edges around the cylinder's axis; Section 4. Bolus formation – two blades are positioned parallel beneath the chewing discs to collect the bolus formed	It closely mimics various key steps and dynamic aspects of oral processing, such as first bite, mastication, bolus formation, and saliva incorporation. The developed *in vitro* system was validated with *in vivo* trials in the case of soft solid foods taking into account dynamic changes that occur during oral processing in bolus particle size, textural attributes, and rheological behavior	

majority of the currently available artificial masticators do not take into account the role that the central nervous system plays in regulating oral mastication performance *via* a feedback mechanism. Figure 5.1 illustrates artificial masticators developed for oral processing research.

5.2.2 Gastrointestinal Systems

The gastrointestinal phase comprises various complex processes such as digestion, absorption, and colonic fermentation, each happening in various regions of the GI tract such as the stomach, small intestine, and gut/colon. In recent years, understanding the fate of food after ingestion in various compartments of the GI tract, food matrix destruction, absorption of macro- and micronutrients, by-products of gut fermentation, and their impact on overall health has become a very popular field of research.[31] The use of human and animal models to investigate food digestion is always more reliable and considered to be the gold standard, but it has various limitations such as involving huge costs, being time consuming and ethical issues.[7] Thus, various *in vitro* gastrointestinal models have been developed to explore food digestion and absorption. The *in vitro* systems can be mainly classified as (1) static/batch methods and (2) dynamic systems, which are further subdivided into (a) monocompartmental systems (*i.e.* simulating a single digestive region), (b) bicompartmental systems (simulating two GI tract regions), and (c) multicompartmental systems, which will specifically imitate from three to several regions of the GI tract such as the stomach, small intestine, and gut.[32]

5.2.2.1 Static/Batch Models

Static models are very popular among researchers owing to their simplicity, fast nature, flexibility to evaluate multiple experimental conditions and a huge number of samples, cheapness, and reproducibility.[33] Nevertheless, they oversimplify the dynamic nature of *in vivo* food digestion by ignoring the dynamic and mechanical aspects of digestion. The most static system consists of a single or a series (two, three, or four) of bioreactors to simulate different digestion stages, namely oral, gastric, small intestinal, and large intestinal digestion.[34] In general, in static systems, the experiments are usually conducted at 37 °C under mixing conditions using devices such as magnetic stirrers, overhead stirrers, blood rotators, and shaking incubators as appropriate.[4] Further, other parameters such as pH and digestive components (*e.g.* water, electrolytes, enzymes, mucins, bile salts) are adjusted to the physiologically appropriate values at the beginning of the specific experimental protocol as in different compartments of the GI tract.[35]

Static gastric and small intestinal models are mainly modeled to mimic the upper GI tract and usually the amount of intermediate products or end products before absorption is measured. In recent years, the pH-stat titration

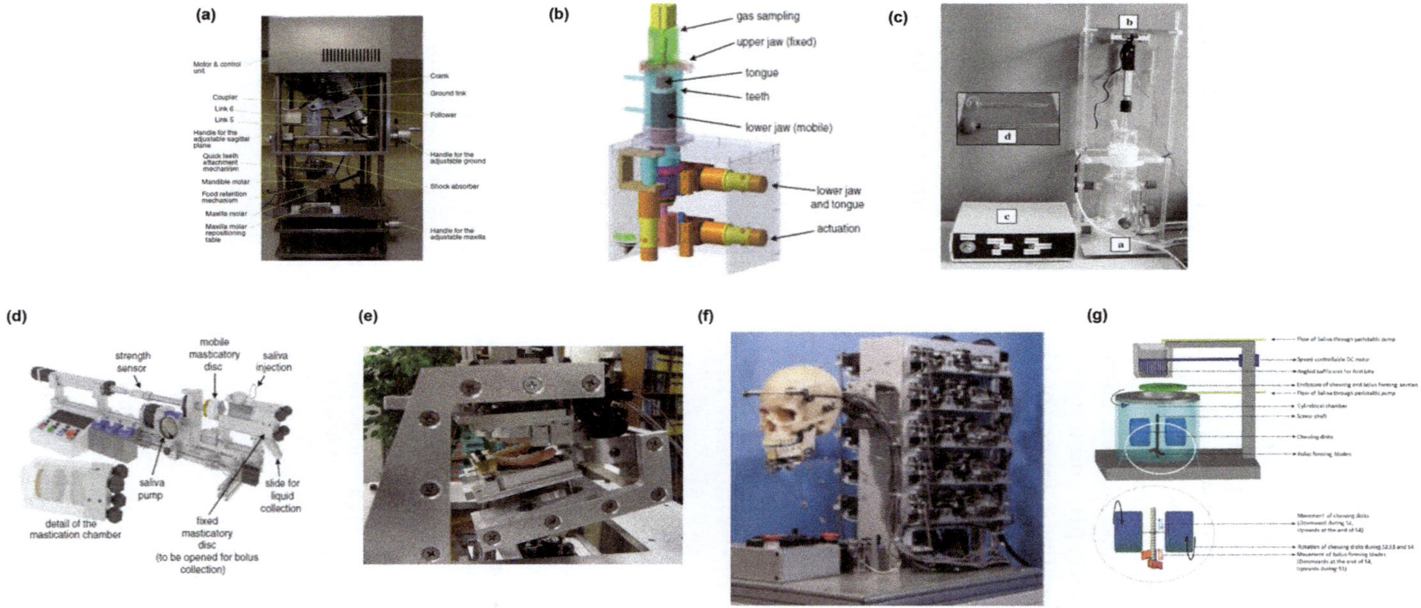

Figure 5.1 Artificial masticators developed for oral processing research: (a) the chewing robot,[77] (b) the artificial mouth,[20] (c) model mouth with artificial tongue,[23] (d) AM2 artificial masticator,[78] (e) Bite Master II,[19] (f) Waseda Jaw robot (WJ-3)[79] (www.takanishi.mech.waseda.ac.jp), and (g) bolus-oriented artificial oral mastication system.[80] (a) Reproduced from ref. 77 with permission from Elsevier, Copyright 2008. (b) Reproduced from ref. 20 with permission from Elsevier, Copyright 2007. (c) Reproduced from ref. 23 with permission from Elsevier, Copyright 2012. (d) Reproduced from ref. 78 with permission from Elsevier, Copyright 2019. (e) Reproduced from ref. 19 with permission from Elsevier, Copyright 2006. (f) Reproduced from ref. 79 with permission from Elsevier, Copyright 2008. (g) Reproduced from ref. 80 with permission from the Royal Society of Chemistry.

method has become a popular and frequently adopted *in vitro* technique in lipid digestion studies. In the case of carbohydrate digestion alone, more than 36 protocols are available for assessing *in vitro* static digestion processes. Several *in vitro* starch hydrolysis protocols are often used to quantify glucose release from different foods and various fractions of starches based on the amount of glucose released in a specific time (such as rapidly digested within 20 min, slowly digested within 120 min, and resistant starches that were not digested after 120 min). Likewise, there is a wide range of *in vitro* procedures to study protein digestion.[10] The main drawback of static systems is that there are considerable variations in key parameters such as the amount of food sample, stirring/agitation speed, amount and type of digestive enzymes, pH, and duration of each digestion step.[36] These differences in key operating parameters hinder the prospect of comparing the findings of individual studies across research groups and drawing conclusions regarding general discoveries. To address this problem, the European COST Action INFOGEST standardized a protocol by defining the key digestion parameters and conditions for *in vitro* static digestion (illustrated in Figure 5.2), and it has shown good interlaboratory reproducibility.[37]

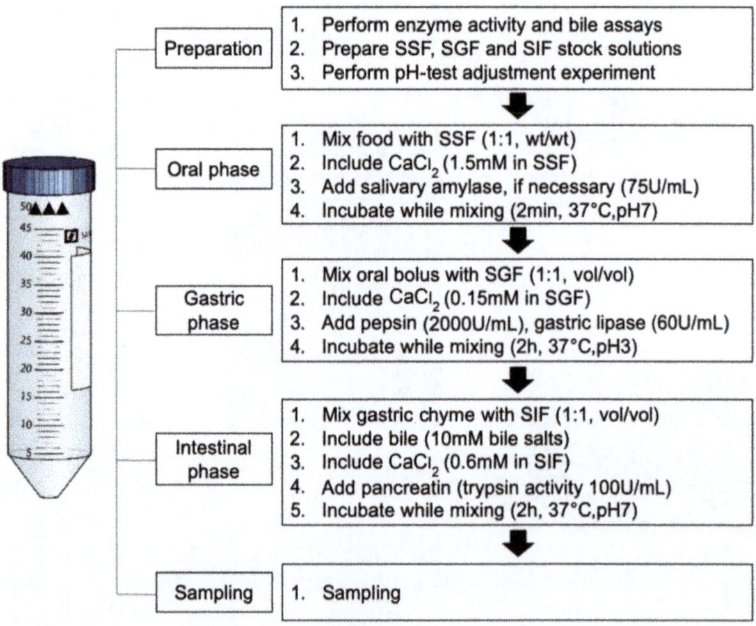

Figure 5.2 The COST INFOGEST static *in vitro* digestion procedure. SSF, SGF, and SIF represent simulated salivary fluid, simulated gastric fluid, and simulated intestinal fluid, respectively. Enzyme activities are in units per mL of the final digestion mixture at each corresponding digestion phase. Reproduced from ref. 37 with permission from Springer Nature, Copyright 2019.

Several other static methods, *e.g.* cell culture models [such as a monolayer of MDCK (Madin–Darby canine kidney) or Caco-2 cells] and membrane models [PAMPA (parallel artificial membrane permeability assay) and Ussing chambers], are frequently used in nutrient absorption studies, for assessment of the bioaccessibility and bioavailability of active molecules, and in drug absorption studies.[4] In the case of *in vitro* colon models, the most basic static models are batch cultures that mimic the gut environment such as temperature (37 °C), specific substrate/growth media, anoxic atmosphere (usually by flushing N_2), microbial growth inhibitor, and particular strains or complex fecal microbial communities.[38] The main purpose of these *in vitro* colonic batch models is to assess the metabolization of specific substrates (prebiotics) by selected gut microbial flora (probiotics and other gut microbes) *via* investigating the various gut metabolites such as short-chain fatty acids (SCFAs) that are released.[39]

5.2.2.2 Dynamic Systems

Various dynamic gastrointestinal systems have been constructed to mimic closely the several dynamic aspects of digestion, absorption, and colonic fermentation along with the aid of cell culture models.[40] The dynamic systems successfully incorporate one or more of the following anatomical and physiological aspects of the GI tract:

(i) Realistic shape and geometric features (*i.e.* J-shaped stomach and tube-like feature of the small intestine and colon) and *in vivo* organ capacity to deal with appropriate portions of food or digesta.
(ii) Some dynamic models can dynamically mimic biochemical features such as pH and enzyme and bile concentrations.
(iii) Some models can realistically replicate the flow and mixing of digesta and the digestive system's motility.
(iv) There are dynamic models with monocompartments [such as the Dynamic Gastric Model (DGM), Human Gastric Simulator (HGS), Dynamic Duodenal Model Artificial Colon (ARCOL)] and also bi-compartmental models [such as the DIDGI® System and ARK®] and multicompartmental models [such as TIM1/2, SHIME, ESIN, and SIMGI].

Figures 5.3, 5.4, 5.5, and 5.6 illustrate dynamic monocompartmental gastric systems, dynamic monocompartmental intestinal systems, dynamic bi-compartmental systems, and dynamic multicompartmental systems, respectively.

Nevertheless, not all dynamic systems mimic all dynamic aspects of human digestion such as mechanical, kinetic, biochemical, and physiological conditions altogether. Table 5.2 summarizes the various dynamic *in vitro* digestive systems used for food applications. In general, the data and information obtained from *in vivo* human and animal studies such as

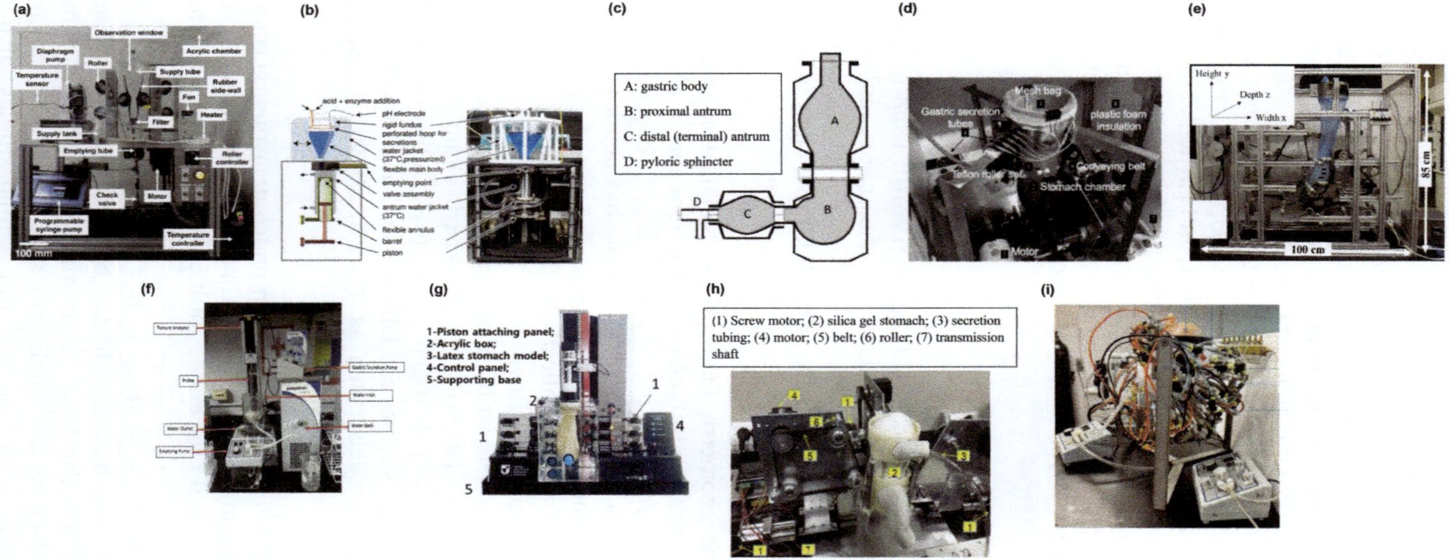

Figure 5.3 Dynamic monocompartmental gastric systems: (a) Gastric Digestion Simulator (GDS);[81] (b) Dynamic Gastric Model (DGM);[82] (c) TIMagc;[83] (d) Human Gastric Simulator Ver. 1;[36,84] (e) Human Gastric Simulator Ver. 2;[85] (f) Dynamic Gastric Simulation Model (DGSM);[86] (g) *In Vitro* Mechanical Gastric System (IMGS);[87] (h) Artificial Gastric Digestive System (AGDS);[82] (i) Gastric Simulation Model (GSM).[83] (a) Reproduced from ref. 81 with permission from Elsevier, Copyright 2017. (b) Reproduced from ref. 88 with permission from Springer Nature, Copyright 2015. (c) Reproduced from ref. 89 with permission from Elsevier, Copyright 2016. (d) Reproduced from ref. 36 with permission from Elsevier, Copyright 2020. (e) Reproduced from ref. 85 with permission from Elsevier, Copyright 2020. (f) Reproduced from ref. 86 with permission from Elsevier, Copyright 2016. (g) Reproduced from ref. 87 with permission from Elsevier, Copyright 2016. (h) Reproduced from ref. 82 with permission from Elsevier, Copyright 2019. (i) Reproduced from ref. 83 with permission from Elsevier, Copyright 2019.

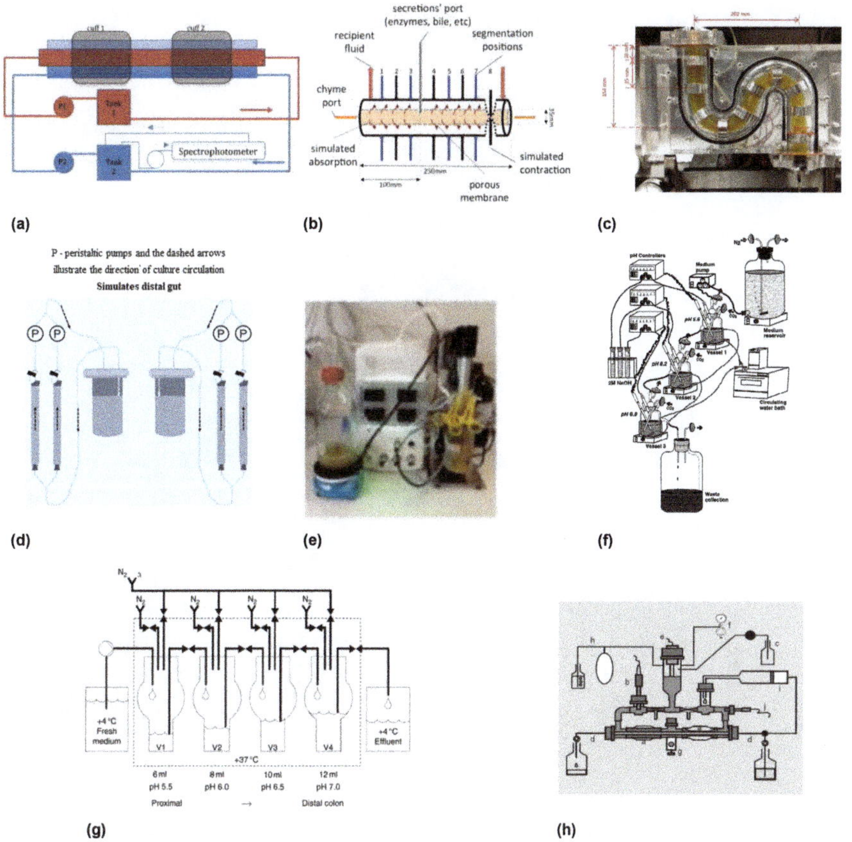

Figure 5.4 Dynamic monocompartmental intestinal systems: (a) Small Intestinal Model (SIM);[90,91] (b) Dynamic Duodenal Model (DDuo);[91] (c) Human Duodenum Model (HDM);[92] (d) Twin-Vessel Single-Stage Chemostats;[93] (e) Artificial Colon (ARCOL);[94,95] (f) Three-Stage Colonic Model System;[96] (g) EnteroMix Four-Stage Colon Simulator;[97,98] (h) TIM-2.[99] (a) and (b) Reproduced from ref. 91 with permission from Elsevier, Copyright 2014. (c) Reproduced from ref. 92 with permission from Elsevier, Copyright 2016. (d) Reproduced from ref. 93 with permission from Elsevier, Copyright 2015. (e) Reproduced from ref. 94 with permission from Elsevier, Copyright 2020. (f) Reproduced from ref. 96 with permission from Springer Nature, Copyright 1998. (g) Reproduced from ref. 97 with permission from Elsevier, Copyright 2009. (h) Reproduced from ref. 99 with permission from Springer Nature, Copyright 2015.

ingested meal transit time, peristaltic movements, pH, contraction type and frequency, temperature, digestive fluid secretion rates, absorption of water and nutrients through the surface of the organs, and microbiota communities and concentration are used in the development of *in vitro* systems, protocols, and simulations.[41] Further, the protocols are specifically designed for particular research interests, food samples (solid, liquid), and

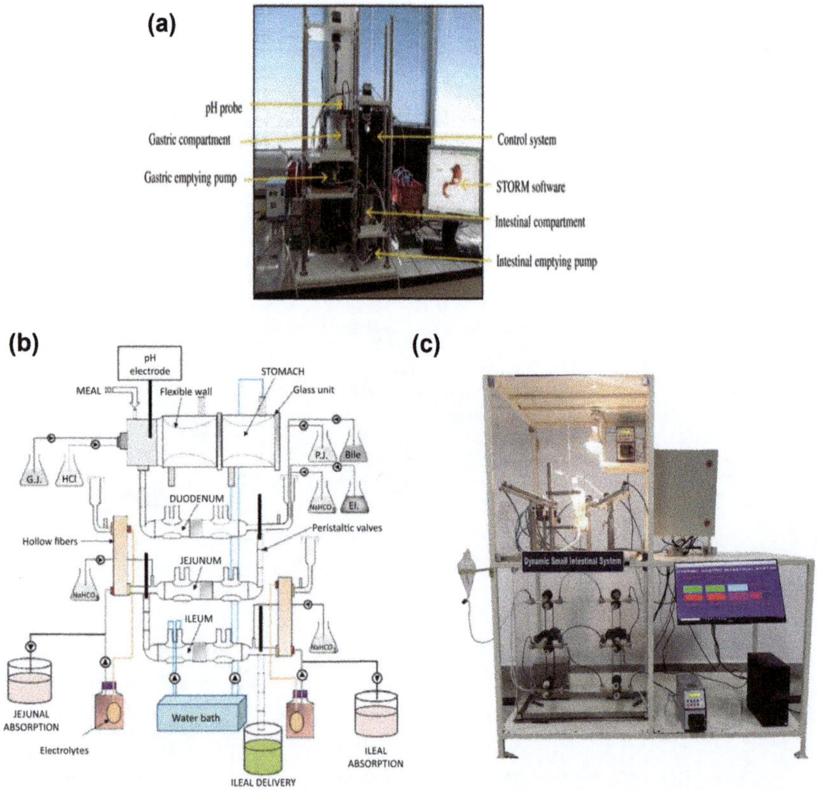

Figure 5.5 Dynamic bicompartmental systems: (a) DIDGI®, gastrointestinal dynamic digestion system;[101] (b) TIM-1;[33,102] (c) ARK®, gastrointestinal dynamic digestion system. (a) Reproduced from ref. 101 with permission from Springer Nature, Copyright 2015. (b) Reproduced from ref. 33 with permission from Elsevier, Copyright 2012.

individuals such as healthy adults, infants, the elderly, or persons with specific needs, and validated against animal or human studies.[42]

5.3 *In Vivo* Approaches

Several specific *in vivo* approaches have been used to investigate the oral processing, gastric and intestinal digestion, and absorption behaviors of foods in humans. Specific *in vivo* approaches are used exclusively to understand various ranges of interrelations amongst food, digestion, and health such as epidemiological and nutritional studies, which explore the link between dietary exposure and health effects.[4] However, the major drawbacks of *in vivo* human studies are that they are time consuming and expensive and have low throughput and ethical issues.[8] Various approaches used in the *in vivo* investigation of food digestion are summarized in Table 5.3.

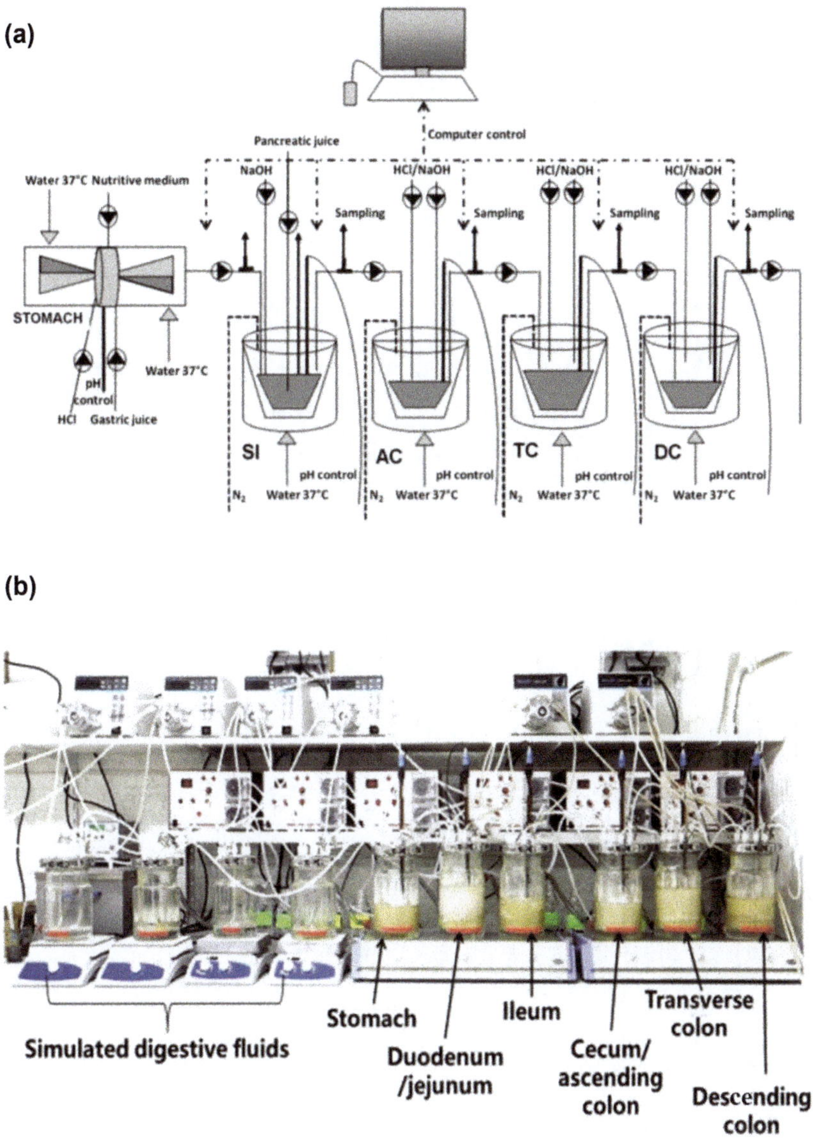

Figure 5.6 Dynamic multicompartmental systems: (a) Simulator of the Gastrointestinal Tract (SIMGI);[103] (b) Simulator of Human Intestinal Microbial Ecosystem (SHIME).[36,104] (a) Reproduced from ref. 103 with permission from Elsevier, Copyright 2015. (b) Reproduced from ref. 36 with permission from Elsevier, Copyright 2020.

5.3.1 Oral Processing Stage

In the oral processing stage, measurements that are made during the consumption of food and/or directly after consumption are considered as *in vivo*

Table 5.2 Summary of static and dynamic *in vitro* digestive systems used for food applications.

In vitro digestion system	Developer and year	Compartment and material	Operating conditions	Advantages	Limitations	Applications	Ref.
Static systems							
COST INFOGEST static model	COST INFOGEST; 2019	Mouth, stomach, and small intestine	*Oral phase* Mastication: mincer Salivation: SSF, $CaCl_2$, salivary amylase Temperature: 37 °C pH: 7 *Gastric phase* Emptying time: 2 h Mixing: rotating mixer Temperature: 37 °C; incubator pH: 3 *Small intestinal phase* Transit time: 2 h Mixing: rotating mixer Temperature: 37 °C pH: 7	• A simple and robust method. • Rhythmic peristaltic contractions and mixing in the upper GI tract are replicated using a revolving mixer, producing good intra- and inter-laboratory reproducibility. • Relatively low-cost and easy assessment of each digestion phase.	• Oversimplifies the physiology of the digestive system and fails to replicate the dynamic features of the digestive process. • In the gastric phase, the pH is kept constant, and there is a lack of the gradual addition of gastric fluid (acid, minerals, and pepsin) and a gradual gastric emptying. • Enzyme activity is kept constant irrespective of food composition. • The intestinal phase is treated as one phase rather than as the sequential duodenal, jejunal, and ileal phases, which exhibit different dilutions, mineral content, pH, enzyme activities, and microbial content.	• Aids in assessing the digestibility of various food components such as proteins, lipids, and starch. • Aids in assessing the release of carotenoids and phenolic compounds from different food matrices.	37

Batch cultures	—	Colon	*Large intestinal phase* Inoculum: human fecal samples Temperature: 37 °C Gas mixture: N_2–CO_2 (80 : 20) Stirring speed: 300 rpm pH: 6.5 Working volume: 100 mL Growth media: Gifu anaerobic broth, pH 6.5, supplemented with fructooligosaccharides, galactooligosaccharides, isomaltooligosaccharides, xylooligosaccharides, raffinose, lactulose, or lactosucrose	• Quick, inexpensive, simple to use, and repeatable. • Sealed and secured system. • Small amounts of test substances; reduced cost. • Contamination risk is low.	• Culture periods are small. • The amount of nutrients decreases as time passes. • Shortage of growth media. • Waste products accumulate. • Absence of environment control.	• They are mainly used to investigate the metabolization of specific substrates by human gut microbiota. • They are useful to evaluate the results of exposure to various sources and doses of compounds and to examine inter-individual variability in response to a bioactive substance or agent.	111
Dynamic monocompartmental systems							
Dynamic Gastric Model (DGM)	Institute of Food Research (Norwich, UK); 2008	Stomach: inverted cone-shaped. Latex	The food is prepared as a bolus with saliva and can be loaded in real time *Gastric phase* Gastric volume: ~800 mL Temperature, pH, and gastric emptying can be controlled. Gastric secretion: dynamic secretion of gastric juice and HCl *via* perforated hoop peristaltic pumps. Gastric mixing: contractions are computer controlled by water-pressure devices with pistons and barrels. Strong mechanical force in the antrum portion and low physical force around the main body	• It has adequate capacity to fill up full meals. • Can be used to study any food matrix. • It is designed in a way to perform *in vivo* correlation studies (*i.e.* the exact meal can be used in clinical studies). • It simulates real-time gastric digestion.	• Antral processing cannot be visually observed. • The stomach's body and antrum are vertically oriented; the fundus is always exposed to air. • It has no *in vivo* satiety signals, which play a crucial role in controlling the rate of digestion.	• It has been utilized in various food-based research studies to investigate the bioaccessibility of nutrients, digestion of various food constituents, structural changes that occur during food digestion, prebiotic delivery, and probiotic survival. • They are also used in pharmaceutical research to examine how different medications dissolve and disintegrate.	88, 112 and 113

Table 5.2 (*Continued*)

In vitro digestion system	Developer and year	Compartment and material	Operating conditions	Advantages	Limitations	Applications	Ref.
Human Gastric Simulator (HGS)	University of California, USA; 2010	*HGS-1* Stomach: inverted cone-shaped. Latex *HGS-2* Stomach: J-shaped. Silicone/plastic	Prepared as a bolus with saliva *Gastric phase* Gastric volume: 1500 mL Temperature and pH can be controlled. Gastric secretions and emptying: dynamically controlled using peristaltic pumps. Gastric mixing: gastric contractions are simulated through roller-driven pulley and variable-speed motor	• It precisely simulates and controls the *in vivo* fluid mechanical forces driving food disintegration during gastric digestion such as the peristaltic activity of the antral contractions waves (motility). • Various gastric digestion parameters, such as secretion rate, pH, and emptying, can be precisely controlled and varied according to different physiological conditions (*e.g.* elderly people, medications that alter gastric secretions, *etc.*).	• Needs more *in vivo* validation on mixing and physical property changes that occur in the system for specific foods. • It does not consider the dynamic nature of oral or small intestinal phases of digestion. • The typical sample size required for HGS is around 150–300 g and it will be expensive to use this system in the case of certain high-value compounds or ingredients.	• The physical and chemical breakdown of food in the stomach can be investigated. • Usually employed to investigate ACW dynamics, motility, and food material characteristics that influence the structural changes and breakdown patterns of different foods during gastric digestion.	84 and 114

Gastric Digestion Simulator (GDS)	Tsukuba University, Japan; 2014	Stomach: inverted funnel-shaped. Latex and acrylic	*Gastric phase* Gastric volume: 550 mL This gastric system models the antrum region of the stomach and has rollers to imitate the ACWs at specific intervals. It has a heating unit to control temperature, an inlet to add gastric fluid, and an outlet to empty the digested samples	• It can stimulate gastric peristalsis. • It is also possible to observe directly the real-time food disintegration due to its transparent design.	• It does not consider a dynamic oral stage. • It also ignores the other regions of the stomach such as the fundus and body. • pH control.	• It has been widely used to study the solid food disintegration and particle size distribution within the stomach due to physical and chemical processes. • It is also used to study intra-gastric flow patterns.	115 and 116
Dynamic Gastric Simulation Model (DGSM)	University of Georgia, USA; 2016	Stomach: cylinder-shaped. Acrylic	Food is prepared as a bolus *Gastric phase* Gastric volume: 300 mL It can control temperature and simulate dynamic pH changes. Continuous gastric secretion and gastric emptying processes are controlled through variable-flow mini-pumps. The gastric mixing action is mimicked using up and down movements of a texture analyzer probe	• It successfully incorporates various dynamic features of gastric digestion such as compressive forces (food disintegration), continuous gastric secretion (simulates pH changes), and continuous gastric emptying (digesta removal).	• It does not take into account the anatomical and geometric features and capacity of the stomach.	• It is chiefly used to evaluate the efficacy of supplementary enzymes in assisting food digestion.	86

Table 5.2 (*Continued*)

In vitro digestion system	Developer and year	Compartment and material	Operating conditions	Advantages	Limitations	Applications	Ref.
TIM Advanced Gastric Compartment[84]	TNO Nutrition and Food Research, Netherlands; 2016	Stomach: L-shaped. Latex	Gastric volume: 300 mL This system consists of three different parts representing specific gastric regions: the gastric body (A), the distal body and proximal antrum (B), and the distal antrum (C). It also has a valve (D), which is placed right after the distal antrum, denoting the pyloric sphincter. Gastric mixing: the gastric contractions of various regions are simulated using water pressure. It also mimics other gastric features such as temperature, digestive secretions (*via* valveless plunger pumps), and gastric emptying (*via* pyloric valve)	• It accurately simulates the dynamic gastric conditions such as motility in various gastric regions and also subsequently mimics mixing liquid velocity, secretion flows, pressure forces in the different regions of the stomach, and gastric emptying without technical complexity. • In addition, it has also adapted the realistic shape of the stomach and compartmentalized different regions of the stomach. • Also, it can be operated as a stand-alone model to study gastric digestion or combined with the TIM system.	• It can only adapt parameters related to specific gastric conditions, based on *in vivo* data. • It also lacks direct feedback mechanisms on meal characteristics such as composition and caloric density.	• It aids in the investigation of realistic gastric behavior of different meal matrices and drug dosage forms.	89

In Vitro Mechanical Gastric System (IMGS)	Universidad Tecnológica Metropolitana, Chile; 2016	Stomach: J-shaped. Latex	Gastric volume: 900 mL The gastric chamber was accurately developed with realistic morphology and dimensions of an adult human stomach using a 3D printed prototype as mold. The mechanical forces of gastric digestion, which are responsible for gastric motility, gastric mixing, and antral contractions, are stimulated using a mechanical system (*i.e.* four pairs of motor-driven acrylic pistons) to generate a frequency of 3 contractions min^{-1} and forces ranging between 0.20 and 1.89 N for liquid and solid meal, respectively. The temperature of the gastric environment is controlled using a water bath to maintain 37 °C	• It incorporates a J-shaped stomach with physiological dimensions. • This system realistically stimulates the mechanical aspects of gastric peristalsis such as peristaltic frequency and force magnitude. • It has precise control of the position where forces are applied and this allows simulation of repeated propulsion, grinding, and retropulsion.	• It does not take into account other dynamic aspects of gastric digestion such as pH control, digestive secretions, or gastric emptying.	• It was used to investigate the impact of gastric peristalsis on the intestinal lipolysis of protein-stabilized oil/water emulsions.	87

Table 5.2 (*Continued*)

In vitro digestion system	Developer and year	Compartment and material	Operating conditions	Advantages	Limitations	Applications	Ref.
Artificial Gastric Digestive System (AGDS)	Zhejiang Gongshang University, China; 2019	Stomach: J-shaped. Silicone	Gastric volume: 1700 mL The stomach model is precisely made using 3D printing to replicate various anatomical features of the adult human stomach such as size, shape, and internal folds of the stomach wall. The gastric mixing and compression forces are mimicked with the aid of mechanical actuators containing compression rollers and a vertically symmetrical pulley system. The system also can dynamically control digestive secretions, gastric emptying temperature, and pH	• The AGDS system incorporates human stomach anatomical features and the dynamic gastric environment, both chemical and mechanical aspects, in the system.	• It cannot automatically modify the digestive secretions such as enzyme concentration and quantity in accordance with different foods with various properties.	• The gastric digestion behavior of α-lactalbumin was investigated. • It can also be used to study the effect of various physiological conditions such as acids, enzyme secretion, and gastric peristalsis on food disintegration and nutrient release during digestion.	82
Gastric Simulation Model (GSM)	University of Georgia, USA; 2019	Stomach: J-shaped. Latex	Gastric volume: 600 mL The GSM precisely reproduces the geometry, motility, gastric secretions, and temperature conditions of the stomach. Peristalsis is produced *via* a series of pneumatically driven syringes that squeeze the inner stomach walls.	• The gastric chamber of the GSM system accurately mimics the human stomach in terms of size, volume, and shape. • The intra-gastric pressures and motility pattern (frequency, amplitude, and	• It does not take the dynamic oral stage and intestinal stage into consideration. • It has no *in vivo* satiety signals, which play a crucial role in controlling the rate of digestion.	• It was used to investigate the effect of contraction force on the breakdown kinetics and size distribution of sausage particles. • Gastric emptying of water-soluble nutrients and non-	83

			The gastric secretions are controlled by a variable-flow mini peristaltic pump, and the temperature of the stomach model is maintained at 37 °C using two light bulbs controlled by a thermostat	mechanical forces) can be altered by programmed control of the syringes to simulate *in vivo* gastric digestion.		digestible solids was examined. • The GSM can be successfully used to investigate the influence of the food matrix and physiological conditions (gastric secretions and contraction forces) on gastric digestion (disintegration kinetics and gastric emptying) of food.	
Small Intestinal Model (SIM)	University of Birmingham, UK; 2010	Small intestine	The model consists of an inner semipermeable flexible dialysis membrane made of cellulose ester and an outer water-impermeable flexible tube made of silicone. The diameter and length of the inner tube where the absorption occurs are 3 and 50 cm, respectively. The segmentation and overall flow of the system are mimicked through the concentric constriction in the tube (cuffs 1 and 2) and by using variable-speed peristaltic pumps (P1 and P2), respectively	• This is a very simple setup that can simulate the key mechanical intestinal conditions such as fluid flow behaviors and mass transfer phenomena of the small intestine.	• It completely ignores various key biological phenomena of the small intestine such as intestinal secretions, temperature, pH, and intestinal wall anatomical features.	• This system demonstrated that the effect of the viscosity of the lumen on mass transfer and absorption of nutrients can be modulated through mass transfer limitations.	90

Table 5.2 (*Continued*)

In vitro digestion system	Developer and year	Compartment and material	Operating conditions	Advantages	Limitations	Applications	Ref.
Dynamic Duodenal Model (DDuo)	University of Birmingham, UK; 2014	Small intestine	This is an improvised version of SIM as it has the same twin-tube concept. In addition, it also as a secretion port, to inject various intestinal secretions such as enzymes and bile. The peristaltic movement and segmentation are mimicked by squeezing the membrane at eight possible positions, and also the motion can be independently controlled	• The secretion port was strategically located 100 mm away from the feeding point to represent physiological conditions. • This improvised system is more automated and flexible than SIM.	• This system also fails to consider the effects of other important physiological conditions such as nutrient transportation through the intestinal membrane and feedback mechanisms.	• It aids in the systematic investigation of the effect of peristaltic and segmentation motions on food digestion. • Especially the effect of mass transfer on food digestibility and nutrient bioaccessibility (glucose) for a range of food hydrocolloids was investigated.	91
Human Duodenum Model (HDM)	University of Georgia, USA; 2016	Duodenum and small intestine	This model mimics the actual sigmoidal geometry of the duodenum of the small intestine comprising ascending and descending sections.Segmentation motion is replicated *via* compression and decompression of the dialysis membrane tube using a series of stainless-steel contraction rings with attached rubber fingers driven by compressed air and vacuum	• Intestinal secretions such as enzymes and salt were supplied using peristaltic pumps. • A buffer fluid was circulated outside the membrane tube simulating the recipient serum and also it was heated using a water bath to help maintain physiological temperature conditions.	• It does not consider the anatomical aspects of the physiological processes that occur over the entire length of the small intestine.	• It was used to study the impact of segmentation, model orientation, flow rate, temperature, and viscosity on the effective permeability of the intestine and the propagation rates of simulated digesta. • The starch digestion efficiency was also investigated.	92

Artificial Colon (ARCOL)	Université Clermont Auvergne, France; 2012	Colon. Glass	This is a fermenter that can stimulate the human colonic conditions. Inoculum: human fecal samples Gas: nitrogen Temperature: 37 °C pH: 6.3 Retention time: 36 h Growth media: mixtures consisting of carbohydrates, proteins, lipids, minerals, and vitamins are supplied to the system to mimic the ileal effluents. A dialysis system is used to simulate the electrolyte and metabolite concentrations *in vivo*	• Quick, inexpensive, and simple to use. • Sealed and secured system. • Small amounts of test substances. • Contamination risk is low.	• Short culture periods. • Nutrients decrease over time. • Anatomy of the intestine is not mimicked. • Nutrient digestion and absorption are not mimicked.	• It can be used to investigate the viability of probiotics and their interactions with gut microbiota.	117
Twin-vessel single-stage chemostat	University of Guelph, Canada, and Wageningen University, Netherlands; 2013	Distal colon	An Infors Multifors bioreactor system was converted into a chemostat by blocking off the condenser vent and displacing oxygen with nitrogen gas. Growth media: peptone water, yeast extract, $NaHCO_3$, $CaCl_2$, pectin, xylan, arabinogalactan, starch, casein, inulin, NaCl, KH_2PO_4, $MgSO_4$, hemin, menadione, bile salts, L-cysteine, porcine gastric mucin.	• Reproducibility. • Lengthy culture period. • Monitoring simultaneous control. • Perturbation studies.	• Daily operation poses a risk of contamination. • The intestine's anatomy is not mimicked. • A lengthy operation raises the expense per run. • There are no host cells.	• It is chiefly used to study the importance and efficacy, bioactivity, stability, bioaccessibility, and bioavailability of functional food ingredients, • It has also been applied to evaluate food safety and potentially hazardous effects of various ingested food additives present in gums, candies, and puddings.	93

Table 5.2 (*Continued*)

In vitro digestion system	Developer and year	Compartment and material	Operating conditions	Advantages	Limitations	Applications	Ref.
			Inoculum: human fecal samples Temperature: 37 °C Working volume: 400 mL pH: 6.9–7.0 Feed flow rate (medium): 400 mL day^{-1} Vessels were allowed to run for 48 days post-inoculation				
Mini Bioreactor Arrays (MBRAs)	Michigan State University, USA; 2014	Distal colon	MBRAs were designed utilizing computer-assisted design software and fabricated with resin *via* stereolithography. Each MBRA consisted of six reactors, with an internal volume of 25 mL. Inoculum: human fecal samples Gas mixture: 5% H_2–5% CO_2 –90% N_2 Stirring: magnetic stir plates Temperature: 37 °C Waste removal flow rate: 1.875 mL h^{-1} pH: 6.8 Working volume: 15 mL Growth media: tryptone, protease peptone, yeast extract, arabinogalactan, maltose, D-cellobiose,	• Minimal processing volumes. • Lower operating costs. • The ability to run up to 48 units simultaneously. • A shorter time to reach the steady state.	• There is no host simulation. • Growth is inhibited by an overabundance of end products. • Only the proximal portion of the colon is simulated.	• It is widely used as a tool to cultivate human fecal communities and to examine the impacts of specific antibiotics and food bioactives (such as polyphenol extracts from fruits) on the dynamics of these communities	118

			NaCl, hemin, $MgSO_4$, $CaCl_2$, KH_2PO_4, phosphate dibasic, Tween 80 pH 6.8, bovine bile, D-glucose, inulin, sodium bicarbonate, and vitamin K_3				
Environment Control System for Intestinal Microbiota (ECSIM)	Université Clermont Auvergne, France; 2010	P-ECSIM: proximal colon; T-ECSIM: transverse colon; D-ECSIM: distal colon; 3S-ECSIM: consists of colon sections such as proximal, transverse, and distal	A versatile system that is made up of three reactors that can be arranged in multiple ways to replicate the relevant nutritional and physicochemical conditions of different zones of the human colon. Inoculum: glycerol stocks of human feces; pH: 5.75 Temperature: 37 °C Gas mixture: N/A, anaerobic conditions maintained by microbial metabolism Working volume: 2 L Retention time: a short one (12.5 h) and a longer one (25 h) Stirring rate: 400 rpm; Growth media: mucin, starch, pectin, guar gum, xylan, arabinogalactan, inulin, L-cysteine, HCl, casein, peptone, tryptone, yeast extract, bile salts, Tween 80, $FeSO_4$, NaCl, KCl, KH_2PO_4, $MgSO_4$, $CaCl_2$, $NaHCO_3$, hemin, $MnSO_4$, $CoSO_4$, $ZnSO_4$,	• Anaerobic conditions are self-maintained. • It aids in simulating different parts of the colon separately or in combination. • It mitigates the sources of fluctuations among experiments, due to strict control of parameters.	• There is no simulation of host behavior, peristaltic movement, digestion, and nutritional absorption.	• It is predominately used to understand the role of the human intestinal microbiota, and also the role and fate of endogenous and exogenous factors (physicochemical, nutrients, prebiotics, probiotics, and synbiotics). • It is also utilized to demonstrate the relationship between the composition and metabolism of the gut microbiota and colonic transit time.	100

Table 5.2 (*Continued*)

In vitro digestion system	Developer and year	Compartment and material	Operating conditions	Advantages	Limitations	Applications	Ref.
			$CuSO_4$, $AlK(SO_4)_2$, H_3BO_3, Na_2MoO_4, $NiCl_2$, Na_2SeO_3, menadione, D-biotin, pantothenate, nicotinamide, vitamin B_{12}, thiamine, and *p*-aminobenzoic acid				
TIM-2	TNO, Netherlands Organization for Applied Scientific Research; 1999	Proximal colon	It has four glass compartments housing flexible membranes that are compressed with pressurized water. Inoculum: human fecal samples with 11% w/w microbiota Gas: nitrogen Temperature: 37 °C pH: 5.8 Working volume: 200 mL Feed flow rate (medium): 4 mL h^{-1} Chyme removal flow rate: 2 mL h^{-1} Dialysis flow rate: 6 mL min^{-1} Retention time: 80 h (12 h microbial adaptation, 2–4 h starvation period) Running time: 3 days Growth media: pectin, xylan, arabinogalactan, amylopectin, starch,	• Peristaltic movements are mimicked by changing the water pressure on the flexible walls and they aid in the mixing and movement of the luminal contents. • Parallel control and treatment. • Water absorption and regulation of the concentration of the metabolites are done with the aid of a dialysis system that prevents microorganism inhibition or death and maintains their physiological concentrations.	• No host simulation.	• It is used to study the gut microbiota dynamics. • It is frequently employed to study fermentation patterns and the impact of prebiotic and probiotic supplementation on microbial composition. • Additionally, it has been utilized to examine how lean and obese populations' gut microbiomes, cholesterol metabolites, and SCFA production are affected by the enrichment dose of food plant sterols.	102 and 119

			Tween 80, bactopeptone, casein, hemin, cysteine, menadione, D-biotin, vitamin B_{12}, pantothenate, nicotinamide, *p*-aminobenzoic acid, thiamine, NaCl, K_2HPO_4, $CaCl_2$, $MgSO_4$, $FeSO_4$, and bile salts				
Three-stage colonic model system	Dunn Clinical Nutrition Centre, Cambridge, UK; 1988	Ascending (V1), transverse (V2), and descending (V3) colon	This system comprises three fermentation vessels (V1, V2, and V3) that are sequentially connected to mimic the microbial activities of ascending, transverse, and descending colon, respectively. Inoculum: human fecal samples Gas: nitrogen Temperature: 37 °C Stirring: magnetically stirred pH: V1 = 6.0, V2 = 6.5, and V3 = 7.0 Working volume: V1 = 220, V2 = 320, and V3 = 320 mL Dilution rates: V1 = 0.14 h^{-1} (retention time = 7 h), V2 = 0.10 h^{-1} (retention time = 10 h), and V3 = 0.10 h^{-1} (retention time = 10 h) Running time: 16 days Growth media: starch, pectin, guar gum, mucin (porcine gastric type III),	• An affordable and trustworthy instrument for modeling colon microbial ecology and activity that simulates the proximal, transversal, and distal colon and permits sampling at various colonic zones.	• No simulation of host cells, nutrition uptake, or digestion. • Contamination risk.	• It was employed to verify that commercial wheat dextrin soluble fiber has positive health effects (such as butyrogenic effects on gut microbiota ecology). • The prebiotic activity and health advantages of orange juice supplemented with a galactooligosaccharide (GOS) mixture with and without Bimuno (B-GOS) were also assessed using this method.	96

Table 5.2 (*Continued*)

In vitro digestion system	Developer and year	Compartment and material	Operating conditions	Advantages	Limitations	Applications	Ref.
			xylan, arabinogalactan, inulin, casein, peptone water, tryptone, bile salts, yeast extract, $FeSO_4$, NaCl, KCl, KH_2PO_4, $MgSO_4$, $CaCl_2$, $NaHCO_3$, cysteine, hemin, and Tween 80				
EnteroMix	Danisco Innovation, Finland; 2006	Ascending (V1), transverse (V2), and descending colon (V3), sigmoidal colon (V4)	This is a semicontinuous culture colon simulator with four parallel units, wherein each unit comprises four glass vessels representing the ascending (V1), transverse (V2), descending (V3), and sigmoid colon/rectum (V4). Inoculum: human fecal samples Gas: O_2-free N_2 Temperature: 37 °C Running time: 48 h Growth media: starch, pectin, guar gum, mucin (porcine gastric type III), xylan, arabinogalactan, inulin, casein, peptone water, tryptone, bile salts, yeast extract, $FeSO_4$, NaCl, KCl, KH_2PO_4, $MgSO_4$, $CaCl_2$, $NaHCO_3$, cysteine, hemin, and Tween 80	• Capacity to conduct four tests in parallel utilizing the same fecal sample as an inoculum.	• Short-term tests are possible. • No simulation of the host cell, nutrient digestion, and absorption.	• Created specifically to study the colonic microbiota's fermentation of carbohydrates. • It was used to investigate various prebiotic, probiotic, and symbiotic combinations and their positive health effects.	98 and 120

Bicompartmental systems							
DIDGI	Institute for Agricultural Research (INRA), France; 2014	Stomach and small intestine. Glass	This is a simple three-compartmental system (stomach, duodenum and jejunum + ileum) that concentrates on the upper regions of the GI tract. *Gastric conditions:* 150 mL (ingested volume) and 30 mL (fasted volume – pepsin 625 U mL^{-1} diluted in 150 mM NaCl) pH: 3 Secretions: pepsin 1250 U mL^{-1} and 0.25 mL min^{-1} and lipase 60 U mL^{-1} and 0.25 mL min^{-1} Transit time: $t_{1/2} = 70$ min and $\beta = 1.23$ *Intestinal conditions:* Fasted volume: 5 mL of bile solution (1%) and 5 mL of pancreatin solution (10%) Secretions: bile 1% and 0.5 mL min^{-1} and pancreatin flow 10% and 0.25 mL min^{-1} pH: 6.5 Transit time: $t_{1/2} = 200$ min and $\beta = 2.2$ It is computer controlled and equipped with temperature, pH, and redox sensors, variable-speed pumps, and peristaltic pumps	• The system can simulate a variety of realistic gastro-intestinal conditions, including the flow of food and chyme, transit times, flow rates, temperature, pH, sequential addition of digestive secretions, and mixing. • Nitrogen gas is used to purge the air, and Teflon membranes with 2 mm holes are placed in front of the transfer pumps that connect the stomach and intestinal compartments, to replicate anaerobic conditions and the sieving function of the pylorus, respectively. • The meal capacity of the system is 200 mL and it can handle any type of meal. • The transparent structure of the system encourages visual observations during gastric processing. • The model is validated against *in vivo* data.	• Biomechanical processes involved in food digestion (*i.e.* gastric contractions) and nutrient absorption are not mimicked. • Vertical orientation of the gastric compartments such as the main body and antrum. • The rate of digestion is not regulated by *in vivo* satiety signals.	• It is mainly used to investigate the digestion of human milk, dairy gels, emulsions, and cheese. • It is also used to study the survival of probiotic food microorganisms in the GI tract.	101

Table 5.2 (*Continued*)

In vitro digestion system	Developer and year	Compartment and material	Operating conditions	Advantages	Limitations	Applications	Ref.
Engineered Stomach and Small Intestinal System (ESIN)	Université Clermont Auvergne, France; 2016	Stomach and small intestine	This is made up of six successive compartments: i. A meal reservoir – feeds the gastric compartment progressively with a realistic amount and size of food particles. ii. A salivary ampoule – continuously mixes the saliva with food. iii. Stomach. iv. Duodenum. v. Jejunum. vi. Ileum. It reproduces several crucial parameters of human digestion, such as body temperature, temporal and longitudinal changes in pH, salivary, gastric, pancreatic, and biliary secretions, transit times, chyme mixing, and passive absorption of digestion products	• The gastric compartment's geometry and architecture facilitate it to replicate the biphasic nature of gastric emptying. • It is the only system that has been designed to digest real-sized food particles with gastric and intestinal components. • For pharmacological applications, the model has been verified against human *in vivo* data.	• Digestion of solid foods and nutritional or microbiological applications require additional validation. • It does not take into account the resident microbiota of the small intestine, but it has been created in a way that allows for the inoculation of microbiota and their maintenance in anaerobic circumstances through nitrogen flushing.	• It is widely used in a wide range of applications including nutritional, pharmaceutical, toxicological, and microbiological applications in various fields.	121

TIM-1	TNO, Netherlands Organization for Applied Scientific Research; 1995	Stomach and small intestine	Four sections, namely the stomach, duodenum, jejunum, and ileum, are connected by peristaltic valve pumps to allow the controlled transfer of chime. Mixing and temperature control of each compartment are maintained by regulating the pressure and temperature of the water circulating outside the flexible walls. *Gastric and intestinal parameters:* Meal size: 300 g bolus with artificial saliva Volumes: stomach 300 mL, duodenum 55 mL, jejunum 130 mL, ileum 130 mL Secretions: gastric (1 mL min^{-1}), bile, pancreatin, and electrolytes (0.5 mL min^{-1}) Gastric pH curve (pH, time): (5.2, 0 min), (3.2, 30 min), (2.2, 60 min), (1.7, 120 min) Gastric emptying curve: $t_{1/2} = 80$ min, $\beta = 2$ Ileal emptying curve: $t_{1/2} = 220$ min, $\beta = 2.2$ Small intestine pH: duodenum 6.2, jejunum 6.5, ileum 7.4 Filtration rate: jejunum 4.5 mL min^{-1}, ileum 4.5 mL min^{-1}	• It is broadly validated, and time and cost efficient. • It closely mimics the dynamic GI conditions • In addition to handling whole meals, it can also manage specific food items and medications. • Throughout the transit of the chyme, samples can be taken from different kinds of compartments for analysis. • Experiments are highly reproducible. • It accurately mimics the geometry and other dynamic physiological GI conditions such as pH, temperature, secretions, gastric emptying, and intestinal permeability and absorption of bioactive compounds.	• There is no feedback signal regarding the energy density of foods and GI conditions. • There is no intestinal mucosa to study absorption and bioavailability including metabolism and excretion.	• It has been demonstrated to be an effective *in vitro* technique for examining the bioaccessibility of nutrients, digestibility of food, and the fate and effectiveness of functional ingredients in simulated dynamic human adult and infant GI environments.	32 and 122

Table 5.2 (*Continued*)

In vitro digestion system	Developer and year	Compartment and material	Operating conditions	Advantages	Limitations	Applications	Ref.
ARK®		Stomach: J-shaped, made of silicone rubber; and small intestine	This consists of a soft-elastic stomach model with a continuous secretion and emptying system using a peristaltic pump, electromechanical driving instruments, and a temperature control unit. Three pneumatic cylinders connected to a solenoid valve, programmed by a PID unit and controlled using a computer, carry out peristaltic contractions that provide the mechanical forces required for the disintegration of solid foods in the antrum region of the stomach. Gastric volume: 900–1400 mL Pneumatic cylinder pressure: 5–10 bar using compressed air Gastric chamber dimensions: 8–10 cm (width), 22–24 cm (greater curvature), 1.2 cm (pylorus diameter) Peristaltic contractions: 3 cycles min^{-1}, mimic the tonic contraction	• It is cost-efficient, simple, and validated and experiments are highly reproducible. • The gastric and intestinal systems can be managed separately or in sequence.	• The absorption of nutrients in the engineered small intestinal system is primarily by passive diffusion. • There is no feedback signal regarding the energy density of foods and GI conditions.	• This model is widely used to investigate particle disintegration kinetics and gastric emptying behavior in various foods. • It is also widely used to study the absorption of nutrients and the glycemic response of various foods.	123–125

			frequency of the human stomach Temperature control: maintained at 37 °C using a halogen bulb. The engineered small intestine system consists of a perfusion chamber, peristaltic pump, pH meter, donor, receiver, and buffer circulation compartments. The perfusion chamber was fixed with intestines from male Wistar rats, circulated with modified Krebs–Henseleit buffer, and oxygenated with carbogen (95% O_2, 5% CO_2). pH: maintained at 7.2 ± 0.1. Intestinal flow parameters (such as segment flow rate and circulating buffer flow rate) were maintained using peristaltic pumps				
Multicompartmental systems							
TNO's Intestinal Model (TIM-1/2)	TNO, Netherlands Organization for Applied Scientific Research	Stomach, duodenum, jejunum, ileum, and large intestine	See TIM 1 and TIM 2				119 and 122
Simulator of Human Intestinal	Ghent University, Belgium; 1993	V1: stomach; V2: small intestine;	This is a five-compartment system that encompasses	• Ensures long-term microbiome stability and potential to	• Lack of feedback mechanism.	• It is frequently used to study the gut microbiome's	104 and 126

Table 5.2 (*Continued*)

In vitro digestion system	Developer and year	Compartment and material	Operating conditions	Advantages	Limitations	Applications	Ref.
Microbial Ecosystem (SHIME)		V3: ascending colon; V4: transverse colon; V5: descending colon	the entire GI tract except for the oral cavity. The whole digestive process occurs in double-jacketed glass vessels (V1–V5) that are linked to each other by peristaltic pumps. The digestive slurry in each compartment is mixed using magnetic stir bars. pH: V1 = 2.0, V2 = slightly acidic to neutral, V3 = 5.6–5.9, V4 = 6.15–6.4, and V5 = 6.6–6.9 Stirring rate: 150 rpm Inoculum: 20% human fecal suspension Growth media: arabinogalactan, pectin, xylan, dextrins, starch, glucose, yeast extract, protease peptone, mucin, cysteine, $NaHCO_3$, NaCl, K_2HPO_4, KH_2PO_4, $CaCl_2$, $MgSO_4$, hemin, Tween 80, menadione, biotin, pantothenate, nicotinamide, vitamin B_{12}, thiamine, and *p*-aminobenzoic acid. Temperature: 37 °C Gas mixture: N_2–CO_2 (90 : 10) or O_2-free N_2	track microbiome adaptation. • Integrates the whole GI tract. • Applicable for colon region-specific research and has microbiome inoculation from different target groups: adult *vs.* infant, healthy *vs.* diseased, and animals. • Permits sampling at various GI tract locations. • Mucosal and luminal microbiomes can be distinguished in an M-SHIME system. • Additionally, with TWIN-SHIME settings, parallel control, and treatment can be accomplished.	• It does not consider a dynamic oral stage. • It completely ignores the anatomical characteristics and some physiological aspects of the GI tract such as peristalsis. • The traditional SHIME does not contain host cells. Epithelial or immunological cells coupled to the HMI module to solve the problem. • The conventional SHIME setup lacks dialysis. However, the incorporation of dialysis modules is possible after small intestine digestion and colon digestion.	receptivity to incoming substrates of interest (such as prebiotics, bioactive substances, or medicines), and also the digestibility of prebiotic substrates and their subsequent fermentability in the colon. • It is also used to examine the resilience of the microbial population against colonization by an opportunistic pathogen and the survival of pathogens or probiotics in the upper digestive tract.	

			Working volume: V3 = 500 mL, V4 = 800 mL, V5 = 600 mL Retention time: V3 = 20 h, V4 = 32 h, V5 = 24 h Feed flow rate: 25 mL h^{-1} Four stages: stabilization (2 weeks), basal (2 weeks), treatment (2–4 weeks), and washout (2 weeks); Running time: typically 30 days per cycle				
Dynamic *In Vitro* Human Stomach–Intestine System (DHIS)	Monash University, Australia; Xiamen University, Soochow University, China; 2008	Esophagus, stomach, and small intestine (duodenum). Silicone	It has a J-shaped gastric chamber with an internal volume of 400 mL and a wrinkled internal lumen (5–10 mm wide and 2–4 mm deep). In addition, 20 silicone tubes are connected to the gastric corpus randomly to replicate the roles of gastric secretory glands. Mixing, disintegration, and transport of chyme are ensured *via* peristaltic contractions that are produced by the action of various electromechanical instruments such as a series of motors, rollers, and eccentric wheels. Temperature: 37 °C using heating lamps and a water bath	• It features a distinctive design that alters the system's overall tilting angle from −45° to +45° to adjust the gastric emptying rate. • It replicates the internal folds, dimensions, and morphology of the human stomach. • The pyloric squeezing device carries out the gastric sieve function of the pylorus.	• It does not consider the entire small intestine. • Finding appropriate operational settings to pre-set gastric emptying profiles is time consuming. • Owing to inadequate mixing of the substrates and digestive enzymes, it suffers from reduced digestion rates.	• It is mainly used to investigate particle size distribution, gastric emptying, and digestion of both solid and liquid fractions of mixed foods.	34, 127 and 128

Table 5.2 (*Continued*)

In vitro digestion system	Developer and year	Compartment and material	Operating conditions	Advantages	Limitations	Applications	Ref.
SIMulator of the GastroIntestinal Tract (SIMGI)	Institute of Food Science Research CIAL (CSIC-UAM), Spain; 2015	Stomach: SI, small intestine; and large intestine: AC, ascending colon; TC, transverse colon; DC, descending colon	*Stomach* It is made of two cylindrical transparent and rigid methacrylate plastic modules covering a reservoir of flexible silicone walls. The gastric contents are mixed by peristaltic movements achieved by changing the pressure of the water. It has separate ports to supply food and gastric secretions. The system has provisions for temperature (37 °C) and pH control (HCl) *Small intestine* It is made of a double-jacketed glass reactor vessel that has stirrers and inlet ports for intestinal secretions (such as pancreatic juice and bile) and gastric contents. Digestion time: 2 h Temperature: 37 °C pH: 6.8 Intestinal secretions: pancreatic juice, $NaHCO_3$, oxgall (dehydrated fresh bile), and porcine pancreatin	• Due to the automated control of the working parameters, the system is flexible and easily adaptable to the needs of experiments. • The SIMGI design dynamically simulates the whole gastrointestinal digestion process except for an oral phase. • Peristaltic movements, controlled gastric emptying, and simulation of other dynamic digestive conditions. • A long period of colon culture.	• No host–gut microbiota interactions. • Risk of contamination due to daily operation. • Oversimplifies the GI tract anatomy and physiology. • There is no feedback signal regarding the energy density of foods and GI conditions. • Lacks devices to evaluate the simulation of intestinal absorption.	• It is mainly used to study the gastric digestion of whey proteins. • Additionally, research has been carried out on the effects of dietary interventions including low- and high-calorie diets on the production of SCFAs and ammonium. • Researchers have also assessed how red wine consumption affects the intestinal microbiota's polyphenol metabolic activity.	103 and 129

Large intestine
Three large intestinal stages (AC, TC, DC) are simulated in three individual double-jacketed glass reactors.
Inoculum: human fecal samples
Gas: N_2
Temperature: 37 °C
Stirring rate: 150 rpm
pH: AC = 5.6 ± 0.2, TC = 6.3 ± 0.2, and DC = 6.8 ± 0.2
Working volume: AC = 250 mL, TC = 400 mL, and DC = 300 mL
Running time: 76 h
Feed flow rate = 5 mL min^{-1}
Stabilization period: 14 days
Growth media: arabinogalactan, pectin from apples, xylan, potato starch, glucose, yeast extract, peptone, mucin, and L-cysteine

Table 5.3 Summary of *in vivo* techniques used to investigate food digestion.

Technique	Applications	Benefits	Drawbacks	Ref.
Tongue pressure sensing system	It links tongue movement and pressure with food texture parameters such as consistency and rheological nature such as elastic or plastic. It consists of a T-shaped sensor or pressure sensor bulb that aids in the real-time monitoring of tongue pressure. It senses the temporal and spatial distribution pattern of tongue pressure during oral processing	The sensor sheets are highly flexible and adapt easily to the oral shape. Does not interfere with the occlusion and perception of taste, aroma, or texture	The positioning of the pressure sensor and food sample during real-time oral processing is particularly challenging in the case of the bulb-type pressure sensor	130 and 131
Electromyography (EMG)	It is predominantly used to evaluate the mastication behavior and muscle activity of individual with respect to specific foods with different textural attributes. It permits one to measure and monitor various physiological parameters associated with the electrical activities of the orofacial muscles linked with the oral processing of foods such as chew style and sequence, chewing time and rate, number of swallows and chews, oral clearance, and muscle activity	A non-invasive technique. It is a feasible, efficient, reliable approach for determining the factors associated with human masticatory parameters related to food texture. It can provide both qualitative and quantitative characterization of oral behavior in response to the specific textural properties of foods	Needs skilled personnel. Involves time-consuming data analysis and interpretation. Requires careful selection of subjects with good health, dental state, and facial morphology. Ethically approved	132 and 133

Jaw tracking/ electrognathography	It is mainly used for recording jaw movements during the consumption of food. Jaw movement parameters such as vertical, lateral, and anterior–posterior amplitudes, the velocity of opening and closing, and durations of opening, closing, and occlusal phase are measured during food oral processing and linked to food texture	A non-invasive technique. It can be used in combination with EMG or separately to determine the effect of texture on the oral processing behavior of various foods among different groups of subjects (children, adults, the elderly, and dysphagia patients)	Various jaw tracking systems such as markers or transducers attached to teeth are reported to impact natural chewing behavior compared with less intrusive skin surface markers. The study of quick movements and minor movements near the intercuspal position is constrained by the lower sampling frequency	134 and 135
Bite force analysis (intra-oral) using a multiple-point sheet sensor system	It is mainly used to measure the bite force while chewing foods. In addition, maximum bite force, duration of the force, chewing cycle time and impulse as the integrated force, the location and degree of pressure exerted by various teeth, and bite pressure distribution can be analyzed	It simultaneously measures the load and contact area. High sampling rate. The sensor can work under the wet conditions of the mouth. Spatial sensitivity is comparable to the human oral somatosensory system	The sensor might have some interference in food chewing behavior. Needs careful selection of subjects with good health, and dental state	136 and 137
Video recording	The recorded video of food consumption by subjects with attached facial markers allows the quantification of various mastication parameters such as several bites, chews, and swallows, chewing and clearance time, oro-sensory exposure time, interval between bites, and meal duration. It is also the most popular and frequently used method to determine oral processing parameters	A very simple technique with fewer disturbances in the measurements. It is non-invasive	Generates too much data, hence data processing consumes excessive time. However, this can be mitigated by using software for data analysis	138 and 139

Table 5.3 (*Continued*)

Technique	Applications	Benefits	Drawbacks	Ref.
Proton transfer reaction mass spectrometry (PTR-MS)	This method is mainly used in nose-space analysis to measure the aroma and flavor release during the oral processing of food with different matrices and textures. It also widely used to correlate sensory data	It precisely and quantitatively measures the release of volatile compounds from foods during mastication. Less invasive in nature	Needs ethical approval. Expensive method. Panelists need training to strictly regulate their breathing	140 and 141
Video fluorography	This technique is used to obtain qualitative data on bolus and mastication features during the oral processing of food with the aid of X-ray imaging and barium sulfate as a fluorescent compound mixed with food. Several measurements including position and shape of food bolus, length of feeding sequence stages, chewing cycles and swallows, maximum gape, and mandibular displacement amplitude can be measured	Assists in gathering spatial and temporal information about the movements of the jaw, teeth, tongue, and hyoid (tongue base). It is possible to obtain real-time information about the food inside the mouth	Use of barium sulfate. Radiation exposure. Lack of quantitative data	142 and 143
Magnetic resonance imaging (MRI)	Aids in examining *in vivo* gastric digestion (protein hydrolysis), emptying process, and gastric motility. Provides information about gastric content such as volume, flow, viscosity, layering (intra-gastric appearance), intra-gastric air, phase separation, and clot formation.	It is a non-invasive technique and does not pose any radiation hazard. MRI results correlate well with scintigraphy and can be validated with time-domain nuclear magnetic resonance relaxometry (TD-NMR) experiments.	Expensive equipment. Needs skilled manpower for operation and interpretation of the equipment and scan results. Demands specific and special posture. Specific absorption rate (SAR) limitation for tissues, which is less strict.	144–146

	Aids in visualizing food state, structure, and composition (*i.e.* water content, fat content, molecular migration, structure change)	It has the potential to bridge *in vitro* and *in vivo* gastric digestion research	It requires a long time for imaging	
Scintigraphy	Facilitates direct measurement of gastric emptying. It aids in the assessment of gastric motility. It helps in the determination of intra-gastric distribution to characterize fundal and antral function	It is a non-invasive, gold-standard technique that provides the most accurate results	Radiation exposure. It is technically demanding, expensive, and availability is limited. It contains radioactive isotopes, so it is inappropriate for use with young persons and pregnant women and in situations that require repeated measurements	147–150
Isotope respiration test	Aids in the measurement of gastric emptying. It is a validated method to investigate gastric emptying for both solids and non-nutrient liquids in healthy individuals and also in people with various disease conditions such as diabetes, dyspepsia, respiratory disease, and gastroparesis. However, it does not provide any information about gastric accommodation, contractile activity, and gastric motility	It is a non-invasive technique with no radiation exposure and is cheaper than scintigraphy. It is a satisfactory alternative to scintigraphy in the case of children, pregnant women, and critically ill patients	Gastric emptying is only measured indirectly. There is still a debate about the ideal analytical algorithm. Lack of a typical commercial diet. It assumes normal small intestine, liver, and lung functions, so there is a chance of the results being inaccurate due to illness	148 and 151–153

Table 5.3 (*Continued*)

Technique	Applications	Benefits	Drawbacks	Ref.
Ultrasonography	This is utilized in the measurement of gastric volume and gastric motility. It helps in the determination of gastric emptying in healthy subjects, diabetic individuals, snd persons with functional disorders and motility disorders of the GI tract. 3D ultrasonography can be exploited to investigate the intra-gastric distribution and meal volume	Ultrasonography poses no radiation hazard. An extensive variety of equipment is available and operating costs are moderate. Validated for both low- and high-nutrient liquid gastric emptying. Several studies have used ultrasonography to study the effect of different nutrients, cereals, and cereal-based products on the gastric emptying rate	Demands a skilled operator and greater expertise is needed in case of 3D than 2D ultrasonography. Imaging efficiency may be affected by obesity and/or bowel gas Necessitates geometric assumptions of stomach shape. Commonly used to measure only the liquid gastric emptying	154–157
Gastroduodenal manometry	This technique is used in investigating gastric and small intestinal motor function. It aids in the indirect measurement of gastric emptying, and allows the detailed study of gastrointestinal motility	It provides a greater degree of functional information that helps in differentiating the neuropathic and non-neuropathic causes of gastroparesis	Invasive and expensive technique. Indirectly measures gastric emptying. It merely records occlusive contractions of the lumen. Restricted accessibility	156 and 158–160

Wireless sports capsule technology	This method evaluates gastric motility and regional intestinal transport and movement. It aids the measurement and differentiation of both normal and delayed gastric emptying. It is capable of measuring the gastrointestinal tract's pH, pressure, and temperature	It is a simple, non-radioactive method that facilitates research anywhere. Correlates with scintigraphy	Does not directly measure gastric emptying. Chance of capsule retention	52 and 161–163
Acetaminophen or paracetamol absorption	This aids in the determination of gastric emptying but it is validated only for liquid emptying. Additionally, it only indirectly quantifies the gastric emptying rate	This is a non-invasive technique with no radiation exposure	Requires frequent blood samples. Less sensitive. Errors in the results are possible as the method is affected by several factors. Lacks standardization	164–167
Blood test	A blood test is mainly used to measure blood glucose concentration, which correlates well with the rate of food digestion and absorption. Used to determine the glycemic response of a meal	It is a non-invasive technique that requires only a small number of blood samples, with less pain and discomfort. It is a simple standard method that facilitates research anywhere	Needs repeated blood sampling	50 and 168

parameters that include sensory, mastication, and bolus parameters.[43] Sensory parameters (such as texture, flavor, taste, appearance, sound, and temperature) are investigated using sensory panelists (trained/untrained) through various sensory analyses such as discriminative, descriptive, hedonic, and temporal tests with the aid of specific sensory software (*e.g.* Compusense®, sensCheck™, and FIZZ) or with the aid of in-house software or methodologies.[44] Various *in vivo* oral processing approaches for sensory parameters are summarized in Table 5.4.

Several *in vivo* oral processing parameters, such as mastication features (*e.g.* bite size, mastication time, chewing cycles, and the number of swallows), are measured using panelists (trained or consumer) under a controlled setup and are monitored either by direct observation or using videography.[45,46] In addition, bolus parameters such as particle size distribution, saliva incorporation, microstructure, and extent of starch digestion are analyzed using image analysis/sieving, weight difference on drying, microscopic techniques, and starch/glucose assays, respectively.[47] Further, to evaluate several oral processing parameters specific to chewing behavior and texture perception, various *in vivo* human physiological factors such as biting force, palatal pressure, tongue pressure, muscle activity [using electromyography (EMG)], jaw movements [using jaw tracking (JT)/electrognathography], and velocity through the pharynx (using the ultrasonic pulse Doppler method) are recorded utilizing non-invasive sensors.[17,48] In addition, the aroma release during the consumption of food is evaluated using nose space analysis in combination with proton transfer reaction mass spectrometry (PTR-MS), retronasal trapping device (RATD), and mass spectrometry (MS).[43,49]

5.3.2 Gastric and Intestinal Stages

Specific to gastric and intestinal stages, various specialized invasive approaches involve intubation, where the measuring devices are inserted through the nose or mouth into specific sites (esophagus, stomach, small intestine, gut) of the human body. For example, barostat balloons and catheters with a manometer (intraluminal manometry) are widely used in studying GI motility behavior and exploring the changes in gastrointestinal pressure and volume during the fasting and fed states.[50,51] Nowadays, an advanced approach known as wireless capsules (SmartPill™) is used as a replacement for balloons and catheters, and it can provide details on pressure, force, torque, temperature, and pH of specific GI segments simultaneously.[52] Further, surgical approaches such as ileostomy may be an option for patients with Crohn's disease and ulcerative colitis. In the case of ileostomy, the terminal part of the ileum is attached to the outside of the body and an external pouch is used for collecting the end products of upper GI tract digestion to investigate the digestion and absorption of foods.[53] Nevertheless, the invasive procedures and collection of ileal samples from humans are unpleasant for the patient but provide unique significant information.

Table 5.4 *In vivo* oral processing approaches for sensory parameters.

Descriptive tests	Discriminative tests	Hedonic tests	Temporal tests	Instrumental techniques
Quantitative descriptive analysis	Triangle test	5/7/9-point hedonic scale	Time–intensity	E-nose
Free-choice profiling	Tetrad test	Just-about-right scaling	Temporal dominance of sensations	E-tongue
Flash profiling	Duo–trio test	Ideal profile method	Temporal check-all-that-apply	
Open-ended questions	ABX test	Degree of satisfaction–difference (DOSD)	Temporal liking	
Free sorting and free multiple sorting	A not-A test	Relative preference mapping (RLM)	Temporal dominance of emotions	
Projective mapping and napping	Paired comparison		Hold-down temporal dominance of sensations and emotions	
Flash profiling	Forced choice		Free comment attack evolution finish	
Preferred attribute elicitation			Projective categorization	
Check-all-that-apply and rate-all-that-apply				
Polarized sensory positioning				

In addition, various non-invasive techniques, such as scanning and imaging (MRI, ultrasonography, X-rays, and scintigraphy), blood tests, and breath tests, are widely used to investigate the several phenomena that occur in the GI tract upon consumption of food.[50] Further, gastric physiological processes such as gastric motility and gastric emptying and other gastric parameters such as gastric pH, volume, temperature, and intra-gastric pressure can be measured with the help of various other tools. In general, gastric motility is commonly measured in terms of frequency, velocity, torque pressure, duration, and intensity of the waves using various diagnostic techniques such as a manometer (barostat), MRI, and the wireless capsule technique. Likewise, the gastric emptying of foods is measured by several non-invasive, radiation-free, and repeatable techniques such as MRI, ultrasonography, isotope respiration test, and acetaminophen or paracetamol absorption. However, scintigraphy remains the "gold standard" technique for measuring gastric emptying, and it uses technetium-99m–sulfur colloid- or technetium-99m–diethylenetriaminepentaacetic acid-labeled food for testing.[158] Further, the results of gastric emptying measurements through MRI and wireless sports capsule technology correlate well with scintigraphy results.[145] Hence all of the above-mentioned tools and techniques aid in understanding the gastrointestinal digestion behaviors of foods in a better manner.

5.4 Animal Studies

Various species of animals, including rats, pigs, sheep, horses, cows, chickens, and dogs, are used in food digestion studies as they offer a certain degree of ethical flexibility compared with human trials.[50] However, the first two (rats and pigs) are the most often encountered as they are relatively cheaper, easier to acquire, and less difficult to handle. Pig models are widely used in food digestion studies such as amino acid metabolism, ileal and fecal digestibility of dietary amino acids,[54] digestion of dietary fiber,[55] and the structural breakdown and glycemic response of starch-based foods,[56] as they are considered as one of the animal models that can closely mimic human digestion. Further, small piglets are an excellent choice for studies related to infant digestion and neonatal nutrition.[57] Nevertheless, no one animal species is the same as humans in terms of gastrointestinal physiology or anatomy.

5.4.1 Sample Collection Techniques

In animal digestion studies, various sample collection techniques, such as (i) slaughtering of the test animal at a predefined time after feeding, (ii) surgical introduction of cannulas at essential sites in the GI tract (stomach and small intestine), and (iii) surgical introduction of catheters in the animal's blood circulation system (portal veins or arteries), are used for collecting sample digesta and blood.[50] Subsequently, these collected samples are analyzed using various methods to understand and quantify the digestion and absorption characteristics. Table 5.5 briefly describes

Table 5.5 Summary of differences between the two techniques most commonly used for sample collection in animal trials: cannulation and slaughtering.

Technique	Procedure	Advantages	Limitations
Cannulation	Requires surgical insertion of a cannula into the GI tract. One end of thecannula protrudes outside the skin and it has a removable stopper, through which samples can be collected at specific times	Samples can be collected from the same animal over a long period of time. Reduced variability as there is flexibility to test various experimental treatments on the same animal	Surgery in the GI tract will affect various gasterointestinal functions such as secretions, motility, and flow patterns, which will modify the actual digestion process
Slaughtering	Fed animals are sedated and euthanized after a specfic time to collect samples	Samples can be collected from multiple locations in the GI tract	Repeated sampling from the same animal is not possible. A large number of animals are needed in a study to ensure statistical significance

the differences between the two most common techniques used for sample collection in animal trials: cannulation and slaughtering.

5.4.2 Limitations and Difficulties

The most significant limitations and difficulties associated with animal studies are the differences between animal and human digestive and metabolic systems, the selection of suitable animal species that resemble humans closely in terms of anatomical and physiological conditions, the handling and feeding of animals and sample collection, sample characterization and correlation with humans, and assessing the reliability, validity, reproducibility, and sensitivity of the animal trials, which is laborious. In addition, experimentation with an animal requires approval from the respective ethical committees to guarantee that adequate care and effort are applied to minimize animal suffering, stress, and pain. Furthermore, when selecting an animal, other factors such as the animal's age, gender, and disease condition need to be carefully considered.[8]

5.5 Cell Culture Models

Cell culture models are predominantly used for understanding the functions of the lower GI tract (especially the small intestine), such as nutrient absorption and its mechanisms, bioavailability, bioaccessibility and bioactivity of food components, and permeability, toxicity, and absorption of drugs. As discussed extensively in Chapter 3, the anatomy, histology, morphology, and physiology of the human small intestine are remarkably complex, consisting of unique finger-like villi to maximize the surface area for nutrient absorption, brush border membrane vesicles with intestinal enterocytes, the lumen made with several distinct layers of mucosa, and a collection various specialized cells, namely goblet cells, Paneth cells, stem cells, and Peyer's patches, with unique functions.[58]

Currently, several cell culture-based transport models. such as (i) monoculture models (*e.g.* MDCK, LS174T, Caco-2, IPEC-J2, and HT29), (ii) co-culture models (*e.g.* Caco-2/HT29-MTX), (iii) triple-culture models (*e.g.* Caco-2/HT29-MTX/Raji B cells), and (iv) 3D cell models, are widely used for studying absorption mechanisms and kinetics.[8] These cell culture models are valuable static *in vitro* models as they offer ethical and economic benefits even though they can only partly mimic the physiological features of the intestine.[7] Another advantage of using cell culture systems is their ability to generate consistent results and reproducibility.[59] Table 5.6 briefly details various monoculture cell line models and their applications.

5.6 *Ex Vivo* Methods

Ex vivo methods/experiments are performed under a controlled external environment similar to the *in vivo* natural conditions using tissues or organs

Table 5.6 Monoculture cell line models and their applications.

Monoculture cell line model	Description	Advantages	Limitations	Applications
Caco-2 cells	• Obtained from human colorectal cancer cells. • Spontaneously differentiate (14–21 days) and polarize in culture. • Their characteristic features are the same as those of enterocytes of the small intestine.	• They possess characteristic apical microvilli and form tight junctions (TJs) between two adjacent cells. • Express typical intestinal enzymes such as disaccharidase and peptidase. • Cost-effective, easy to use. • Commercially available. • Extensive literature available.	• Time-consuming procedure (21 days). • Shortened culture durations, *i.e.* 7 or 3 days to modify the cell morphology, and expression of enzymes and transporters, and may not give TJs enough time to develop. • Cancerous origin. • Difficulty in controlling differentiation.	• Predominantly used for assessing the intestinal transport and absorption of food compounds, bioactives, drugs, and formulations. • They are also widely used for interpreting the relative transport mechanisms behind designer food and drug systems (*e.g.* micelles, polymeric and lipid-based nanoparticles). • To investigate the processes through which the intestinal epithelium's barrier function is impacted by the microbiota, microbial metabolites, and food digesta.

Table 5.6 (*Continued*)

Monoculture cell line model	Description	Advantages	Limitations	Applications
HT-29	• Derived from human colorectal adenocarcinoma. • Largely undifferentiated but polarized. • The cells produce mucus and cytokines (such as interleukins, TNF-α).	• Contain mucus-producing goblet cells. • Commercially available.	• They are malignant colonic cells with impaired glucose metabolism. • They only replicate some characteristic features of the small intestine and colon enterocytes.	• They are frequently employed to research host–microbiome interactions, immunological response, and nutrition. • Methotrexate-treated HT-29 cells (HT29-MTX) have mostly been employed to study bacterial adhesion and survival. • Utilized as host models to explore *in vitro* fermentation of prebiotics and fecal fermentation.
T84	• Although extracted from lung metastasized cells, these cells originated from human colon cancer cells. • Express various brush border properties such as enzymes and transporters. • They can achieve tight TJs and also express various TJ proteins and ion transporters.	• Cost-effective. • Easy to use. • Commercially available. • Can be polarized.	• Cancerous origin. • Less utilized model for microbiome research.	• They are often used as a model for studying electrolyte transport and epithelial barrier function. • Commonly used to explore the effect of enterotoxin on intestinal secretions and the protective effects of probiotic bacteria and food compounds.

IEC-6 and IEC-18	• The rat (whole) small intestine is the source of IEC-6. • IEC-18 is derived from the ileum of the rat's small intestine. • Displays Toll-like receptors and cytokines such as interleukin-6.	• Non-cancerous origin. • Excellent for small intestinal studies. • Can be polarized. • Commercially available.	• Do not precisely replicate the colonic morphology and physiology.	• IEC-6 is an excellent tool to study the bacterial adhesion of many species. • IEC-6 has also been used to investigate the transportation of probiotics, products derived from microflora, and small intestine microbiota. • IEC-18 is widely used to explore the protective effects of *Bifidobacteria* spp.
IPEC-J2 and IPEC-1	• The intestinal porcine epithelial cell (IPEC) lines are obtained from the small intestine of neonatal piglets. • IPEC1 and IPEC-J2 are derived from the jejunum and the ileum and jejunum, respectively. • They express TJ proteins, inflammatory pathway markers, mucins, Toll-like receptors, and cytokines.	• Physiological relevance to humans is greater than that for rats. • Polarizable and non-cancerous origin. • Commercially available. • Beneficial for studying the small intestine. • Consist of a variety of cell types.	• Do not precisely mimic the human colon.	• Utilized for studying bacterial infections, nutrient and ion transport, proliferation of cells, pathogens, and viral and fungal infections. • IPEC-J2 is increasingly used for probiotic research and innate immune responses to many pathogens.

obtained from a living organism, usually animals such as rats, pigs, and chickens.[7] In addition, tissues are less frequently obtained from other animal sources such as mice, rabbits, dogs, and monkeys. They are regarded as the middle ground between *in vivo* and *in vitro* methods. Although human tissues are the most applicable, their use in research is constrained by several problems, including limited availability, the inability to conduct regional studies, ethical considerations, and the lack of standardization of human tissue samples (with respect to gender, age, medication, and illness conditions).[8] In food digestion research, *ex vivo* experiments are commonly used in (1) oral texture analysis to predict fatty mouthfeel using a tribometer attached to a pig tongue to investigate phase inversion of fat droplets in emulsions[60] and (2) intestinal absorption and transport studies. Various *ex vivo* models such as intestinal rings/slices, intestinal segments, InTESTine™, everted sac, diffusion chamber (Ussing chamber and Franz cell), enteroids/organoids model, gut loops, and intestinal perfusion model are frequently used to perform intestinal absorption and transport studies.[59] Table 5.7 details the salient features of various *ex vivo* methods.

The significant advantage of *ex vivo* techniques is that they offer anatomical and physiological characteristics of the tissues that are difficult to reproduce otherwise. Further, they also closely mimic the *in vivo* conditions providing ethical flexibility to carry out experiments that cannot be executed in live organisms, and also provide deeper knowledge and understanding of the test processes for sample foods or compounds. However, various experimental aspects, such as (i) source, age, species, state (fasting/fed) of animal tissue used, (ii) morphological similarities (in terms of physiology, anatomy, and intestinal environment) with respect to the human equivalent, (iii) use of anesthesia and animal euthanasia methods and tissue harvesting time, and (iv) postharvest handling and preparation of tissue for experiments, need to be given careful consideration to ensure the validity and reliability of the results.

5.7 Microfluidic Systems: Organ-on-a-chip Models

Microfluidic systems are emerging in food digestion and absorption research specifically to mimic the dynamic and active microenvironment of the human intestine (small intestine and gut) as present *in vivo* in a small-scale fluid flow system, similar to a chip.[31] Several microfluidic-based microbiome–host models, such as gut-on-a-chip, host–microbiota interaction (HMI™), human–microbial crosstalk (HuMiX), and multi-organ tissue-flow (MOTiF) intestine-on-chip, are now widely used to study nutrient absorption and metabolism, host–microbiota response to probiotics and prebiotics, bacteria–host interactions, and bacterial overgrowth with inflammatory bowel disease (IBD).[38] A small intestine-on-a-chip model has recently been created using grown organoids generated from human biopsy alone or in combination with intestinal microvascular endothelial cells to mimic accurately the complicated *in vivo* intestinal epithelial tissue.[61] Figure 5.7 illustrates various microfluidic systems: organ-on-a-chip models.

Table 5.7 Salient features of *ex vivo* methods.

Ex vivo model	System description	Advantages	Limitations	Applications	Ref.
Intestinal rings/ slices	• Intestinal tissues from humans or animals are removed freshly, everted, chopped into rings or slices (2–5 mm wide), and then immersed in an oxygenated buffer comprising the test substance. • Intestinal tissues such as rings and segments are viable for 30–60 min and 2 h, respectively.	• Relatively simple, fast, and reliable. • High throughput. • Multiple compounds can be tested simultaneously. • Excellent agreement with *in vivo* bioavailability findings.	• Both luminal and serosal sides are exposed. • Limited viability of the tissue. • Difficulty in differentiating between binding and absorption. • Applicable analytical methods are scarce.	• Used to examine regional variations in intestinal absorption, drug metabolism, and the modulation of enzymes, GI hormones, and transporters involved in absorption mechanisms. • Used to analyze the kinetics of carrier-mediated transport and the accumulation of drugs in enterocytes.	7 and 8
Everted sac	• This system uses fresh *ex vivo* intestinal tissues from humans and animals (rats, hamsters, and chickens). • Small tubes (2–3 cm long) made from the intestine segments are inverted on glass rods and fastened at both ends. • Intestinal tissues are viable for 2 h under optimal conditions and handling.	• Moderately fast and easy. • Low cost. • Either the luminal or mucosal side can be exposed *via* an individual compartment.	• Intestinal tissue structure might be damaged during eversion. • Poor stirring conditions. • Low throughput.	• Frequently used for studies involving drug interactions, multidrug resistance, efflux transport, and pharmacokinetics. • Frequently used for investigating regional absorption mechanisms, the function of intestinal enzymes and transporters during the absorption of medication in the intestine, and glucose permeability.	169 and 170

Table 5.7 (*Continued*)

Ex vivo model	System description	Advantages	Limitations	Applications	Ref.
Ussing chamber	• This is a widely used model that utilizes the fresh intestinal epithelial tissues from humans and various animal species (mice, rats, rabbits, dogs, rats, and monkeys). • The intestinal segments are split apart to form planar epithelial sheets, which are then installed between two half-cells filled with buffer solutions. • The buffer solution is oxygenated using carbogen (95% O_2–5% CO_2) to guarantee the buffer's pH stabilization and sufficient oxygenation of the tissue and to reduce the unstirred water layer. • The intestinal tissue is viable for 2–3 h.	• Commercially available. • Directed transport. • Various sections of the small and large intestine can be used.	• The equipment required is costly. • Long and laborious. • Only a few experiments per animal are possible. • Requires special skill.	• Widely used for measuring apparent permeability and restoration of barrier, regional differences in absorption, and absorption mechanism. • It is also used for exploring translocation proteins/bacteria, GI hormone release, transcriptomics, and early immune response.	171

InTESTine™	• This model was developed by TNO and precisely mimics intestinal tract conditions. • It utilizes freshly removed intestinal tissue from various regions of healthy pigs. • It is available in 24- to 96-well plate formats.	• Commercially available. • Cost-effective. • Simple to use. • Moderate throughput.	• Issues with tissue integrity during transport. • Difficult to customize and modify. • Lack of published data using this technology. • Poor stirring conditions. • Potential for non-specific binding.	• It can be utilized to study the intricate physiology of the intestine and also absorption and metabolism and the complex physiology of the intestine.	172
Intestine perfusion model	• This is a versatile technique that requires the administration of anesthesia and animal surgery. • Rodents are widely used in this approach. • The small intestinal segment can be perfused with test solution in an open-loop or closed-loop setup.	• Effective stirring. • Executable with or without a blood supply. • Coordinated transport.	• Time consuming and expensive. • Complex process and requires a skilled team. • Low throughput per animal.	• It is predominantly used in investigating GI physiology, mucoadhesion, drug transport, permeability studies, and the biochemical barrier function of the small intestine.	173 and 174

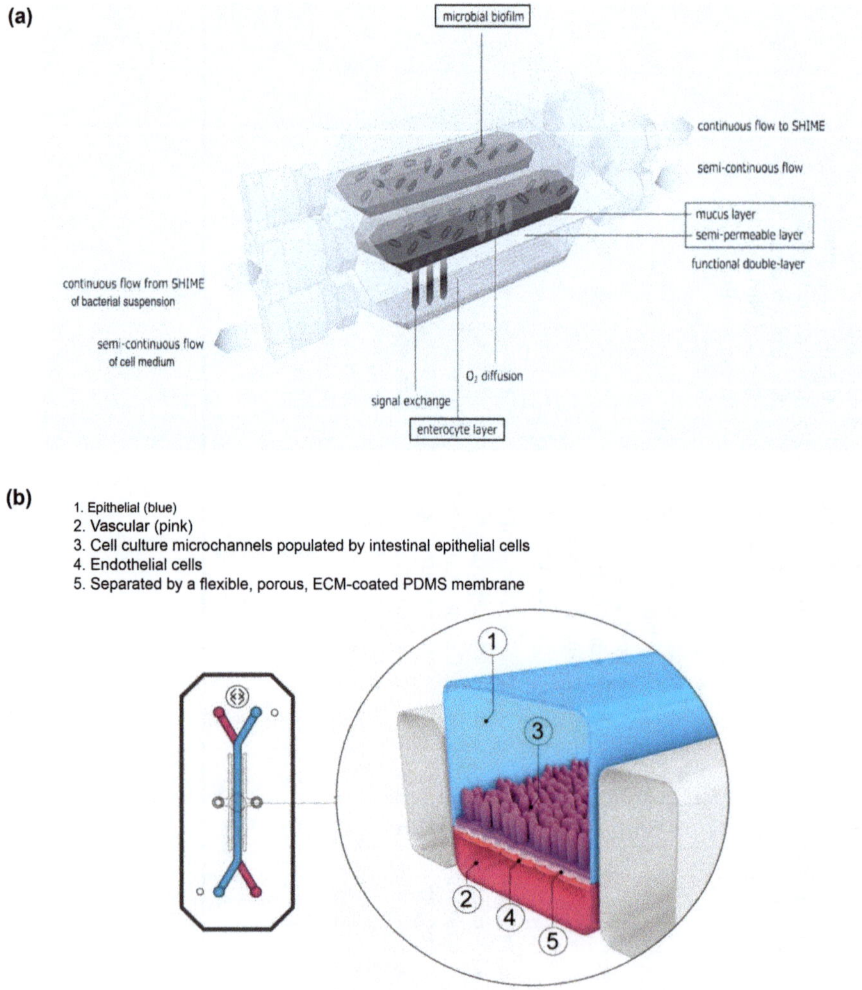

Figure 5.7 Microfluidic systems: organ-on-a-chip models: (a) HMI™,[105] and (b) duodenum–intestine–chip.[31,106] (a) Reproduced from ref. 105, https://doi.org/10.1186/1471-2180-14-133, under the terms of the CC BY 2.0 license https://creativecommons.org/licenses/by/2.0. (b) Reproduced from ref. 31 with permission from Elsevier, Copyright 2021.

An intestine-on-a-chip model known as the NutriChip has been used to measure the immunomodulatory properties of dairy products.[62] Various other intestine-on-a-chip and organ-on-a-chip models have also been extensively involved in studies to evaluate toxicity and drug efficacy, assess calcium transport through the epithelial layer, and analyze the transport of lipophilic dioxin, the flavonoid apigenin, and soy isoflavone.[63] Most of the microfluidic models use Caco-2 cells to simulate intestinal epithelium to assess permeability, even though Caco-2 cells are colon cancer-derived cell lines that may potentially behave differently to normal epithelial cells.[64] Nevertheless, these systems are considered an upgrade to the static cell culture models as they replicate the 3D topology, dynamic environment, intestinal cellular morphology and physiology, intestinal microenvironment (nutrients, microbiome, and gas composition), and physiology (barrier and permeability, blood flow, shear stress, and peristalsis) with the aid of microfluidic technology, bioengineering approaches, and cell cultures.[8,65] The main limitations of this technology are that they are time consuming, expensive, and complex in nature.

5.8 *In Silico* Approaches

In silico approaches involve mathematical equations and computational modeling methods, and they are predominantly used as a predictive tool in digestion and absorption research. They are utilized for simulating various aspects of digestion to provide a mechanistic understanding and significant insights into complex processes that occur during food oral processing and gastric and intestinal digestion and absorption. Further, the modeling of food digestion must follow certain basic rules such as conservation of momentum, mass, and energy, physicochemical and biological reactions, anatomical features, and physiological characteristics of various areas of the GI tract.[66] In general, an *in silico* digestion model typically consists of the following:

- (i) Computational model (*i.e.* a geometric mesh of the section to be modeled).
- (ii) Numerical modeling (usually included in the template to represent the movements of the mesh).
- (iii) Governing equations associated with the physiochemical properties of food in the GI tract.
- (iv) Boundary conditions can be added in cases where kinetic equations are dependent on the geometry of the model. Nevertheless, in specific conditions where movement and deformation of the boundary may occur as in biomechanical systems, meshless models are preferred.[67]

Further, in the case of mathematical models, the purely empirical models do not take mechanisms that control the system behavior into consideration as in mechanistic models. The predominant advantage of mechanistic mathematical models is their ability to extrapolate beyond the data with which they were constructed.[68] Two interesting approaches to modeling food digestion and absorption in the GI tract are continuum and compartmental models. The continuum model considers the GI tract as a tubular construct with starting and ending points and use a mathematical modeling approach with partial differential equations to describe simultaneously various aspects of digestion, including transit, digestion, physical breakdown, enzymatic degradation, and nutrient absorption. However, in the case of the compartmental modeling approach, the GI tract is represented as a series of compartments such as the mouth, stomach, small intestine (duodenum, jejunum, and ileum), and/or large intestine with distinct functionalities and digestive conditions.[69] A complete GIT illustration in a computational or mathematical model is extremely uncommon. Nevertheless, in recent years, many interesting studies have been carried out in this area, as extensively reviewed in the literature.[68–71] *In silico* approaches widen our understanding of complex processes, complement experimental studies, and lessen the requirement for time-consuming and costly *in vivo* studies. However, to enhance the power of computational predictions, correlations among *in vitro*, *in vivo*, and *in silico* approaches need to be established and validated. Table 5.8 gives an overview of significant phenomena modeled and challenges in modeling of the GI tract for food digestion and absorption research.

5.9 Advanced Tools and Techniques Involved in Understanding Digestion and Absorption

The digestion behavior of foods can be understood by characterizing the structural changes that the foods undergo during *in vitro* and *in vivo* digestion research. Various imaging techniques, such as 3D time-series microcomputed tomography, magnetic resonance imaging (MRI), time-lapse synchrotron deep-UV fluorescence microscopy, transmission electron microscopy (TEM), confocal laser scanning microscopy (CLSM), scanning electron microscopy (SEM), and light microscopy, are widely used to study the structure of foods and the microstructural changes that occur in foods due to oral processing[71,72] and gastric and intestinal digestion.[10,73] Further, techniques involving fluorescence, such as fluorescence recovery after photobleaching (FRAP) and fluorescence correlation spectroscopy (FCS), are also used to determine the diffusion rates and mobility of the enzyme pepsin in different food matrices, which aid in understanding the effect of pepsin on gastric digestion.[74,75] However, the main disadvantage of both FRAP and FCS techniques is the inactivation of the pepsin due to the fluorescent labeling reaction that occurs at a denaturing pH for the protein.[75,76] Table 5.9 presents various advanced tool/techniques involved in understanding the digestion and absorption of foods.

5.10 Conclusion and Future Perspectives

Overall, the entire gastrointestinal digestion process comprising various mechanical movements (grinding, mixing, sieving, mass transfer, peristalsis), biochemical processes [digestive secretions (saliva, HCl, mucin, water, bile, enzymes), enzymatic hydrolysis, changes in pH conditions], and other physiological processes (absorption of nutrients, water, and bile salts, microbial growth, residence time, and metabolism) is very complex to

Table 5.8 Overview of significant phenomena modeled and challenges in modeling of the GI tract for food digestion and absorption research.

Physiological compartment of the GI tract	Significant phenomena that are modeled	Challenges in modeling	Salient features that should be considered and included in modeling approaches
Oral cavity/mouth	• Movements of jaw and tongue during mastication. • Mechanical breakdown behavior of foods. • Taste and aroma released into saliva.	• Establishing a feedback system and inter-individual tongue and jaw motion variation. • Understanding how the state of ingested foodstuffs affects the movements and deformation of tissues. • Assessing the physical, chemical, and enzymatic reactions that take place inside the mouth. • Flow behavior of saliva and its consequence on softening and lubrication. • Modeling representations of anatomical movements, soft-tissue deformations, measured release and transport rates of tastant and aroma compounds.	• Saliva incorporation (amylase, water, mucus). • Enzymatic hydrolysis of starch • Changes to the material properties of solid foods due to mixing and saliva incorporation. • Thermal effects on the properties of the food (melting of ice-cream or chocolate).

Table 5.8 (*Continued*)

Physiological compartment of the GI tract	Significant phenomena that are modeled	Challenges in modeling	Salient features that should be considered and included in modeling approaches
Gastric chamber/stomach	• Mixing and sieving. • Advective flow performance. • Food disintegration. • Gastric emptying. • Role antral contraction waves on gastric flow and mixing. • Impact on gastric flow dynamics, mixing, and emptying due to food viscosity.	• Precise characterization of complex geometry. • Mechanical and contractile properties of the stomach wall with various stomach contents (solids, liquids, and gases), in relation to the timing of meals and emptying. • Influence of enzymatic and chemical reactions on material responses and flow behavior. • Kinetics of protein digestion.	• Gastric secretion (enzymes, HCl, mucus). • pH decrease. • Enzymatic hydrolysis of starch, proteins, and lipids. • Phase separation (sedimentation, creaming).
Small intestine	• Starch hydrolysis. • Enzymatic hydrolysis. • Mixing. • Mass transfer. • Nutrient and water absorption. • Lipid digestion. • Bioavailability of bioactive compounds. • Prediction of glycemic response. • Peristaltic flow patterns. • Impact of lumen and villi motility on transport, mixing, and absorption.	• Individual differences. • Intestinal anatomy is often ignored in mathematical models. • Feedback control from the brain. • Processes occur at different length scales of the small intestine.	• Multiple secretions (pancreatic enzymes, bile, bicarbonate, water, mucus). • pH increase. • Enzymatic hydrolysis of starch, proteins, and lipids. • Residence time. • Absorption of bile salts in the ileum.

Large intestine (ascending, transverse, and descending colon)	• Absorption of water. • Fermentation of carbohydrates. • Breakdown of undigested food and nutrients by microbes. • Colonic motility. • Mass transfer between the mucus and lumen. • Impact of dietary fiber on the synthesis of SCFAs. • Microbial growth, gut microbiome variation through life. • Effect of food entering the colon on microbial growth and metabolism. • Elastic deformations of the intestinal wall. • Mucus layer and bacterial metabolism.	• An enormous number of bacterial species. • Fermentation and gas generation.	• Accurate geometric features of the colon. • Fermentation and gas production. • Motor patterns and diffusion processes accountable for mass transfer, mixing, and absorption. • Intestinal secretions (water, bicarbonate, mucus). • Absorption of water and microbial metabolites. • Residence time.

Table 5.9 Advanced tools and techniques employed in monitoring and understanding human digestion.

Technique	Applications	Benefits	Drawbacks	Ref.
Time-domain nuclear magnetic resonance relaxometry (TD-NMR)	It measures the relaxation times of protons and provides insights into the state of water in foods and the degree of exchange between water and solutes (proteins) or particle surfaces. Variations in relaxation rate and proton population could be exploited to demonstrate the changes in macromolecule concentration, water migration, and structure in food matrices. It helps in investigating the accessibility of gastric juices to various compartments within a matrix, hydrolysis of proteins, and *in situ* swelling/shrinking of molecules during the digestion process	No need for any pretreatment such as filtration or centrifugation. Simple and rapid technique. TD-NMR results can be compared and validated with MRI results for digestion research	The phase separation and identification results obtained from TD-NMR are less accurate than those from MRI image analysis. Professionals are needed for the operation of sophisticated equipment. High equipment cost	144, 175 and 176
3D time-series microcomputed tomography (micro-CT)	It helps in the evaluation of changes together with comprehensive structural characterization of solid foods (porous, high-moisture food)	Non-destructive technique High-quality images, quantitative, and interpretable data. Higher image resolution capacity than in MRI.	Long time required for image acquisition Expensive equipment	177

	during *in vitro* gastric digestion. The density map aids in differentiating between air and liquid and in assessing the subtle variations in density	It exhibits similar trends to those observed in destructive analysis, including hardness and moisture content		
Time-lapse synchrotron deep-UV fluorescence microscopy	It allows the real-time quantification and visualization of proteins with high resolution using the natural fluorescence of the proteins (such as tryptophan and tyrosine at 275 nm). It permits *in situ* monitoring of the microstructural changes of protein gels during simulated gastric digestion. It aids in the visualization of the adsorption and diffusion of enzymes (amylase) during starch granule degradation in real time	It allows the real-time and variable hydrolysis-time 2D and 3D tracking of enzymes within specific structures (such as amylase in single starch granules). Uses natural fluorescence. High resolution. Specific location tracing of enzymes in 3D mapping without any probe	Does not require any labeling. No tedious sample preparation required before visualizing, such as staining and sectioning	178–180
Fluorescence recovery after photobleaching (FRAP)	It helps with determining the diffusion coefficients of solutes in both homogeneous and heterogeneous matrices and exploring the dynamics of diffusion. It helps in the estimation of effective diffusion coefficients of fluorescently labeled molecules.	Advanced non-destructive technique. It is a well-established and versatile technique. Frequently used along with microscopic techniques. Provides a deeper understanding of structure–interaction–diffusion relationships	Protein labeling and bleaching with light are mandatory. Quantitative FRAP necessitates careful experimental design, solid controls, data collection, and analysis. Experienced and skilled professionals are required for operation and for choosing	75, 76 and 181–183

Table 5.9 (*Continued*)

Technique	Applications	Benefits	Drawbacks	Ref.
	In recent times, it has been used to study the mobility of the digestive enzymes within the food matrices, *e.g.* the diffusion of pepsin in food protein gels was measured		specific FRAP models for a particular application	
Fluorescence correlation spectroscopy (FCS)	It measures spontaneous fluctuations in fluorescence intensity for a group of fluorescent molecules diffusing in the microscopic detection volume. Fluorescence intensity fluctuations brought on by the Brownian motion of particles make it easier to measure particle size and concentration. It is used to characterize the diffusion of pepsin in different food matrices to understand the effect of pepsin on digestion (dairy-based gel microstructures) by investigating various parameters such as diffusion coefficient, diffusivity, and diffusion rates of the fluorescently labeled pepsin	Non-destructive technique. High spatial and temporal resolution. Short analysis time. Low sample consumption. Largely a non-perturbing technique. Compared with FRAP it has a larger detection range of diffusion time (μs to s), works at low fluorophore concentrations, and distinguishes multicomponent diffusion rates	Suffers from poor signal-to-noise ratios when performed alone. Low detection efficiency. Frequently combined with confocal or two-photon microscopy to improve the sensitivity	74, 184 and 185

understand and very challenging to simulate or mimic completely. In addition, the above-stated parameters are greatly influenced by food properties (structure, composition, length scale, texture, rheological properties, and energy density), individual variations, health and disease conditions, and feedback control from the brain. Hence the various *in vitro*, *in vivo*, *ex vivo*, and *in silico* approaches and advanced techniques outlined in this chapter play a very crucial role in improving our knowledge and quantitation of gastrointestinal digestion processes. This area is still novel and emerging in the case of food-related research compared with pharmacology and drug studies. There is still a long way to go to gain greater insights that will facilitate personalized, health- and disease-targeted food product development with optimized and controlled functional food properties, oro-sensory perception, digestibility, and bioavailability.

References

1. G. M. Bornhorst, O. Gouseti, M. S. Wickham and S. Bakalis, Engineering digestion: multiscale processes of food digestion, *J. Food Sci.*, 2016, **81**(3), R534–R543.
2. P. Wu and X. D. Chen, Validation of *in vitro* bioaccessibility assays—A key aspect in the rational design of functional foods towards tailored bioavailability, *Curr. Opin. Food Sci.*, 2021, **39**, 160–170.
3. A. A. Xavier and L. R. Mariutti, Static and semi-dynamic *in vitro* digestion methods: State of the art and recent achievements towards standardization, *Curr. Opin. Food Sci.*, 2021, **41**, 260–273.
4. S. Muttakin, T. E. Moxon and O. Gouseti, *In vivo*, *in vitro*, and *in silico* studies of the GI tract, in *Interdisciplinary approaches to food digestion*, Springer, 2019, pp. 29–67.
5. J. Chen and L. Engelen, *Food oral processing: fundamentals of eating and sensory perception*, John Wiley & Sons, 2012.
6. D. H. Baker, Animal models in nutrition research, *J. Nutr.*, 2008, **138**(2), 391–396.
7. K. Verhoeckx, P. Cotter and I. López-Expósito, *et al.*, *The impact of food bioactives on health: in vitro and ex vivo models*, 2015.
8. Y. Xu, N. Shrestha, V. Préat and A. Beloqui, An overview of *in vitro*, *ex vivo* and *in vivo* models for studying the transport of drugs across intestinal barriers, *Adv. Drug Delivery Rev.*, 2021, **175**, 113795.
9. S. Le Feunteun, S. Verkempinck and J. Floury, *et al.*, Mathematical modelling of food hydrolysis during *in vitro* digestion: From single nutrient to complex foods in static and dynamic conditions, *Trends Food Sci. Technol.*, 2021, **116**, 870–883.
10. O. Gouseti, G. M. Bornhorst, S. Bakalis and A. Mackie, *Interdisciplinary approaches to food digestion*, Springer, 2019.
11. P. Puerta, L. Laguna, B. Villegas, A. Rizo, S. Fiszman and A. Tárrega, Oral processing and dynamics of texture perception in commercial gluten-free breads, *Food Res. Int.*, 2020, **134**, 109233.

12. D. P. Bolhuis and C. G. Forde, Application of food texture to moderate oral processing behaviors and energy intake, *Trends Food Sci. Technol.*, 2020, **106**, 445–456.
13. K. Petrowski, G.-B. Wintermann, P. Joraschky and S. Päßler, Chewing after stress: psychosocial stress influences chewing frequency, chewing efficacy, and appetite, *Psychoneuroendocrinology*, 2014, **48**, 64–76.
14. A. Van Der Bilt, Oral physiology, mastication and food perception, in *Designing functional foods*, Elsevier, 2009, pp. 3–37.
15. J. R. Stokes, M. W. Boehm and S. K. Baier, Oral processing, texture and mouthfeel: From rheology to tribology and beyond, *Curr. Opin. Colloid Interface Sci.*, 2013, **18**(4), 349–359.
16. S. Panda, J. Chen and O. Benjamin, Development of model mouth for food oral processing studies: Present challenges and scopes, *Innovative Food Sci. Emerging Technol.*, 2020, 102524.
17. P. Morell, I. Hernando and S. M. Fiszman, Understanding the relevance of in-mouth food processing. A review of *in vitro* techniques, *Trends Food Sci. Technol.*, 2014, **35**(1), 18–31.
18. M.-A. Peyron and A. Woda, An update about artificial mastication, *Curr. Opin. Food Sci.*, 2016, **9**, 21–28.
19. J.-F. Meullenet and R. Gandhapuneni, Development of the BITE Master II and its application to the study of cheese hardness, *Physiol. Behav.*, 2006, **89**(1), 39–43.
20. C. Salles, A. Tarrega and P. Mielle, *et al.*, Development of a chewing simulator for food breakdown and the analysis of *in vitro* flavor compound release in a mouth environment, *J. Food Eng.*, 2007, **82**(2), 189–198.
21. R. A. de Wijk, A. M. Janssen and J. F. Prinz, Oral movements and the perception of semi-solid foods, *Physiol. Behav.*, 2011, **104**(3), 423–428.
22. L. Lvova, S. Denis and A. Barra, *et al.*, Salt release monitoring with specific sensors in "*in vitro*" oral and digestive environments from soft cheeses, *Talanta*, 2012, **97**, 171–180.
23. O. Benjamin, P. Silcock, J. Kieser, J. Waddell, M. Swain and D. Everett, Development of a model mouth containing an artificial tongue to measure the release of volatile compounds, *Innovative Food Sci. Emerging Technol.*, 2012, **15**, 96–103.
24. C. De Loubens, M. Panouillé, A. Saint-Eve, I. Déléris, I. C. Tréléa and I. Souchon, Mechanistic model of *in vitro* salt release from model dairy gels based on standardized breakdown test simulating mastication, *J. Food Eng.*, 2011, **105**(1), 161–168.
25. M.-A. Peyron, V. Santé-Lhoutellier, O. François and M. Hennequin, Oral declines and mastication deficiencies cause alteration of food bolus properties, *Food Funct.*, 2018, **9**(2), 1112–1122.
26. W. M. Qazi and M. Stading, *In vitro* models for simulating swallowing, in *Dysphagia*, Springer, 2017, pp. 549–562.
27. I. Kato, A. Takanisi, K. Asari and T. Tani, Development of artificial mastication system. Construction of one degree of freedom antagonistic muscle model WJ-O, *Anat. Anz.*, 1988, **165**(2–3), 197–203.

28. P. Mielle, A. Tarrega and E. Sémon, *et al.*, From human to artificial mouth, from basics to results, *Sens. Actuators, B*, 2010, **146**(2), 440–445.
29. A. Mishellany-Dutour, M.-A. Peyron and J. Croze, *et al.*, Comparison of food boluses prepared *in vivo* and by the AM2 mastication simulator, *Food Qual. Preference*, 2011, **22**(4), 326–331.
30. W. L. Xu, D. Lewis, J. E. Bronlund and M. P. Morgenstern, Mechanism, design and motion control of a linkage chewing device for food evaluation, *Mech. Mach. Theory*, 2008, **43**(3), 376–389.
31. C. O'Farrell, K. Stamatopoulos, M. Simmons and H. Batchelor, In vitro models to evaluate ingestible devices: Present status and current trends, *Adv. Drug Delivery Rev.*, 2021, **178**, 113924.
32. D. Dupont, M. Alric and S. Blanquet-Diot, *et al.*, Can dynamic *in vitro* digestion systems mimic the physiological reality?, *Crit. Rev. Food Sci. Nutr.*, 2019, **59**(10), 1546–1562.
33. A. Guerra, L. Etienne-Mesmin, V. Livrelli, S. Denis, S. Blanquet-Diot and M. Alric, Relevance and challenges in modeling human gastric and small intestinal digestion, *Trends Biotechnol.*, 2012, **30**(11), 591–600.
34. J. Wang, P. Wu and M. Liu, *et al.*, An advanced near real dynamic *in vitro* human stomach system to study gastric digestion and emptying of beef stew and cooked rice, *Food Funct.*, 2019, **10**(5), 2914–2925.
35. M. Minekus, M. Alminger and P. Alvito, *et al.*, A standardised static *in vitro* digestion method suitable for food–an international consensus, *Food Funct.*, 2014, **5**(6), 1113–1124.
36. C. Li, W. Yu, P. Wu and X. D. Chen, Current *in vitro* digestion systems for understanding food digestion in human upper gastrointestinal tract, *Trends Food Sci. Technol.*, 2020, **96**, 114–126.
37. A. Brodkorb, L. Egger and M. Alminger, *et al.*, INFOGEST static *in vitro* simulation of gastrointestinal food digestion, *Nat. Protocols*, 2019, **14**(4), 991–1014.
38. D. Roupar, P. Berni, J. T. Martins, A. C. Caetano, J. A. Teixeira and C. Nobre, Bioengineering approaches to simulate human colon microbiome ecosystem, *Trends Food Sci. Technol.*, 2021, **112**, 808–822.
39. A. N. Payne, A. Zihler, C. Chassard and C. Lacroix, Advances and perspectives in *in vitro* human gut fermentation modeling, *Trends Biotechnol.*, 2012, **30**(1), 17–25.
40. A. Mackie, A.-I. Mulet-Cabero and A. Torcello-Gómez, Simulating human digestion: developing our knowledge to create healthier and more sustainable foods, *Food Funct.*, 2020, **11**(11), 9397–9431.
41. I. Sensoy, A review on the food digestion in the digestive tract and the used *in vitro* models, *Curr. Res. Food Sci.*, 2021, **4**, 308–319.
42. C. Shani-Levi, P. Alvito and A. Andrés, *et al.*, Extending *in vitro* digestion models to specific human populations: Perspectives, practical tools and bio-relevant information, *Trends Food Sci. Technol.*, 2017, **60**, 52–63.
43. S. Adams, A. J. Taylor, Oral processing and flavour sensing mechanisms, in *Food oral processing: fundamentals of eating and sensory perception*, Blackwell Publishing Ltd, Oxford, 2012, pp. 177–202.

44. C. Marques, E. Correia, L.-T. Dinis and A. Vilela, An Overview of Sensory Characterization Techniques: From Classical Descriptive Analysis to the Emergence of Novel Profiling Methods, *Foods*, 2022, **11**(3), 255.
45. A. Slyper, Oral processing, satiation and obesity: overview and hypotheses, *Diabetes, Metab. Syndr. Obes.: Targets Ther.*, 2021, **14**, 3399.
46. P. Sethupathy, S. Priyadarshini, J. A. Moses and C. Anandharamakrishnan, Matrix-dependent oral processing, oro-sensory perception, and glycemic index of chocolate bars, *J. Food Process. Preserv.*, 2021, **45**(12), e16067.
47. Q. He, F. Bramante, A. Davies, C. Elleman, K. Fourtouni and B. Wolf, Material properties of *ex vivo* milk chocolate boluses examined in relation to texture perception, *Food Funct.*, 2018, **9**(6), 3532–3546.
48. T. Funami, In vivo and rheological approaches for characterizing food oral processing and usefulness of polysaccharides as texture modifiers-A review, *Food Hydrocolloids*, 2017, **68**, 2–14.
49. C. Muñoz-González, J. J. Rodríguez-Bencomo, M. V. Moreno-Arribas and M. Á. Pozo-Bayón, Feasibility and application of a retronasal aroma-trapping device to study *in vivo* aroma release during the consumption of model wine-derived beverages, *Food Sci. Nutr.*, 2014, **2**(4), 361–370.
50. G. M. Bornhorst and R. Paul Singh, Gastric digestion *in vivo* and *in vitro*: how the structural aspects of food influence the digestion process, *Annu. Rev. Food Sci. Technol.*, 2014, **5**, 111–132.
51. W. Schwizer, A. Steingötter and M. Fox, *et al.*, Non-invasive measurement of gastric accommodation in humans, *Gut*, 2002, **51**(suppl 1), i59–i62.
52. A. D. Farmer, S. M. Scott and A. R. Hobson, Gastrointestinal motility revisited: the wireless motility capsule, *United Eur. Gastroenterol. J.*, 2013, **1**(6), 413–421.
53. O. Livny, R. Reifen, I. Levy, Z. Madar, R. Faulks, S. Southon and B. Schwartz, β-carotene bioavailability from differently processed carrot meals in human ileostomy volunteers, *Eur. J. Nutr.*, 2003, **42**, 338–345.
54. P. J. Moughan and R. R. Wolfe, Determination of dietary amino acid digestibility in humans, *J. Nutr.*, 2019, **149**(12), 2101–2109.
55. P. Rojas-Bonzi, C. T. Vangsøe, K. L. Nielsen, H. N. Lærke, M. S. Hedemann and K. E. B. Knudsen, The relationship between *in vitro* and *in vivo* starch digestion kinetics of breads varying in dietary fibre, *Foods*, 2020, **9**(9), 1337.
56. J. Nadia, J. Bronlund, R. P. Singh, H. Singh and G. M. Bornhorst, Structural breakdown of starch-based foods during gastric digestion and its link to glycemic response: In vivo and *in vitro* considerations, *Compr. Rev. Food Sci. Food Saf.*, 2021, **20**(3), 2660–2698.
57. L. R. Brink, K. Matazel and B. D. Piccolo, *et al.*, Neonatal diet impacts bioregional microbiota composition in piglets fed human breast milk or infant formula, *J. Nutr.*, 2019, **149**(12), 2236–2246.

58. J. Campbell, J. Berry and Y. Liang, Anatomy and physiology of the small intestine, in *Shackelford's Surgery of the Alimentary Tract, 2 Volume Set*, Elsevier, 2019, pp. 817–841.
59. S. C. Pearce, H. G. Coia, J. P. Karl, I. G. Pantoja-Feliciano, N. C. Zachos and K. Racicot, Intestinal *in vitro* and *ex vivo* models to study host-microbiome interactions and acute stressors, *Front. Physiol.*, 2018, 1584.
60. D. Dresselhuis, E. De Hoog, M. C. Stuart and G. Van Aken, Application of oral tissue in tribological measurements in an emulsion perception context, *Food Hydrocolloids*, 2008, **22**(2), 323–335.
61. M. Kasendra, A. Tovaglieri and A. Sontheimer-Phelps, *et al.*, Development of a primary human Small Intestine-on-a-Chip using biopsy-derived organoids, *Sci. Rep.*, 2018, **8**(1), 1–14.
62. Q. Ramadan, H. Jafarpoorchekab and C. Huang, *et al.*, NutriChip: nutrition analysis meets microfluidics, *Lab Chip*, 2013, **13**(2), 196–203.
63. C. A. M. Fois, T. Y. L. Le and A. Schindeler, *et al.*, Models of the Gut for Analyzing the Impact of Food and Drugs, *Adv. Healthcare Mater.*, 2019, **8**(21), 1900968.
64. K. Kulthong, L. Duivenvoorde and H. Sun, *et al.*, Microfluidic chip for culturing intestinal epithelial cell layers: Characterization and comparison of drug transport between dynamic and static models, *Toxicology in Vitro*, 2020, **65**, 104815.
65. H. J. Kim, D. Huh, G. Hamilton and D. E. Ingber, Human gut-on-a-chip inhabited by microbial flora that experiences intestinal peristalsis-like motions and flow, *Lab Chip*, 2012, **12**(12), 2165–2174.
66. K. Kristensen, N. David-Rogeat and N. Alshammari, *et al.*, Food Digestion Engineering, in *Sustainable Food Processing and Engineering Challenges*, Elsevier, 2021, pp. 343–368.
67. M. J. Ferrua, F. Kong and R. P. Singh, Computational modeling of gastric digestion and the role of food material properties, *Trends Food Sci. Technol.*, 2011, **22**(9), 480–491.
68. S. Le Feunteun, A. R. Mackie and D. Dupont, In silico trials of food digestion and absorption: how far are we?, *Curr. Opin. Food Sci.*, 2020, **31**, 121–125.
69. S. Le Feunteun, A. Al-Razaz, M. Dekker, E. George, B. Laroche and G. Van Aken, Physiologically based modeling of food digestion and intestinal microbiota: State of the art and future challenges. An INFOGEST review, *Annu. Rev. Food Sci. Technol.*, 2021, **12**, 149–167.
70. P. W. Cleary, M. D. Sinnott, B. Hari, S. Bakalis and S. M. Harrison, Modelling food digestion, *Modeling Food Processing Operations*, Woodhead Publishing, 2015, pp. 255–305.
71. M. W. Boehm, S. K. Baier and J. R. Stokes, Capturing changes in structure and rheology of an oily brittle snack food during *in vitro* oral processing, *Food Res. Int.*, 2013, **54**(1), 544–551.
72. M. Stieger and F. van de Velde, Microstructure, texture and oral processing: New ways to reduce sugar and salt in foods, *Curr. Opin. Colloid Interface Sci.*, 2013, **18**(4), 334–348.

73. Q. Guo, A. Ye, H. Singh and D. Rousseau, Destructuring and restructuring of foods during gastric digestion, *Compr. Rev. Food Sci. Food Saf.*, 2020, **19**(4), 1658–1679.
74. Q. Luo, J. W. Borst, A. H. Westphal, R. M. Boom and A. E. Janssen, Pepsin diffusivity in whey protein gels and its effect on gastric digestion, *Food Hydrocolloids*, 2017, **66**, 318–325.
75. J. Thevenot, C. Cauty, D. Legland, D. Dupont and J. Floury, Pepsin diffusion in dairy gels depends on casein concentration and microstructure, *Food Chem.*, 2017, **223**, 54–61.
76. G. Somaratne, F. Nau and M. J. Ferrua, *et al.*, Characterization of egg white gel microstructure and its relationship with pepsin diffusivity, *Food Hydrocolloids*, 2020, **98**, 105258.
77. W. Xu, D. Lewis, J. Bronlund and M. Morgenstern, Mechanism, design and motion control of a linkage chewing device for food evaluation, *Mech. Mach. Theory*, 2008, **43**(3), 376–389.
78. M.-A. Peyron, V. Sante-Lhoutellier and D. Dardevet, *et al.*, Addressing various challenges related to food bolus and nutrition with the AM2 mastication simulator, *Food Hydrocolloids*, 2019, **97**, 105229.
79. W. Xu, J. Bronlund and J. Potgieter, *et al.*, Review of the human masticatory system and masticatory robotics, *Mech. Mach. Theory*, 2008, **43**(11), 1353–1375.
80. V. Raja, S. Priyadarshini, J. Moses and C. Anandharamakrishnan, A dynamic *in vitro* oral mastication system to study the oral processing behavior of soft foods, *Food Funct.*, 2022, **13**(20), 10426–10438.
81. H. Kozu, I. Kobayashi and M. Nakajima, *et al.*, Mixing characterization of liquid contents in human gastric digestion simulator equipped with gastric secretion and emptying, *Biochem. Eng. J.*, 2017, **122**, 85–90.
82. W. Liu, D. Fu, X. Zhang, J. Chai, S. Tian and J. Han, Development and validation of a new artificial gastric digestive system, *Food Res. Int.*, 2019, **122**, 183–190.
83. Y. Li, L. Fortner and F. Kong, Development of a Gastric Simulation Model (GSM) incorporating gastric geometry and peristalsis for food digestion study, *Food Res. Int.*, 2019, **125**, 108598.
84. F. Kong and R. P. Singh, A human gastric simulator (HGS) to study food digestion in human stomach, *J. Food Sci.*, 2010, **75**(9), E627–E635.
85. S. Keppler, S. O'Meara, S. Bakalis, P. Fryer and G. Bornhorst, Characterization of individual particle movement during *in vitro* gastric digestion in the Human Gastric Simulator (HGS), *J. Food Eng.*, 2020, **264**, 109674.
86. D. H. T. Do, F. Kong, C. Penet, D. Winetzky and K. Gregory, Using a dynamic stomach model to study efficacy of supplemental enzymes during simulated digestion, *LWT – Food Sci. Technol.*, 2016, **65**, 580–588.
87. L. Barros, C. Retamal, H. Torres, R. N. Zúñiga and E. Troncoso, Development of an *in vitro* mechanical gastric system (IMGS) with realistic peristalsis to assess lipid digestibility, *Food Res. Int.*, 2016, **90**, 216–225.

88. E. C. Thuenemann, G. Mandalari, G. T. Rich and R. M. Faulks, Dynamic gastric model (DGM), *The impact of food bioactives on health*, 2015, pp. 47–59.
89. S. Bellmann, J. Lelieveld, T. Gorissen, M. Minekus and R. Havenaar, Development of an advanced *in vitro* model of the stomach and its evaluation *versus* human gastric physiology, *Food Res. Int.*, 2016, **88**, 191–198.
90. A. Tharakan, I. Norton, P. Fryer and S. Bakalis, Mass transfer and nutrient absorption in a simulated model of small intestine, *J. Food Sci.*, 2010, **75**(6), E339–E346.
91. O. Gouseti, M. R. Jaime-Fonseca, P. Fryer, C. Mills, M. Wickham and S. Bakalis, Hydrocolloids in human digestion: Dynamic *in vitro* assessment of the effect of food formulation on mass transfer, *Food Hydrocolloids*, 2014, **42**, 378–385.
92. N. D. Wright, F. Kong, B. S. Williams and L. Fortner, A human duodenum model (HDM) to study transport and digestion of intestinal contents, *J. Food Eng.*, 2016, **171**, 129–136.
93. J. A. McDonald, S. Fuentes and K. Schroeter, *et al.*, Simulating distal gut mucosal and luminal communities using packed-column biofilm reactors and an *in vitro* chemostat model, *J. Microbiol. Methods*, 2015, **108**, 36–44.
94. P. Wu and X. D. Chen, On designing biomimic *in vitro* human and animal digestion track models: ideas, current and future devices, *Curr. Opin. Food Sci.*, 2020, **35**, 10–19.
95. S. Blanquet-Diot, S. Denis, S. Chalancon, F. Chaira, J.-M. Cardot and M. Alric, Use of artificial digestive systems to investigate the biopharmaceutical factors influencing the survival of probiotic yeast during gastrointestinal transit in humans, *Pharm. Res.*, 2012, **29**, 1444–1453.
96. G. T. Macfarlane, S. Macfarlane and G. Gibson, Validation of a three-stage compound continuous culture system for investigating the effect of retention time on the ecology and metabolism of bacteria in the human colon, *Microb. Ecol.*, 1998, **35**, 180–187.
97. A. Ouwehand, K. Tiihonen, H. Mäkeläinen, N. Rautonen, O. Hasselwander and G. Sworn, Non-starch polysaccharides in the gastrointestinal tract, in *Designing Functional Foods*, Elsevier, 2009, pp. 126–147.
98. H. A. Makivuokko, M. T. Saarinen, A. C. Ouwehand and N. E. Rautonen, Effects of lactose on colon microbial community structure and function in a four-stage semi-continuous culture system, *Biosci., Biotechnol., Biochem.*, 2006, **70**(9), 2056–2063.
99. K. Venema, The TNO *in vitro* model of the colon (TIM-2), in *The Impact of Food Bioactives on Health: in vitro and ex vivo models*, 2015, pp. 293–304.
100. J.-F. Brugère, D. Féria-Gervasio, Z. Popse, W. Tottey and M. Alric, The ECSIM concept (Environmental Control System for Intestinal

Microbiota) and its derivative versions to help better understand human gut biology, *Appl. Biomed. Eng.*, 2011, **4**, 63–82.
101. O. Ménard, D. Picque and D. Dupont, The DIDGI® system, in *The impact of food bioactives on health*, 2015, pp. 73–81.
102. M. Minekus, *The TNO gastro-intestinal model (TIM)*, Springer International Publishing, 2015.
103. E. Barroso, C. Cueva, C. Peláez, M. C. Martínez-Cuesta and T. Requena, Development of human colonic microbiota in the computer-controlled dynamic SIMulator of the GastroIntestinal tract SIMGI, *LWT – Food Sci. Technol.*, 2015, **61**(2), 283–289.
104. T. Van de Wiele, P. Van den Abbeele, W. Ossieur, S. Possemiers and M. Marzorati, The simulator of the human intestinal microbial ecosystem (SHIME®), in *The impact of food bioactives on health*, 2015, pp. 305–317.
105. M. Marzorati, B. Vanhoecke and T. De Ryck, *et al.*, The HMI™ module: a new tool to study the Host-Microbiota Interaction in the human gastrointestinal tract *in vitro*, *BMC Microbiol.*, 2014, **14**, 1–14.
106. M. Kasendra, R. Luc and J. Yin, *et al.*, Duodenum Intestine-Chip for preclinical drug assessment in a human relevant model, *eLife*, 2020, **9**, e50135.
107. H. Takanobu, N. Kuchiki and A. Takanishi, Control of rapid closing motion of a robot jaw using nonlinear spring mechanism, *IEEE*, 1995, 372–377.
108. F. Winquist, P. Wide, T. Eklöv, C. Hjort and I. Lundström, Crispbread quality evaluation based on fusion of information from the sensor analogies to the human olfactory, auditory and tactile senses, *J. Food Process Eng.*, 1999, **22**(5), 337–358.
109. J. S. Pap, W. L. Xu and J. Bronlund, A robotic human masticatory system: kinematics simulations, *Int. J. Intell. Syst. Technol. Appl.*, 2005, **1**(1–2), 3–17.
110. A. Woda, A. Mishellany-Dutour and L. Batier, *et al.*, Development and validation of a mastication simulator, *J. Biomech.*, 2010, **43**(9), 1667–1673.
111. R. Takagi, K. Sasaki and D. Sasaki, *et al.*, A single-batch fermentation system to simulate human colonic microbiota for high-throughput evaluation of prebiotics, *PLoS One*, 2016, **11**(8), e0160533.
112. M. Vardakou, A. Mercuri, S. A. Barker, D. Q. Craig, R. M. Faulks and M. S. Wickham, Achieving antral grinding forces in biorelevant *in vitro* models: comparing the USP dissolution apparatus II and the dynamic gastric model with human *in vivo* data, *AAPS PharmSciTech*, 2011, **12**(2), 620–626.
113. M. Wickham, R. Faulks, J. Mann and G. Mandalari, The design, operation, and application of a dynamic gastric model, *Dissolution Technol.*, 2012, **19**(3), 15–22.
114. M. J. Ferrua and R. P. Singh, Human gastric simulator (Riddet model), *The impact of food bioactives on health*, 2015, pp. 61–71.

115. H. Kozu, Y. Nakata and M. Nakajima, *et al.*, Development of a human gastric digestion simulator equipped with peristalsis function for the direct observation and analysis of the food digestion process, *Food Sci. Technol. Res.*, 2014, **20**(2), 225–233.
116. I. Kobayashi, H. Kozu, Z. Wang, H. Isoda and S. Ichikawa, Development and fundamental characteristics of a human gastric digestion simulator for analysis of food disintegration, *Jpn. Agric. Res. Quart.*, 2017, **51**(1), 17–25.
117. S. Blanquet-Diot, S. Denis, S. Chalancon, F. Chaira, J.-M. Cardot and M. Alric, Use of artificial digestive systems to investigate the biopharmaceutical factors influencing the survival of probiotic yeast during gastrointestinal transit in humans, *Pharm. Res.*, 2012, **29**(6), 1444–1453.
118. J. M. Auchtung, C. D. Robinson and R. A. Britton, Cultivation of stable, reproducible microbial communities from different fecal donors using minibioreactor arrays (MBRAs), *Microbiome*, 2015, **3**(1), 1–15.
119. M. Minekus, M. Smeets-Peeters and A. Bernalier, *et al.*, A computer-controlled system to simulate conditions of the large intestine with peristaltic mixing, water absorption and absorption of fermentation products, *Appl. Microbiol. Biotechnol.*, 1999, **53**(1), 108–114.
120. H. Makivuokko, J. Nurmi, P. Nurminen, J. Stowell and N. Rautonen, In vitro effects on polydextrose by colonic bacteria and caco-2 cell cyclooxygenase gene expression, *Nutr. Cancer*, 2005, **52**(1), 94–104.
121. A. Guerra, S. Denis and O. le Goff, *et al.*, Development and validation of a new dynamic computer-controlled model of the human stomach and small intestine, *Biotechnol. Bioeng.*, 2016, **113**(6), 1325–1335.
122. M. Minekus, P. Marteau, R. Havenaar and J. H. J. Huis In't Veld, A multicompartmental dynamic computer-controlled model simulating the stomach and small intestine, *Altern. Lab. Anim.*, 1995, **23**(2), 197–209.
123. S. Ranganathan, E. M. Vasikaran, A. Elumalai, J. A. Moses and C. Anandharamakrishnan, Gastric emptying pattern and disintegration kinetics of cooked rice in a 3D printed *in vitro* dynamic digestion model ARK®, *Int. J. Food Eng.*, 2021, **17**(5), 385–393.
124. A. Payal, A. Elumalai, J. Moses and C. Anandharamakrishnan, An investigation on gastric emptying behavior of apple in the dynamic digestion model ARK® and its validation using MRI of human subjects–A pilot study, *Biochem. Eng. J.*, 2021, **175**, 108134.
125. S. Parthasarathi, J. A. Bhushani and C. Anandharamakrishnan, Engineered small intestinal system as an alternative to *in-situ* intestinal permeability model, *J. Food Eng.*, 2018, **222**, 110–114.
126. K. Molly, M. V. Woestyne and W. Verstraete, Development of a 5-step multi-chamber reactor as a simulation of the human intestinal microbial ecosystem, *Appl. Microbiol. Biotechnol.*, 1993, **39**(2), 254–258.
127. J. Y. Yoo, *Development and application of an in vitro physicochemical upper gastrointestinal system (IPUGS) simulating the human digestive processes*, PhD thesis, Monash University, 2008.

128. L. Chen, Y. Xu and T. Fan, *et al.*, Gastric emptying and morphology of a 'near real' *in vitro* human stomach model (RD-IV-HSM), *J. Food Eng.*, 2016, **183**, 1–8.
129. E. Barroso, C. Cueva, C. Peláez, M. C. Martínez-Cuesta and T. Requena, The computer-controlled multicompartmental dynamic model of the gastrointestinal system SIMGI, in *The Impact of Food Bioactives on Health*, 2015, p. 319.
130. W. A. Alsanei, J. Chen and R. Ding, Food oral breaking and the determining role of tongue muscle strength, *Food Res. Int.*, 2015, **67**, 331–337.
131. K. Hori, H. Hayashi and S. Yokoyama, *et al.*, Comparison of mechanical analyses and tongue pressure analyses during squeezing and swallowing of gels, *Food Hydrocolloids*, 2015, **44**, 145–155.
132. Y. G. Espinosa and J. Chen, 13 Applications of Electromyography (EMG) Technique for Eating Studies, in *Food oral processing: Fundamentals of eating and sensory perception*, 2012, p. 289.
133. A. C. Mosca and J. Chen, *Objective and Subjective Aspects of Food Oral Texture Assessment*, 2016
134. H. Koç, C. Vinyard, G. Essick and E. Foegeding, Food oral processing: conversion of food structure to textural perception, *Annu. Rev. Food Sci. Technol.*, 2013, **4**, 237–266.
135. M. Devezeaux De Lavergne, A. K. Young, J. Engmann and C. Hartmann, Food Oral Processing—An Industry Perspective, *Front. Nutr.*, 2021, **8**, 634410.
136. K. Kohyama, Oral sensing of food properties, *J. Texture Stud.*, 2015, **46**(3), 138–151.
137. H. Dan, F. Hayakawa and K. Kohyama, Modulation of biting procedures induced by the sensory evaluation of cheese hardness with different definitions, *Appetite*, 2008, **50**(1), 158–166.
138. R. A. de Wijk, F. Wulfert and J. F. Prinz, Oral processing assessed by M-mode ultrasound imaging varies with food attribute, *Physiol. Behav.*, 2006, **89**(1), 15–21.
139. M. G. Aguayo-Mendoza, E. C. Ketel, E. van der Linden, C. G. Forde, B. Piqueras-Fiszman and M. Stieger, Oral processing behavior of drinkable, spoonable and chewable foods is primarily determined by rheological and mechanical food properties, *Food Qual. Prefer.*, 2019, **71**, 87–95.
140. D. Frank, I. Appelqvist, U. Piyasiri, T. J. Wooster and C. Delahunty, Proton transfer reaction mass spectrometry and time intensity perceptual measurement of flavor release from lipid emulsions using trained human subjects, *J. Agric. Food Chem.*, 2011, **59**(9), 4891–4903.
141. E. Aprea, F. Biasioli, F. Gasperi, T. D. Märk and S. van Ruth, In vivo monitoring of strawberry flavour release from model custards: effect of texture and oral processing, *Flavour Fragrance J.*, 2006, **21**(1), 53–58.
142. A. Okada, M. Honma, S. Nomura and Y. Yamada, Oral behavior from food intake until terminal swallow, *Physiol. Behav.*, 2007, **90**(1), 172–179.

143. K. Hiiemae, Mechanisms of food reduction, transport and deglutition: how the texture of food affects feeding behavior, *J. Texture Stud.*, 2004, **35**(2), 171–200.
144. R. Deng, A. E. Janssen and F. J. Vergeldt, *et al.*, Exploring *in vitro* gastric digestion of whey protein by time-domain nuclear magnetic resonance and magnetic resonance imaging, *Food Hydrocolloids*, 2020, **99**, 105348.
145. W. Liu, Y. Jin, P. J. Wilde, Y. Hou, Y. Wang and J. Han, Mechanisms, physiology, and recent research progress of gastric emptying, *Crit. Rev. Food Sci. Nutr.*, 2021, **61**(16), 2742–2755.
146. R. Spiller and L. Marciani, Intraluminal impact of food: new insights from MRI, *Nutrients*, 2019, **11**(5), 1147.
147. J.-P. Guo, A. H. Maurer, R. S. Fisher and H. P. Parkman, Extending gastric emptying scintigraphy from two to four hours detects more patients with gastroparesis, *Dig. Dis. Sci.*, 2001, **46**(1), 24–29.
148. N. Q. Nguyen, L. K. Bryant and C. M. Burgstad, *et al.*, Gastric emptying measurement of liquid nutrients using the 13 C-octanoate breath test in critically ill patients: a comparison with scintigraphy, *Intensive Care Med.*, 2013, **39**(7), 1238–1246.
149. J. Punkkinen, I. Konkka, O. Punkkinen, T. Korppi-Tommola, M. Färkkilä and J. Koskenpato, Measuring gastric emptying: comparison of 13 C-Octanoic acid breath test and Scintigraphy, *Dig. Dis. Sci.*, 2006, **51**(2), 262–267.
150. W. Schwizer, M. Fox and A. Steingötter, Non-invasive investigation of gastrointestinal functions with magnetic resonance imaging: towards an "ideal" investigation of gastrointestinal function, *Gut*, 2003, **52**(suppl 4), iv34–iv39.
151. B. Braden, B. Lembcke, W. Kuker and W. Caspary, 13C-breath tests: current state of the art and future directions, *Dig. Liver Dis.*, 2007, **39**(9), 795–805.
152. G. Camps, M. Mars, B. Witteman, C. de Graaf and P. Smeets, Indirect *vs* direct assessment of gastric emptying: A randomized crossover trial comparing C-isotope breath analysis and MRI, *Neurogastroenterol. Motil.*, 2018, **30**(7), e13317.
153. S. Hellmig, F. Von Schöning and C. Gadow, *et al.*, Gastric emptying time of fluids and solids in healthy subjects determined by 13C breath tests: influence of age, sex and body mass index, *J. Gastroenterol. Hepatol.*, 2006, **21**(12), 1832–1838.
154. T. Mazzawi, E. Bartsch and S. Benammi, *et al.*, Gastric emptying of low- and high-caloric liquid meals measured using ultrasonography in healthy volunteers, *Ultrasound Int. Open*, 2019, **5**(01), E27–E33.
155. C. Muresan, T. Surdea Blaga, L. Muresan and D. L. Dumitrascu, Abdominal Ultrasound for the Evaluation of Gastric Emptying Revisited, *J. Gastrointest. Liver Dis.*, 2015, **24**(3), 329–338.
156. L. K. Phillips, C. K. Rayner, K. L. Jones and M. Horowitz, Measurement of gastric emptying in diabetes, *J. Diabetes Complications*, 2014, **28**(6), 894–903.

157. M. Sugita, M. Matsumoto, Y. Tsukano, C. Fukunaga and T. Yamamoto, Gastric emptying time after breakfast in healthy adult volunteers using ultrasonography, *J. Anesth.*, 2019, **33**(6), 697–700.
158. R. K. Goyal, Y. Guo and H. Mashimo, Advances in the physiology of gastric emptying, *Neurogastroenterol. Motil.*, 2019, **31**(4), e13546.
159. A. C. Meyer-Gerspach, J. R. Biesiekierski and E. Deloose, *et al.*, Effects of caloric and noncaloric sweeteners on antroduodenal motility, gastrointestinal hormone secretion and appetite-related sensations in healthy subjects, *Am. J. Clin. Nutr.*, 2018, **107**(5), 707–716.
160. L. A. Szarka and M. Camilleri, Methods for measurement of gastric motility, *Am. J. Physiol.-Gastrointest. Liver Physiol.*, 2009, **296**(3), G461–G475.
161. D. Cassilly, S. Kantor and L. Knight, *et al.*, Gastric emptying of a nondigestible solid: assessment with simultaneous SmartPill pH and pressure capsule, antroduodenal manometry, gastric emptying scintigraphy, *Neurogastroenterol. Motil.*, 2008, **20**(4), 311–319.
162. B. Kuo, R. McCallum and K. Koch, *et al.*, Comparison of gastric emptying of a nondigestible capsule to a radio-labelled meal in healthy and gastroparetic subjects, *Aliment. Pharmacol. Ther.*, 2008, **27**(2), 186–196.
163. S. Maqbool, H. P. Parkman and F. K. Friedenberg, Wireless capsule motility: comparison of the SmartPill® GI monitoring system with scintigraphy for measuring whole gut transit, *Dig. Dis. Sci.*, 2009, **54**(10), 2167–2174.
164. P. Kar, K. L. Jones, M. Horowitz, M. J. Chapman and A. M. Deane, Measurement of gastric emptying in the critically ill, *Clin. Nutr.*, 2015, **34**(4), 557–564.
165. D.-Y. Kim, S.-J. Myung and M. Camilleri, Novel testing of human gastric motor and sensory functions: rationale, methods, and potential applications in clinical practice, *Am. J. Gastroenterol.*, 2000, **95**(12), 3365–3373.
166. R. B. Raffa, J. V. Pergolizzi Jr, R. Taylor Jr, J. F. Decker and J. T. Patrick, Acetaminophen (paracetamol) oral absorption and clinical influences, *Pain Pract.*, 2014, **14**(7), 668–677.
167. Y. Zhu, W. H. Hsu and J. H. Hollis, The impact of food viscosity on eating rate, subjective appetite, glycemic response and gastric emptying rate, *PLoS One*, 2013, **8**(6), e67482.
168. S. Priyadarshini, J. Moses and C. Anandharamakrishnan, Prediction of *in vitro* glycemic responses of biscuits in an engineered small intestine system, *Food Res. Int.*, 2021, **147**, 110459.
169. S. Priyadarshini, E. Arunkumar, J. Moses and C. Anandharamakrishnan, Predicting human glucose response curve using an engineered small intestine system in combination with mathematical modeling, *J. Food Eng.*, 2021, **293**, 110395.
170. M. A. Alam, F. I. Al-Jenoobi and A. M. Al-Mohizea, Everted gut sac model as a tool in pharmaceutical research: limitations and applications, *J. Pharm. Pharmacol.*, 2012, **64**(3), 326–336.

171. K. L. Hamilton, Ussing's "Little Chamber": 60 Years+ old and counting, *Front. Physiol.*, 2011, **2**, 6.
172. E. Abeln, M. Posno and K. Nishide, *InTESTine™ Physiologically relevant intestinal tissue model*, 2013
173. J. Stappaerts, J. Brouwers, P. Annaert and P. Augustijns, *In situ* perfusion in rodents to explore intestinal drug absorption: challenges and opportunities, *Int. J. Pharm.*, 2015, **478**(2), 665–681.
174. D. Dahlgren, C. Roos and K. Peters, *et al.*, Evaluation of drug permeability calculation based on luminal disappearance and plasma appearance in the rat single-pass intestinal perfusion model, *Eur. J. Pharm. Biopharm.*, 2019, **142**, 31–37.
175. A. Bordoni, L. Laghi and E. Babini, *et al.*, The foodomics approach for the evaluation of protein bioaccessibility in processed meat upon *in vitro* digestion, *Electrophoresis*, 2014, **35**(11), 1607–1614.
176. L. Laghi, G. Picone and F. Capozzi, Nuclear magnetic resonance for foodomics beyond food analysis, *TrAC, Trends Anal. Chem.*, 2014, **59**, 93–102.
177. A. G. Olenskyj, I. R. Donis-González and G. M. Bornhorst, Nondestructive characterization of structural changes during *in vitro* gastric digestion of apples using 3D time-series micro-computed tomography, *J. Food Eng.*, 2020, **267**, 109692.
178. J. Floury, T. Bianchi and J. Thévenot, *et al.*, Exploring the breakdown of dairy protein gels during *in vitro* gastric digestion using time-lapse synchrotron deep-UV fluorescence microscopy, *Food Chem.*, 2018, **239**, 898–910.
179. F. Jamme, D. Bourquin, G. Tawil, A. Viksø-Nielsen, A. Buleon and M. Réfrégiers, 3D imaging of enzymes working *in situ*, *Anal. Chem.*, 2014, **86**(11), 5265–5270.
180. G. Tawil, F. Jamme, M. Réfrégiers, A. Viksø-Nielsen, P. Colonna and A. Buléon, *In situ* tracking of enzymatic breakdown of starch granules by synchrotron UV fluorescence microscopy, *Anal. Chem.*, 2011, **83**(3), 989–993.
181. Q. Guo, N. Bellissimo and D. Rousseau, Role of gel structure in controlling *in vitro* intestinal lipid digestion in whey protein emulsion gels, *Food Hydrocolloids*, 2017, **69**, 264–272.
182. N. Lorén, J. Hagman and J. K. Jonasson, *et al.*, Fluorescence recovery after photobleaching in material and life sciences: putting theory into practice, *Q. Rev. Biophys.*, 2015, **48**(3), 323–387.
183. J. V. C. Silva, P. Peixoto, S. Lortal and J. Floury, Transport phenomena in a model cheese: The influence of the charge and shape of solutes on diffusion, *J. Dairy Sci.*, 2013, **96**(10), 6186–6198.
184. Q. Luo, E. Sewalt, J. W. Borst, A. H. Westphal, R. M. Boom and A. E. Janssen, Analysis and modeling of enhanced green fluorescent protein diffusivity in whey protein gels, *Food Res. Int.*, 2019, **120**, 449–455.
185. Y. Tian, M. M. Martinez and D. Pappas, Fluorescence correlation spectroscopy: a review of biochemical and microfluidic applications, *Appl. Spectrosc.*, 2011, **65**(4), 115A–124A.

CHAPTER 6

Food Structure and the Complexity of Food Matrices

6.1 Introduction

Foods are complex heterogeneous materials, made up of various structural elements that (co-)exist in various states (solids, liquids, and/or gases) at different length scales. Food technology has been defined as *a controlled attempt to preserve, transform, create, or destroy a structure that has been imparted by nature or processing*.[1] Food structure can be defined as *the spatial organization of different food components arising from the assembly and interactions of macronutrients (i.e. proteins, lipids, and carbohydrates) at various length scales ranging from the molecular scale to macroscale*.[2] The micronutrients (vitamins and minerals) and other bioactive compounds present in foods (polyphenols, antioxidants, enzymes, hormones) are commonly present within the food structure/matrix.[3] Food structure and matrix effects play a very crucial and significant role in the perception of sensory properties (texture, flavor), stability and shelf life, and digestion and absorption behavior of foods and nutrients.[4,5]

During the 20th century, major advances in the field of food technology were dominated by chemical and mechanical engineering concepts and mainly impacted food processing at a macroscopic level, primarily to transform and preserve foods through the adaptation of unit operations and design of food processing equipment. Interestingly, in the 21st century, the main advances and interventions in foods were at the micro- and nanostructural levels, to enhance the quality of already existing food products and to design and develop novel food products to satisfy expanding consumer needs in terms of nutritional, textural, and sensorial aspects. Figure 6.1 illustrates the structural elements in foods and relevant length scales. Most of

Food Chemistry, Function and Analysis No. 42
Food Digestion and Absorption: Its Role in Food Product Development
By C. Anandharamakrishnan, Jeyan Arthur Moses and S. Priyanka

Published by the Royal Society of Chemistry, www.rsc.org

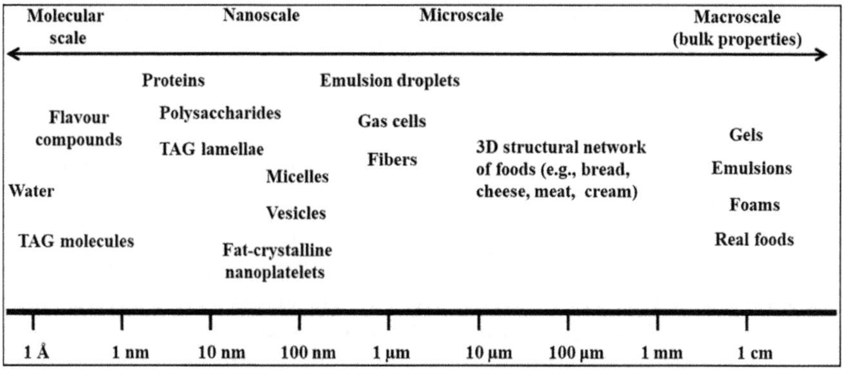

Figure 6.1 Structural elements in foods and relevant length scales. Reproduced from ref. 5 with permission from Elsevier, Copyright 2017.

the food structural elements at the micro- and nanostructural levels (such as meat fibers, plant cells, cell walls, food polymer networks, powders, small particulate material in protein assemblies, oil droplets, gas bubbles, crystals of several types, starch granules, and colloidal particles, *etc.*) play a crucial role in the physical and transport properties, textural properties, rheological behavior, and sensorial traits of foods.[1] Another critical factor that assisted in the in-depth research and investigations on food structure is the availability of advanced analytical tools and experimental techniques such as microscopy, scattering methods, spectroscopy, sensitive rheometers, reliable particle sizing equipment, zeta sizers, and differential scanning calorimetry.[6,7] Further, along with the advances in measurement science, image analysis techniques and numerical simulations and modeling have been used to quantify and predict microstructural information.[8–10]

This chapter summarizes the multidisciplinary aspects of food structure, food structure and matrix classification, food matrix effects, the impact of food structure and the matrix on food texture and other sensory aspects, and analytical techniques used in the characterization of food structure.

6.2 Multidisciplinary Aspects of Food Structure

In various research investigations, "food structure" was correlated with the disciplines of biology, engineering, chemistry, and physics.[11] Also, food structure was substantially related to texture, taste, rheology, digestion, and absorption behavior. The meaning and understanding of food structure vary depending on the perspective of the observer (physicist, chemist, biologist, or engineer), and the type of food (natural or processed/structured). For instance, a biological perspective can be effectively adopted when considering natural food materials such as encapsulated embryos (*e.g.* pulses and grains), fibrous structures (*e.g.* meat), and fleshy materials (*e.g.* vegetables and fruits), which are considered hierarchical fibrous composites made of a few basic components and structural units (such as cells, fibers, membranes, *etc.*) that contribute to their structure and functionality (food texture).[12]

However, foods consumed by humans on a day-to-day basis in the modern world are structurally complex. Foods processed either in industry or at home consist of multiple ingredients and components, they undergo various processes (mechanical, biochemical, and thermal) at different stages, and they are structurally complex. Hence, in the case of processed food products, the food structure is usually approached from an engineering or technological perspective.[13] Processed foods are investigated experimentally for their structure through a physicochemical perspective, and they are conveniently described as "food colloids" as they are a multiphase food system with a wide range of spatial length scales.[14] Further, food structure has mostly been described and expressed by many authors in physicochemical language using the terminology of colloid science, *e.g.* gel, particle, emulsion, interface, and foam.[11]

6.3 Food Matrix: Concept and Classification

In recent times, the term "food matrix" has been widely used in food science and nutrition research mainly to indicate the complex assembly (structuring/restructuring) of food components (nutrients and non-nutrients) and their physical and chemical interactions that affect their behavior in the gastrointestinal (GI) tract during oral processing, digestion, and absorption in terms of mass transfer, stability, digestibility, release, and accessibility.[15] The food matrix is a part of the food microstructure; it is both scale sensitive and compound specific and plays a significant role in the bioaccessibility and bioavailability of food nutrients.[15]

Let us consider the example of bread to understand the scale-sensitive nature of the food matrix: the protein–starch walls surrounding the air cells (scale: a few hundred microns) contribute to the porous crumb texture of bread;[16] during baking, the starch granules undergo gelatinization, which may be viewed as an additional feature in the continuous gluten matrix (scale: 10 μm);[17] further, during digestion, the enzyme degradation of starch granules by α-amylases to release glucose molecules occurs at the nanoscale.[18] To understand the constituent specificity (compound-specific nature) of the food matrix, let us consider the example of milk. During the heating of milk, milk fat (solid fraction) melts within the fat globules and whey protein undergoes denaturation.[19]

Figure 6.2 summarizes the concept of the food matrix and its role in bioaccessibility and bioavailability.[15] Recently, an initiative was taken by Aguilera[15] to classify food matrices into several types based on several case studies and literature references with respect to food science and nutrition, and other related sciences. The food matrices are broadly classified into 10 types, namely aqueous liquid, liquid emulsions, gel matrices, cellular matrices, network exocellular matrices, fibrous extracellular matrices, viscoelastic matrices, dense matrices, porous matrices, and artificial matrices. Table 6.1 outlines the classification and characteristic features of different types of matrices.

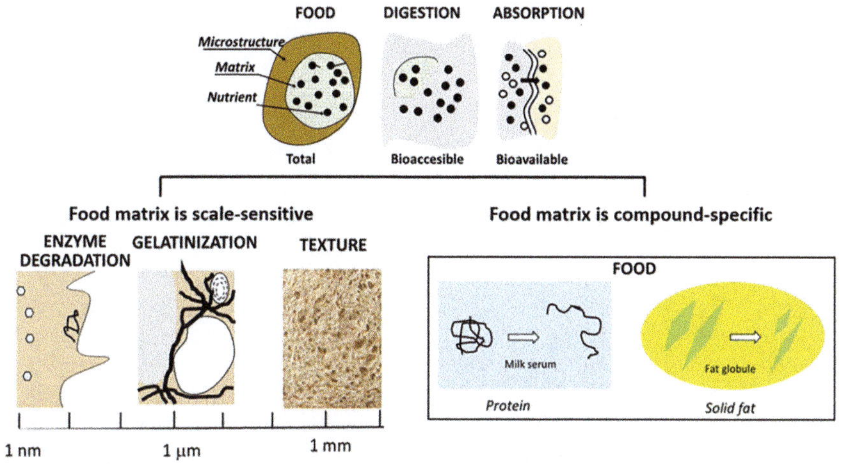

Figure 6.2 The concept of the food matrix (scale sensitivity in bread and compound specificity in milk) and its role in bioaccessibility and bioavailability. Reproduced from ref. 15 with permission from Taylor & Francis, Copyright 2021.

6.4 Impact of Food Structure on Food Functionality

In recent times, consumer expectations regarding food have been very critical, such as desirable sensory properties, affordability, and personalized food composition and functionality (reduced salt, sugar, fat, and calories and improved nutritional benefits). Various studies have revealed that food microstructural features such as the type of structural units and their interactions significantly influence the physical, sensorial, and nutritional quality of food products. Thus, to understand completely the behavior of food products (physical, sensorial, and nutritional), knowledge about their structure/microstructure, composition, and molecular dynamics through the three-dimensional arrangement of the several structural elements and their interactions is necessary. The food structure/microstructure is a combination of different structural elements [such as low molecular weight ingredients, biopolymers (proteins and polysaccharides), oil droplets, granules, fat crystals, micelles, strands, water, and interfaces] with different spatial arrangements and various interactions. Further, all of these structural elements can coexist in different states (glassy/rubbery/crystalline and solubilized) in various combinations and proportions even under uniform temperature conditions and water activities, and inevitably impact the macroscopic food quality attributes and behavior throughout the storage period. The microstructure of food affects its stability, quality, and safety as it influences the availability of water for the growth of microorganisms, water–solute interactions, and degradation reactions. Further, changes in microstructural elements can affect the shelf life of products, for instance the formation of fat bloom (hazy white appearance) on chocolates due to the

Table 6.1 Classification and characteristic features of different types of matrices. Reproduced from ref. 15 with permission from Taylor & Francis, Copyright 2021.

Type of matrix	Description	Characteristic features	Examples	Ref.
Cellular	They are mainly used to refer to plant-based foods such as fruits and vegetables as the hierarchical structure of plants is mainly made of cells. The cell walls are responsible for their mechanical properties and strength and also aid in the protection of cell contents from damage	Cell walls offer dietary fiber, enclose nutrients, impact bioaccessibility during digestion, and are responsible for texture and turgor	Majority of edible fresh fruits and vegetables	51 and 52
Fibrous extracellular	Usually used to describe animal tissues such as meat. Long muscle fibers, encased in layers of connective tissue, and fat tissue (marbling) are the building blocks of meat	Impacts textural properties of meat such as tenderness and toughness	Connective tissue collagen surrounds and binds meat muscle fiber	53 and 54
Aqueous liquids	Commonly used for fluids that contain other components/bioactives in minor quantities	They play a key role in the release of aromas and taste sensations	Aroma compounds of wines and aqueous phase of fruit juices. Aqueous/ethanolic medium with trace amounts of proteins, carbohydrates, tannins, and polyphenolic compounds	55 and 56

Liquid emulsions	There are two categories based on the scale: 1. continuous phase of oil-in-water (O/W) emulsions at the macroscale; 2. the architecture of the interface at the submicron level.	Affect rheological stability and characteristics. Serve as a vehicle for bioactives and help to protect and transport them to specific target sites. The interface may restrict the digestion of lipids and alter their absorption kinetics, induce satiety, and increase the bioavailability of nutrients	Mayonnaise, salad dressings, and emulsion-based beverages	57 and 58
Gels	They consist of a biopolymer network having the capacity to contain sizable volumes of water in addition to other dispersed substances, giving them a semi-solid texture and viscoelastic behavior. Gel biopolymer networks can be particulate (protein aggregates) or fine stranded (gelatin, pectin gels). Based on dispersed elements, the gel matrices can be classified as filled gels (particles), emulsion gels (oil droplets), and aerated gels (air bubbles)	Provide structure and texture to foods. Entrap fat droplets and act as a carrier and protect bioactives. Modulate flavor intensity and prolonged perception	Proteins and polysaccharides interact together to generate 3D networks. Examples: processed meats, yogurt, and desserts	59 and 60

Table 6.1 (*Continued*)

Type of matrix	Description	Characteristic features	Examples	Ref.
Viscoelastic	They refer to food products that resume their original form following repeated cycling under significant deformations	Helps bread and cakes acquire a porous texture by trapping gas bubbles, and also controls their expansion during baking. Limits the starch in pasta from gelatinizing. Serves as a continuous matrix for flavorings, glycerol, and sugars (sweeteners)	Wheat dough produced a 3D network of gluten proteins loaded with starch (*e.g.* baked and pasta products). Elastic network of chewing gum	61 and 62
Porous	Many different foods are porous in nature and made of a continuous solid, viscoelastic, or liquid matrix usually enclosing a dispersed phase in the form of open or closed gas cells. Porous matrices can be produced by freeze-drying, extrusion, aeration, gas release from chemical reactions, fermentation, and baking	Imparts a light texture. Modifies the appearance and the mouthfeel. Facilitates reconstitution and rehydration	Bread, marshmallows, and whipped egg white. Extrusion-based low-density food products and snacks. Aerated chocolate and instant coffee powder	63 and 64
Dense	Most frequently they have low-moisture, glassy, semicrystalline, or crystalline structures	Provides stability and convenience in use as ingredients; easy to store and long shelf life. Spray-dried powders encapsulate bioactive compounds and flavor	Sugar-based confections – amorphous (ungrained caramel), glassy (hard candy), crystalline (rock candy), or partially crystalline (fondants).	65 and 66

			Compact and brittle structures of flour, and dry powders (skim milk powder, instant coffee, cereal flours, ground dry spices)	
Network extracellular	Exopolysaccharides (EPS) are released by microbes, primarily *Lactobacillus* species	Provides rheological properties to some fluid food matrices, *e.g.* enhances viscosity, enhances mouthfeel, and lessens syneresis. Provides beneficial impacts on gut health. Has therapeutic properties such as antitumor, antimutagenic, anticancer, and cholesterol-lowering effects in addition to immunostimulatory activity	Exopolysaccharides in some fermented vegetables and fermented dairy products (yogurt, kefir, and fermented cream)	67 and 68
Artificial	Food matrices that are artificially engineered or structured, such as those that include a bioactive ingredient or a microbe that has been encapsulated in a thin shell for protection	Encapsulation of microorganisms and other functional substances (flavors, bitter peptides, nutrients, bioactive molecules) aids in their protection and targeted delivery of the substance	Bioactives, flavors, or microorganisms encapsulated in gels or solid walls	69 and 70

crystallization of cocoa butter and phase separation in the case of emulsion-based foods such as mayonnaise or spreads due to various instability mechanisms.

6.4.1 Influence of Food Microstructure on the Sensory Properties of Foods

The microstructure of food significantly affects the texture, rheological, and other sensorial properties of food products such as stiffness, snap, and hardness, and imparts specific organoleptic properties (melting and creaminess).[20] The sensory properties of a food are established when our senses such as vision, smell, touch, hearing, and taste interact with the food during consumption. Various qualitative, quantitative, and analytical tools and techniques are used in the evaluation of the sensory properties of food.[21]

The texture is one of the important sensory properties that is perceived during the oral processing of food and is also an important factor that determines the acceptability of food. In a recent study by Czajkowska-González *et al.*,[22] a wheat bread formulation was partially replaced with phenolic compounds through food by-products, flours, and extracts to improve the health properties of the bread (antioxidants). However, sensory modifications of the bread texture were predominant owing to the interactions of phenolic compounds with gluten protein by reducing the disulfide bonds that are an essential part of the gluten matrix, which plays a crucial role in bread structure.[22] Creaminess is a sensory attribute and is an important indicator of quality and richness in the case of many food products such as creams and chocolates. The perceived creaminess is a complex attribute in a multiphase food system as it is dependent on both the flavor and texture characteristics of the product. Additionally, the microstructural features of food, especially the size and nature (solid, liquid, and gaseous) of the dispersed particles in the food, also significantly affect the perceived creaminess.[23]

Food microstructure also plays a crucial role in the retention and release of flavor during the processing and consumption of food. For instance, in the case of solids such as hard candy, flavor becomes trapped in the hydrophilic or lipid phase during processing and is released when the matrix softens, swells, or dissolves during consumption.[24] Another study with lemon-flavored oil/water model emulsions demonstrated that fat content and thickener type affected both rheological behavior and structural properties, and also perceived flavor and texture.[25]

6.4.2 Effects of the Food Matrix on Nutritional Performance and Health Potential

The effects of the food matrix are of particular interest as they play a key role in determining the health potential and nutritional performance of food products having identical chemical compositions. The key terms used in food matrix research related to nutritional/health effects include bioaccessibility, bioavailability, bioconversion, bioactivity, and bioefficacy.[15] Table 6.2 gives a summary of key terminology used in food matrix studies and associated with nutritional/health effects.

Various studies have elucidated that food matrices and their interactions among food components have a key role in the release and bioavailability of numerous nutrients. For instance, consumption of raw carrots decreased the bioavailability of carotenoids fivefold compared with carotenoid supplements dissolved in oil, and this was due to the entrapment of carotenoids within plant cell structures and also the complexing of carotene with proteins in chloroplasts, which made them inaccessible for digestion and absorption.[26] Further, food matrix components such as fiber can hinder the absorption of carotenoids but increase the bioavailability of ferulic acid.[27,28] The mechanisms of several key food matrix effects, namely entrapment within food matrices, confinement within the plant cell walls, attachment to specific cellular organelles, complexes with other food components such as macromolecules, and molecular conformation of the released nutrients, affect the digestion and absorption of nutrients and bioactive compounds.

Various other factors, such as food processing, fermentation, and germination, influence food matrix effects and their possible nutrition/health outcomes. Food processing modifies food structures through changes in the physical state, chemical condition, and state of aggregation or dispersion, among others. Owing to various food processing approaches, there are significant changes in food matrices and their effects, *e.g.* cooking modifies the texture and increases digestibility,[29,30] bioavailability of functional bioactives from fruits and vegetables can be enhanced by homogenizing them into juices,[31,32] and all-*trans* isomers of lycopene are converted to more bioavailable and bioactive *cis* isomers during processing.[33] Fermentation by

Table 6.2 Terminology used in food matrix studies and associated with nutritional/health effects. Reproduced from ref. 15 with permission from Taylor & Francis, Copyright 2021.

Term	Definition
Bioaccessibility	Fraction of an ingested compound (nutrient, bioactive) that is released or liberated from the food matrix in the GI tract
Bioavailability	Fraction of a given compound or its metabolite that reaches the systemic circulation
Bioconversion	Fraction of a bioavailable nutrient that is converted to its active form from an absorbed precursor (*e.g.* retinol from provitamin A)
Bioactivity	Specific effect of a compound in the body. It includes tissue uptake and the consequent physiological response (*e.g.* antioxidant, anti-inflammatory, *etc.*)
Bioefficacy/ bioefficiency	Fraction of an ingested nutrient converted to the active form after biotransformation in the body that produces desirable (or undesirable) human health outcomes in target populations

microbes brings about favorable changes in the natural food matrices of various foods such as milk and dairy products, cereal doughs, meats, fruits, cereals, and grains through the creation of new textures, flavors, and metabolites. Additionally, fermentation improves the bioaccessibility of various nutrients by breaking down complex structures into simpler absorbable forms (*e.g.* starch to simple sugars and proteins to peptides and amino acids) and also breaks down anti-nutrients such as phytates complexed with minerals, thereby enhancing the bioavailability of numerous minerals.[34,35] Further, in the case of cereals and legumes, germination/sprouting modifies the original matrix by partly hydrolyzing the cell walls and other storage constituents, thereby resulting in the enhancement of nutrient content (such as B-group vitamins, minerals, certain essential amino acids, and total sugars) and a reduction in anti-nutritional factors.[36]

The food matrix also undergoes significant changes during oral processing and digestion. Destruction of the food matrix during oral processing releases various food components such as flavors and aroma compounds.[37] In addition, some of the released components may react with oral components such as saliva and enzymes. For instance, polyphenols (from coffee, tea, wine, and some fruits) interact with salivary protein (proline rich) to form insoluble complexes accountable for astringency perception; α-amylase hydrolyzed some starches to glucose.[38] Further, the food matrix effect also plays a significant role in satiety (the inhibitory effect of dietary consumption on appetite), *i.e.* in the case of foods with equivalent caloric value solid foods have a greater effect on satiety than liquid food matrices, and satiation (end of eating), *i.e.* a high-viscosity food matrix containing gums and gelling fiber produces a satiation response through delaying gastric emptying and/or by hindering the digestive enzyme's activity.[39,40]

6.5 Advanced Tools and Experimental Techniques Used in the Characterization of Food Structure

In the last few decades, the demand for foods with superior nutritional quality, improved functionality, and satisfying sensory properties has been increasing among consumers. Accordingly, food scientists and manufacturers are attempting to cater to the needs of consumers by focusing on the development and creation of novel products, improving the quality of existing products, and online monitoring of quality attributes during food production using reliable, rapid, robust technologies.[41] Nowadays, food scientists are working with foods at micro- and nanostructural levels (such as colloidal assemblies and biological macromolecules) to manipulate and control their functional properties such as physicochemical properties, rheological behavior, texture and sensory properties, nutritional properties, and digestion behavior.[42] The microstructures of foods are either broken down or built up during food processing. Therefore, a good knowledge of food microstructure is critical as there is a strong relationship between food structure and functionality.[43] Further, a more in-depth understanding of physiochemical aspects of food structure and its connection with macroscopic food properties will play a crucial role in the design and engineering of foods with modified and controlled physical and organoleptic properties.

The investigation of food microstructure is a tedious task owing to the complex multiphase and multicomponent nature of food systems. Therefore, various advanced analytical tools and techniques have been adapted from various fields, such as spectroscopic approaches (mass spectrometry, nuclear magnetic resonance spectroscopy, and ultrasonic spectroscopy), microscopy techniques (atomic force microscopy, confocal laser scanning microscopy, scanning electron microscopy, transmission electron microscopy, and scanning transmission electron microscopy),[44,45] tomographic techniques (X-ray tomography, magnetic resonance imaging, and neutron tomography),[46,47] and scattering techniques (such as X-ray diffraction, neutron scattering/reflection, and diffusing wave spectroscopy),[48,49] for the characterization of food microstructure. In addition, computational tool packages and simulations are used in combination with image analysis algorithms to investigate dynamic processes with microscopic techniques.[9,50] Table 6.3 presents a summary of various advanced analytical tools and experimental techniques used in food structure characterization.

6.6 Conclusion and Future Perspectives

The subject of food structure is very complex, and the terms food structure/food microstructure and food matrix are used interchangeably based on the context. In this chapter, we have tried to simplify and summarize various key concepts of food structure and food matrices. In addition, various advanced analytical tools and techniques that have been advantageously exploited for examining food microstructure have been discussed. The strong growth in food microstructural research over the last century is based on the realization that changes in microstructure impact the macroscopic properties of food products, such as nutritional aspects, physical and chemical properties, and sensorial attributes. Further, the health potential of food is mainly dependent on the matrix structure and nutrient composition. Food structure plays a key role when the chemical composition of products is identical and the health potential of a food varies widely depending on its structure. Hence it is necessary to understand structure–property relationships when designing novel food products with improved functionality, acceptable sensory properties, and controlled delivery, digestion, and absorption profile to provide intended health benefits and consumer satisfaction. Further in-depth research investigations are warranted to achieve a deeper understanding of the relationships among food microstructure, matrix effects, processing, and the digestion and absorption behavior of different food systems.

Table 6.3 Summary of advanced analytical tools and experimental techniques used in food structure characterization.

Tool/technique	Description/principle	Advantages	Drawbacks	Applications	Ref.
Optical/light microscopy	Visible light (photons) acts as an illumination source. They produce 2D contrast images of microstructural phases due to the differences in their absorbance of transmitted light. Approximate resolution of 200 nm laterally and 600 nm axially. There are different types of bright field (dark field viewing, phase contrast, oil immersion microscopy, differential interference contrast), polarizing, and fluorescence microscopy	Functions well at room temperature and pressure. Permits the characterization of a sample in its natural and hydrated state. Provides full-color information. Allows precise staining of various chemical components. Low initial and ongoing costs. Polarized light and temperature/pressure/shear-controlled stages allow for customization	Diffraction-limited resolution power and poor depth of focus. Requires flat, thin samples	Used to study the supramolecular structure of starch granules, the microstructure of cereals, grains, dough, and bread, emulsion microstructure (droplet flocculation), and morphological features of fruits and vegetables (cell wall structures). Polarized light microscopy was used to study microstructural features (shape, size, and state of agglomeration) of crystals such as fat crystal networks (palm oil, cocoa butter, oleogels) and the crystallinity of carbohydrate particles (starch and cellulose) in multicomponent systems	41 and 71–73

Confocal laser scanning electron microscopy (CLSM)	Through pinholes, a beam of light containing four laser lines encompassing the UV to IR spectral region is directed exclusively at a tiny sample volume. Resolution 200 nm; enhanced depth of focus	3D reconstruction of images. Simple sample preparation. Minimal disturbance. Enhanced resolution compared with optical microscopy	The sample needs to be stained with an appropriate fluorochrome. A longer image acquisition time is needed for dynamic liquid samples. To guarantee equal illumination, the sample surface must be flat	Microstructures and interfacial morphology of various multiphasic colloidal systems can be studied. Employed to determine the concentrations and locations of certain macromolecules in a sample. Microstructural and viscoelastic characterization of soft matter systems can be performed in tandem when connected with a rheometer	74–76
Scanning electron microscopy (SEM) and cryoSEM	Image is created by gathering the scattered electrons from the sample surface as the electron beam sweeps the specimen line by line over a rectangular cross-section. It can achieve a resolution of 1 nm	Image interpretation is simple. A broad range of focus depth (mm to nm). Elements can be analyzed using an energy-dispersive X-ray spectrometry (EDS) detector	The sample preparation chamber needs a vacuum. A sample might need coating, chemical fixation, and/or cryogenic conditions. Unable to examine the internal structure	Employed to visualize the morphology of particle-stabilized droplets, ice crystals, and the interface of gas bubbles in a variety of food systems, including emulsions, foams, and ice-cream. Dynamic studies at various temperatures are possible, as demonstrated by the observation of chocolate bloom over time and the spatial	77–79

Table 6.3 (*Continued*)

Tool/technique	Description/principle	Advantages	Drawbacks	Applications	Ref.
				distribution of various components (such as fats and sugars) on the sample surface. Further analysis of SEM images through fractal dimension and lacunarity analysis can provide information on the descriptive parameters of a food microstructure, namely surface roughness and porosity	
Transmission electron microscopy (TEM) and cryoTEM	An electron spread beam is directed through the specimen, changing the phase and intensity of the incident electrons. As a result, contrast is produced and monitored by a detector that is located on the other side of the beam generator. The approximate resolution is 0.1 nm	Can examine a sample's internal structure. The highest resolution available in microscopy	The sample requires coating and cutting (100 nm thickness). Compared with SEM, complex image interpretation and operational problems	Employed to calculate the difference in thickness between the shell and core of nanoparticles. Particles size can be interpreted from TEM images. Emulsion microstructures (O/W) were investigated with TEM to observe the surface covering of proteins on oil droplets and the	80–83

				impact of thermal treatment on the system. Utilized to study the casein protein network in milk and the caking mechanism in cocoa powder particles	
Atomic force microscopy (AFM)	An image of a sample is created by measuring the interaction of attractive and repulsive forces between a nanometer-sized probe and the surface of the specimen. Provides high-resolution images up to 1 nm. AFM has three main operating methods, namely the contact mode, non-contact mode, and tapping mode	3D topographical images of samples can be obtained. Mechanical behavior of samples can be mapped. Sample preparation is minimal. Images can be generated in both air and liquid environments. Detects particular intermolecular forces on surfaces	Small scan size. Scanning speeds are low. The internal structure cannot be detected	Used for investigating the surface topology of soft food materials, *e.g.* • Polysaccharides were studied at molecular levels. • The complex hierarchical structure of starch granules was characterized. • The structure–function relationship of a bioactive carbohydrate was elucidated. • Shape and size between modified and native casein micelles were assessed. • The morphology fat crystals growing on the surface of chocolate-based confectionery products was studied.	44, 84 and 85

Table 6.3 (*Continued*)

Tool/technique	Description/principle	Advantages	Drawbacks	Applications	Ref.
				The interfacial behavior of protein and surfactant film in food emulsions and foams can be elucidated	
Magnetic resonance imaging (MRI)	This is based on how a sample exposed to a magnetic field absorbs and emits energy. An image from an MRI scan incorporates spatially resolved information from both relaxometry and combined spectroscopy	Internal structures of opaque samples can be detected. Non-invasive and non-destructive technique. Many different contrast methods are available	Long acquisition times. Limited spatial resolution (up to a few hundred microns)	MRI helps in the study of multiphasic food systems because it can use various contrast techniques to discriminate between aqueous phases and lipid, protein, and solid and liquid states. Utilized to examine the size and distribution of bubbles in porous hydrogel foams. Employed to examine the local porosity of dough both qualitatively and quantitatively as it is being proved. Utilized to evaluate concentrated emulsion flow and microstructure	86–88

X-ray tomography	X-rays penetrate sample materials and the differences in attenuation of X-radiation by the materials are exploited to generate transmission images (radiographs) of samples. It is a 3D imaging technique with sub-micron resolution and a field of view ranging from a few centimeters to a few millimeters	Captures intact internal microstructural data. Simple sample preparation. A non-destructive and non-invasive technique	Expensive and relatively time consuming. Extensive data analysis. Restricted access to synchrotron facilities. Limitations (requires a special setup) when studying semi-liquids (such as emulsions and foams) and temperature-sensitive samples (such as ice-creams)	With the aid of phase contrast, used for investigating the microstructural features of meat such as fat distribution, myofibril diameter and length, and connective tissues. By utilizing a synchrotron X-ray source, microstructural details of O/W emulsions and ice-cream samples have been investigated. Used for assessing the microstructures of bread and chocolates such as their pores, voids, and cracks, the effect of new ingredients on their microstructure, and the spatial distribution of fat and sugar, and can also capture dynamic changes during baking	46 and 89–91

References

1. J. M. Aguilera, Why food microstructure?, *J. Food Eng.*, 2005, **67**(1–2), 3–11.
2. A. Acevedo-Fani, Q. Guo, N. Nasef and H. Singh, Aspects of food structure in digestion and bioavailability of LCn-3PUFA-rich lipids, *Omega-3 Delivery Systems*, Elsevier, 2021, pp. 427–448.
3. J. Parada and J. Aguilera, Food microstructure affects the bioavailability of several nutrients, *J. Food Sci.*, 2007, **72**(2), R21–R32.
4. C. Mengucci, P. Ferranti, A. Romano, P. Masi, G. Picone and F. Capozzi, Food structure, function and artificial intelligence, *Trends Food Sci. Technol.*, 2022, **123**, 251–263.
5. Q. Guo, A. Ye, N. Bellissimo, H. Singh and D. Rousseau, Modulating fat digestion through food structure design, *Prog. Lipid Res.*, 2017, **68**, 109–118.
6. D. J. Mcclements, Critical review of techniques and methodologies for characterization of emulsion stability, *Crit. Rev. Food Sci. Nutr.*, 2007, **47**(7), 611–649.
7. J. Ubbink, A. Burbidge and R. Mezzenga, Food structure and functionality: a soft matter perspective, *Soft Matter*, 2008, **4**(8), 1569–1581.
8. R. Van der Sman, Soft matter approaches to food structuring, *Adv. Colloid Interface Sci.*, 2012, **176**, 18–30.
9. S. R. Euston, Modelling and computer simulation of food structures, *Food microstructures*, Elsevier, 2013, pp. 336–385.
10. J. Aguilera and J. Germain, Advances in image analysis for the study of food microstructure, *Understanding and controlling the microstructure of complex foods*, 2007, pp. 261–287.
11. E. Dickinson, Understanding food structures: the colloid science approach, *Food structures, digestion and health*, Elsevier, 2014, pp. 3–49.
12. J. F. Vincent, The composite structure of biological tissue used for food, *Food Materials Science: Principles and Practice*, 2008, pp. 11–20.
13. J. M. Aguilera, Seligman lecture 2005 food product engineering: building the right structures, *J. Sci. Food Agric.*, 2006, **86**(8), 1147–1155.
14. M. Boland, M. Golding and H. Singh, *Food structures, digestion and health*, Academic Press, 2014.
15. J. M. Aguilera, The food matrix: implications in processing, nutrition and health, *Crit. Rev. Food Sci. Nutr.*, 2019, **59**(22), 3612–3629.
16. Z. Liu and M. Scanlon, Predicting mechanical properties of bread crumb, *Food Bioprod. Process.*, 2003, **81**(3), 224–238.
17. T. Maeda, M. Kokawa and M. Miura, *et al.*, Development of a novel staining procedure for visualizing the gluten–starch matrix in bread dough and cereal products, *Cereal Chem.*, 2013, **90**(3), 175–180.
18. S. Dhital, F. J. Warren, P. J. Butterworth, P. R. Ellis and M. J. Gidley, Mechanisms of starch digestion by α-amylase—Structural basis for kinetic properties, *Crit. Rev. Food Sci. Nutr.*, 2017, **57**(5), 875–892.
19. U. Kulozik, Structuring dairy products by means of processing and matrix design, *Food Materials Science: Principles and Practice*, 2008, pp. 439–473.

20. J. F. Fundo and C. L. Silva, Microstructure, composition and their relationship with molecular mobility, food quality and stability, *Food microstructure and its relationship with quality and stability*, Elsevier, 2018, pp. 29–41.
21. S. Ray, Sensory Properties of Foods and Their Measurement Methods, *Techniques to Measure Food Safety and Quality: Microbial, Chemical, and Sensory*, 2021, 345–381.
22. Y. A. Czajkowska-González, E. Alvarez-Parrilla, N. del Rocío Martínez-Ruiz, A. A. Vázquez-Flores, M. Gaytán-Martínez and L. A. de la Rosa, Addition of phenolic compounds to bread: antioxidant benefits and impact on food structure and sensory characteristics, *Food Prod., Process. Nutr.*, 2021, **3**(1), 1–12.
23. D. Kilcast and S. Clegg, Sensory perception of creaminess and its relationship with food structure, *Food Qual. Prefer.*, 2002, **13**(7–8), 609–623.
24. K. B. de Roos, Effect of texture and microstructure on flavour retention and release, *Int. Dairy J.*, 2003, **13**(8), 593–605.
25. C. Arancibia, L. Jublot, E. Costell and S. Bayarri, Flavor release and sensory characteristics of o/w emulsions. Influence of composition, microstructure and rheological behavior, *Food Res. Int.*, 2011, **44**(6), 1632–1641.
26. J. J. Castenmiller and C. E. West, Bioavailability and bioconversion of carotenoids, *Annu. Rev. Nutr.*, 1998, **18**(1), 19–38.
27. L. Rondini, M.-N. Peyrat-Maillard and A. Marsset-Baglieri, *et al.*, Bound ferulic acid from bran is more bioavailable than the free compound in rat, *J. Agric. Food Chem.*, 2004, **52**(13), 4338–4343.
28. E. D. Brown, S. R. Blakely, U. Babu, E. Grundel and G. V. Mitchell, Vegetable concentrates interact with canthaxanthin to affect carotenoid bioavailability and superoxide dismutase activity but not immune response in rats, *Nutr. Res.*, 1997, **17**(6), 989–998.
29. J. Singh, A. Dartois and L. Kaur, Starch digestibility in food matrix: a review, *Trends Food Sci. Technol.*, 2010, **21**(4), 168–180.
30. L. Kaur, E. Maudens, D. R. Haisman, M. J. Boland and H. Singh, Microstructure and protein digestibility of beef: The effect of cooking conditions as used in stews and curries, *LWT – Food Sci. Technol.*, 2014, **55**(2), 612–620.
31. A. E. Quirós-Sauceda, C.-Y. O. Chen, J. B. Blumberg, H. Astiazaran-Garcia, A. Wall-Medrano and G. A. González-Aguilar, Processing 'ataulfo'mango into juice preserves the bioavailability and antioxidant capacity of its phenolic compounds, *Nutrients*, 2017, **9**(10), 1082.
32. V. Raikos, Food matrix: Natural barrier or vehicle for effective delivery of carotenoids from processed foods?, *Insights Nutr. Metab.*, 2017, **1**(1), 1–6.
33. K. N. Honest, H. W. Zhang and L. Zhang, Lycopene: Isomerization effects on bioavailability and bioactivity properties, *Food Rev. Int.*, 2011, **27**(3), 248–258.
34. M. Samtiya, R. E. Aluko, A. K. Puniya and T. Dhewa, Enhancing micronutrients bioavailability through fermentation of plant-based foods: A concise review, *Fermentation*, 2021, **7**(2), 63.

35. M. L. Marco, D. Heeney and S. Binda, *et al.*, Health benefits of fermented foods: microbiota and beyond, *Curr. Opin. Biotechnol*, 2017, **44**, 94–102.
36. K. Lorenz and B. D'Appolonia, Cereal sprouts: Composition, nutritive value, food applications, *CRC Crit. Rev. Food Sci. Nutr.*, 2009, **13**(4), 353–385.
37. E. Guichard and C. Salles, Retention and release of taste and aroma compounds from the food matrix during mastication and ingestion, *Flavor*, Elsevier, 2016, pp. 3–22.
38. P. Sethupathy, J. A. Moses and C. Anandharamakrishnan, Food oral processing and tribology: Instrumental approaches and emerging applications, *Food Rev. Int.*, 2021, **37**(5), 538–571.
39. S. Fiszman and P. Varela, The role of gums in satiety/satiation. A review, *Food Hydrocolloids*, 2013, **32**(1), 147–154.
40. L. Chambers, K. McCrickerd and M. R. Yeomans, Optimising foods for satiety, *Trends Food Sci. Technol.*, 2015, **41**(2), 149–160.
41. L. Metilli, M. Francis and M. Povey, *et al.*, Latest advances in imaging techniques for characterizing soft, multiphasic food materials, *Adv. Colloid Interface Sci.*, 2020, **279**, 102154.
42. U. Lesmes and D. J. McClements, Structure–function relationships to guide rational design and fabrication of particulate food delivery systems, *Trends Food Sci. Technol.*, 2009, **20**(10), 448–457.
43. I. Sensoy, A review on the relationship between food structure, processing, and bioavailability, *Crit. Rev. Food Sci. Nutr.*, 2014, **54**(7), 902–909.
44. Y. Wen, Z. Xu, Y. Liu, H. Corke and Z. Sui, Investigation of food microstructure and texture using atomic force microscopy: A review, *Compr. Rev. Food Sci. Food Saf.*, 2020, **19**(5), 2357–2379.
45. J. A. R. Cardona, C. H. Iriart and M. L. Herrera, Applications of confocal laser scanning microscopy (CLSM) in foods, *Confocal laser microscopy-principles and applications in medicine, biology, and the food sciences*, InTechOpen, 2013, pp. 203–234.
46. Z. Wang, E. Herremans, S. Janssen, D. Cantre, P. Verboven and B. Nicolaï, Visualizing 3D food microstructure using tomographic methods: Advantages and disadvantages, *Annu. Rev. Food Sci. Technol.*, 2018, **9**, 323–343.
47. P. Verboven, T. Defraeye and B. Nicolai, Measurement and visualization of food microstructure: Fundamentals and recent advances, *Food microstructure and its relationship with quality and stability*, 2018, pp. 3–28.
48. T. Nicolai, Food characterisation using scattering methods, *Understanding and controlling the microstructure of complex foods*, 2007, pp. 288–310.
49. E. P. Gilbert, Small-angle X-Ray and neutron scattering in food colloids, *Curr. Opin. Colloid Interface Sci.*, 2019, **42**, 55–72.
50. M. Greiner, B. Sonnleitner, M. Mailänder and H. Briesen, Modeling complex and multi-component food systems in molecular dynamics

simulations on the example of chocolate conching, *Food Funct.*, 2014, **5**(2), 235–242.
51. Y. Ogawa, N. Donlao and S. Thuengtung, *et al.*, Impact of food structure and cell matrix on digestibility of plant-based food, *Curr. Opin. Food Sci.*, 2018, **19**, 36–41.
52. G. Mandalari, R. M. Faulks and G. T. Rich, *et al.*, Release of protein, lipid, and vitamin E from almond seeds during digestion, *J. Agric. Food Chem.*, 2008, **56**(9), 3409–3416.
53. E. Tornberg, Engineering processes in meat products and how they influence their biophysical properties, *Meat Sci.*, 2013, **95**(4), 871–878.
54. T. Nishimura, The role of intramuscular connective tissue in meat texture, *Anim. Sci. J.*, 2010, **81**(1), 21–27.
55. R. R. Villamor and C. F. Ross, Wine matrix compounds affect perception of wine aromas, *Annu. Rev. Food Sci. Technol.*, 2013, **4**, 1–20.
56. H. Caswell, The role of fruit juice in the diet: an overview, *Nutr. Bull.*, 2009, **34**(3), 273–288.
57. D. J. McClements and E. A. Decker, Lipid oxidation in oil-in-water emulsions: Impact of molecular environment on chemical reactions in heterogeneous food systems, *J. Food Sci.*, 2000, **65**(8), 1270–1282.
58. E. Dickinson, Interfacial structure and stability of food emulsions as affected by protein–polysaccharide interactions, *Soft Matter*, 2008, **4**(5), 932–942.
59. S. Banerjee and S. Bhattacharya, Food gels: gelling process and new applications, *Crit. Rev. Food Sci. Nutr.*, 2012, **52**(4), 334–346.
60. M. Corredig, N. Sharafbafi and E. Kristo, Polysaccharide–protein interactions in dairy matrices, control and design of structures, *Food Hydrocolloids*, 2011, **25**(8), 1833–1841.
61. M. Jekle and T. Becker, Wheat dough microstructure: The relation between visual structure and mechanical behavior, *Crit. Rev. Food Sci. Nutr.*, 2015, **55**(3), 369–382.
62. R. V. Potineni and D. G. Peterson, Mechanisms of flavor release in chewing gum: cinnamaldehyde, *J. Agric. Food Chem.*, 2008, **56**(9), 3260–3267.
63. K. Niranjan and S. Silva, Bubble-containing foods, *Food Materials Science: Principles and Practice*, 2008, pp. 281–303.
64. A. Marabi and I. S. Saguy, Rehydration and reconstitution of foods, *Advances in food dehydration*, CRC Press, 2008, pp. 255–302.
65. B. R. Bhandari, N. Bansal, M. Zhang and P. Schuck, *Handbook of food powders: Processes and properties*, Elsevier, 2013.
66. A. B. D. Nandiyanto and K. Okuyama, Progress in developing spray-drying methods for the production of controlled morphology particles: From the nanometer to submicrometer size ranges, *Adv. Powder Technol.*, 2011, **22**(1), 1–19.
67. A. Patel and J. Prajapat, Food and health applications of exopolysaccharides produced by lactic acid bacteria, *Adv. Dairy Res.*, 2013, 1–8.
68. P. Singh and P. Saini, Food and health potentials of exopolysaccharides derived from Lactobacilli, *Microbiol. Res. J. Int.*, 2017, **22**, 1–14.

69. K. M. Crowe, Designing functional foods with bioactive polyphenols: highlighting lessons learned from original plant matrices, *J. Hum. Nutr. Food Sci.*, 2013, **1**(3), 1018.
70. D. J. McClements, E. A. Decker, Y. Park and J. Weiss, Structural design principles for delivery of bioactive components in nutraceuticals and functional foods, *Crit. Rev. Food Sci. Nutr.*, 2009, **49**(6), 577–606.
71. W. Błaszczak and G. Lewandowicz, Light microscopy as a tool to evaluate the functionality of starch in food, *Foods*, 2020, **9**(5), 670.
72. K. Autio and M. Salmenkallio-Marttila, Light microscopic investigations of cereal grains, doughs and breads, *LWT – Food Sci. Technol.*, 2001, **34**(1), 18–22.
73. S. S. Narine and A. G. Marangoni, Fractal nature of fat crystal networks, *Phys. Rev. E: Stat. Phys., Plasmas, Fluids, Relat. Interdiscip. Top.*, 1999, **59**(2), 1908.
74. R. Besseling, L. Isa, E. R. Weeks and W. C. Poon, Quantitative imaging of colloidal flows, *Adv. Colloid Interface Sci.*, 2009, **146**(1–2), 1–17.
75. K. Ako, D. Durand, T. Nicolai and L. Becu, Quantitative analysis of confocal laser scanning microscopy images of heat-set globular protein gels, *Food Hydrocolloids*, 2009, **23**(4), 1111–1119.
76. M. Auty, Confocal microscopy: principles and applications to food microstructures, *Food Microstructures*, Elsevier, 2013, pp. 96–P8.
77. X. Zhuang, M. Han and X. Jiang, *et al.*, The effects of insoluble dietary fiber on myofibrillar protein gelation: Microstructure and molecular conformations, *Food Chem.*, 2019, **275**, 770–777.
78. J. Rahimi and M. O. Ngadi, Structure and irregularities of surface of fried batters studied by fractal dimension and lacunarity analysis, *Food Struct.*, 2016, **9**, 13–21.
79. P. D. Tran, D. Van de Walle and M. Hinneh, *et al.*, Controlling the stability of chocolates through the incorporation of soft and hard StOSt-rich fats, *Eur. J. Lipid Sci. Technol.*, 2015, **117**(11), 1700–1713.
80. A. Sarkar, B. Murray, M. Holmes, R. Ettelaie, A. Abdalla and X. Yang, In vitro digestion of Pickering emulsions stabilized by soft whey protein microgel particles: influence of thermal treatment, *Soft Matter*, 2016, **12**(15), 3558–3569.
81. W. D. Pyrz and D. J. Buttrey, Particle size determination using TEM: a discussion of image acquisition and analysis for the novice microscopist, *Langmuir*, 2008, **24**(20), 11350–11360.
82. J. V. Silva, D. Legland, C. Cauty, I. Kolotuev and J. Floury, Characterization of the microstructure of dairy systems using automated image analysis, *Food Hydrocolloids*, 2015, **44**, 360–371.
83. J. Petit, F. Michaux and C. Jacquot, *et al.*, Storage-induced caking of cocoa powder, *J. Food Eng.*, 2017, **199**, 42–53.
84. H. Yang, Y. Wang, S. Lai, H. An, Y. Li and F. Chen, Application of atomic force microscopy as a nanotechnology tool in food science, *J. Food Sci.*, 2007, **72**(4), R65–R75.

Food processing techniques can be broadly classified into conventional (thermal processing, milling, grinding, drying, freezing, fermentation, *etc.*) and emerging technologies (high-pressure homogenization and pulsed electric field, microwave, and ultrasound processing). Based on the type of initial food product/constituents, each processing technique imparts various structural modifications to develop a specific final product. Major structural modifications obtained through food processing are summarized in Table 7.1.

Additionally, various food processing techniques are used for the preservation, processing, and structuring of foods with different origins (plant and animal) into different states (solid, dispersions, and fluids) and at various length scales (molecular, nano, micro and macro levels). Different food processing approaches uniquely impact the physical and sensorial properties of food, food structure, digestion behavior, and nutritional aspects (nutrient release and availability). Figure 7.1 illustrates the interconnections between food processing, structure modification, and food properties.

It is worth mentioning that it is difficult to generalize the effects of processing on food structure and its effect on digestion behavior owing to the complex nature of the subject. Hence a proper understanding of the effects of different food processing approaches on food structure and nutrients is essential for the selection of specific processes and process parameters to manipulate the food digestion behavior and nutrient absorption. In this chapter, we discuss the influence of various food processing methods on food structure, the impact of processing treatments on the gastrointestinal fate of different foods through case studies, and the influence of food processing on nutrient release, bioaccessibility, and bioavailability.

7.2 Effects of Processing Treatments on the Gastrointestinal Fate of Different Food Structures

In this section, we summarize the fate of natural food structures and their processed versions in the gastrointestinal (GI) tract. For liquids, we have considered milk as it is nature's most complex emulsion system. In the case of solid foods, we chose almond nuts (cellular matrix) as they can be consumed in their raw form or with minimal processing. Table 7.2 summarizes recent studies exploring the effects of various food processing treatments on food structure and digestion behavior.

7.2.1 Case Study 1: Milk and Dairy Products

Milk is considered a balanced food and it is consumed by people across all age groups. The major components of milk are proteins, fats, lactose, and minerals. Milk from various species such as cows, goats, sheep, and others is commonly consumed by humans. The composition of milk is highly variable within the same species and across species. Milk and milk protein ingredients play an integral role in the development of various dairy-based

Table 7.1 Major structural modifications resulting from food processing and their impact on digestion and nutrient absorption. Reproduced from ref. 44 with permission from Elsevier, Copyright 2023.

Food processing	Specific food products/ constituents	Structural changes	Final food products	Significance	Ref.
Grinding, milling, and refining	Grains, cereals, and pulses	Tissue disintegration. Cell wall disruption. Size reduction	Flours and powders	Increased digestibility of starch and protein due to an increase in surface area and improved enzyme accessibility. Increased nutrient availability due to cell wall disruption. Loss of certain nutrients and fiber due to the removal of bran	46 and 47
Cutting, chopping, and grinding	Fruits and vegetables	Disruption of membrane and cell wall. Breakdown of intracellular organelles. Change in shape and decrease in size	Purée, pulp, juice, pieces	Nutrients such as carotenoids become more accessible and easier to absorb. Exposure to air, light, and enzymes during processing can lead to loss of nutrients such as vitamin C and phenolic compounds due to oxidative degradation	33, 48 and 49

Thermal processing, cooking, and microwave, pulsed electric field, ultrasonication, and high-pressure processing	Meat, grains, fruits, and vegetables	Compaction and softening of tissues. Formation of channels. Alteration of structure. Shrinkage, debonding, and disintegration of cells. Alterations to intercellular porosity. Breakage of the cell wall and membrane	Foods that have been minimally processed, ready-to-eat foods, snacks, and pretreated products for further processing	Denaturation of proteins in meat and other plant sources improves accessibility to digestive enzymes, thereby improving protein digestibility. Processing can improve the bioaccessibility and bioavailability of certain nutrients such as β-carotene, lycopene, and polyphenols. However, overprocessing and boiling in excess water will lead to nutrient degradation and leaching of water-soluble nutrients such as vitamin C and phenolic compounds	50–52
Fermentation	Various food products consisting of milk, fruits, vegetables, and grains	Breakdown of complex constituents such as starch and protein into simple sugars, peptides, and amino acids. Degradation of the food matrix, antinutritional factors such as	Fermented products	Enhanced protein and starch digestibility. Improved antioxidative attributes. Enhanced bioaccessibility and bioavailability of minerals and vitamins	30 and 31

Table 7.1 (*Continued*)

Food processing	Specific food products/ constituents	Structural changes	Final food products	Significance	Ref.
		phytates and tannins, and release of bound nutrients such as minerals. Prominent structural and textural changes			
High-shear mixing, colloidal mill, homogenization, thermal processing, and ultrasonication	Lipids, proteins, and polysaccharides	Decrease in droplet/ particle size. Alterations in interfacial structural design. Structuring of the layer/ multilayer	Various emulsions, coated drops, and hydrogel structures	Improved bioaccessibility and bioavailability of a range of encapsulated ingredients such as bioactive compounds, vitamins, minerals, and probiotics	53–55

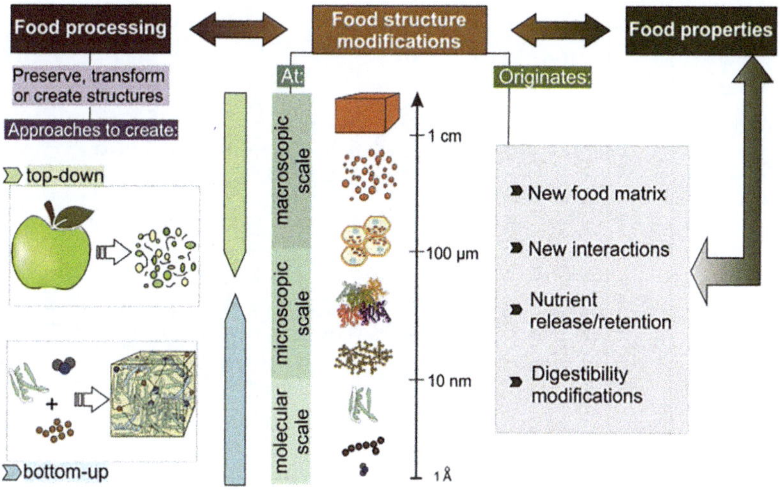

Figure 7.1 Interconnections among food processing, structure modification, and food properties. Reproduced from ref. 44 with permission from Elsevier, Copyright 2023.

products such as yogurt, cheese, and other artificially structured products such as emulsions. Milk is a complex oil-in-water emulsion system from nature owing to its exclusive production from mammary glands, complex structure, and assembly of nutrients such as colloidal casein protein and fat globules, and it has a complex digestion behavior compared with artificially structured products (coagulation behavior in the stomach).[1]

Milk remains relatively unaltered during oral processing owing to its shorter residence time in the mouth as it is liquid in nature and does not possess any macroscopic structure to be broken down before swallowing. Thus, the gastric coagulation behavior of milk was reported to have a critical role and it influences the digestion, delivery, and absorption of nutrients. The digestion behavior of different protein fractions in milk varies owing to their susceptibility to coagulation. Casein micelles are coagulated in the stomach because of the acidic conditions and the action of the enzyme pepsin. Hence they are relatively slowly hydrolyzed by pepsin, which results in their slow gastric emptying and delivery to the small intestine. In contrast, whey proteins in milk are soluble globular proteins and are not coagulated by either proteolysis or low pH, resulting in rapid gastric emptying and a fast postprandial rate of release of amino acids into the bloodstream.[2]

Several studies have investigated the gastric coagulation behavior of various milk types (whole milk, skim milk, and milk from different species) and the effects of different processing conditions such as heat treatments (pasteurization and UHT milk) and homogenization using different digestion models such as infants, humans, animals, and dynamic models.[1,3–5,45] In brief, raw unheated milk forms a firm coagulum (*i.e.* a tight and dense clot network), whereas processed milk forms a fragmented coagulum (*i.e.* open

Table 7.2 Summary of recent studies exploring the effects of various food processing treatments on food structure and digestion behavior.

Type of test food and processing conditions	Objective of the study	Digestion parameters	Significance of the study	Ref.
Starchy foods with different structures: • Bakery products (biscuits, crackers, bread) – baking and leavening. • Breakfast cereals – flaking and puffing. • Pastas – extrusion. • Porridge, muesli, barley – minimally processed. • Starchy root vegetables and pulses.	To explore the pattern of carbohydrate digestion through glycemic response in differently structured foods rich in carbohydrates	Oral processing, *in vitro* gastric digestion, sugar analysis, and glycemic impact. Rapidly available carbohydrate (RAC). Slowly digested starch (SDS). Inaccessible digestible starch (IDS)	High RAC, low SDS, and minimal IDS were characteristics of highly processed cereal-based foods, particularly morning cereals and baked goods with less resilient structures. Both highly processed pasta with dense starch structures and minimally processed whole grain cereal items showed roughly identical amounts of RAC and SDS but provided relatively little IDS. Leguminaceous pulses with preserved cellular morphology digested more linearly and produced a very small amount of RAC. Further, the majority of their readily available carbohydrates were SDS and also they had higher quantities of IDS	56

Granny Smith apples: Three different processing conditions: • Freezing (using liquid N_2). • Freeze-drying (using a freeze-drier at −50 °C). • Convective drying (using a hot air dryer at 60 °C and 2 m s^{-1} air velocity)	To assess the impact of various processing methods on microstructure, chemical properties, and release of bioactive substances	*In vitro* oral and gastric digestion. Total polyphenol content and antioxidant activity. Microstructure analysis (cell walls). Mathematical modeling	The initial composition and microstructure of Granny Smith apples were altered by processing in comparison with raw apples. As a result of processing and 180 min of *in vitro* stomach digestion, there was significant cellular damage and changes in cell wall composition. Processing of apples resulted in decreases in total polyphenol content and antioxidant activity	57
Durum wheat porridge meals with different particle sizes: Milling and size reduction	To evaluate the starch bioaccessibility of a variety of products made from the wheat endosperm with different particle sizes	*In vitro* oral, gastric, and small intestinal digestion. Starch and glucose analysis	The release of starch and protein and the glycemic response were enhanced with a decrease in the size of wheat because the finely milled wheat endosperm lacks intact cells compared with coarsely milled wheat (contains intact cells)	58
Bovine longissimus dorsi muscle meat: High-pressure processing (HPP) at 175 and 600 MPa	To study the effect of HPP of meat on muscle protein breakdown during digestion	*In vitro* gastrointestinal digestion. Microstructure. Protein digestibility	HPP resulted in structural changes in the myofibrillar protein and protein denaturation, which increased the susceptibility to pepsin action, resulting in increased protein breakdown and protein solubility.	59

Table 7.2 (*Continued*)

Type of test food and processing conditions	Objective of the study	Digestion parameters	Significance of the study	Ref.
			HPP-treated meat had better protein digestibility than cooked meat	
Cold-boned beef, semimembranosus: Pulsed electric field (PEF) [T1, 5.0 kV (90 Hz); T2, 10 kV (20 Hz)]	To assess the impact of PEF processing on the kinetics of beef digestion	*In vitro* gastrointestinal digestion. Microstructure. Protein digestibility and solubility	PEF processing resulted in structural changes in proteins and increased the membrane permeability through electroporation, which led to a positive impact on the enzymatic hydrolysis of meat proteins. Thereby, PEF processing resulted in greater protein digestibility and soluble protein-free amino acids	60
Plant-based shakes: mixing and high-pressure homogenization	To evaluate the influence of processing on the digestion kinetics of lipids and proteins	Static *in vitro* digestion. Particle size. Microstructure. Lipid and protein digestion kinetics	High-pressure homogenization treatment resulted in small and homogeneous particles, which enhanced protein solubility and also resulted in a faster rate and a higher extent of both lipid and protein digestion compared with plant-based shakes prepared from only mixing. However, plant-based shakes prepared from only mixing had protein aggregates that slowed the lipid digestion	61

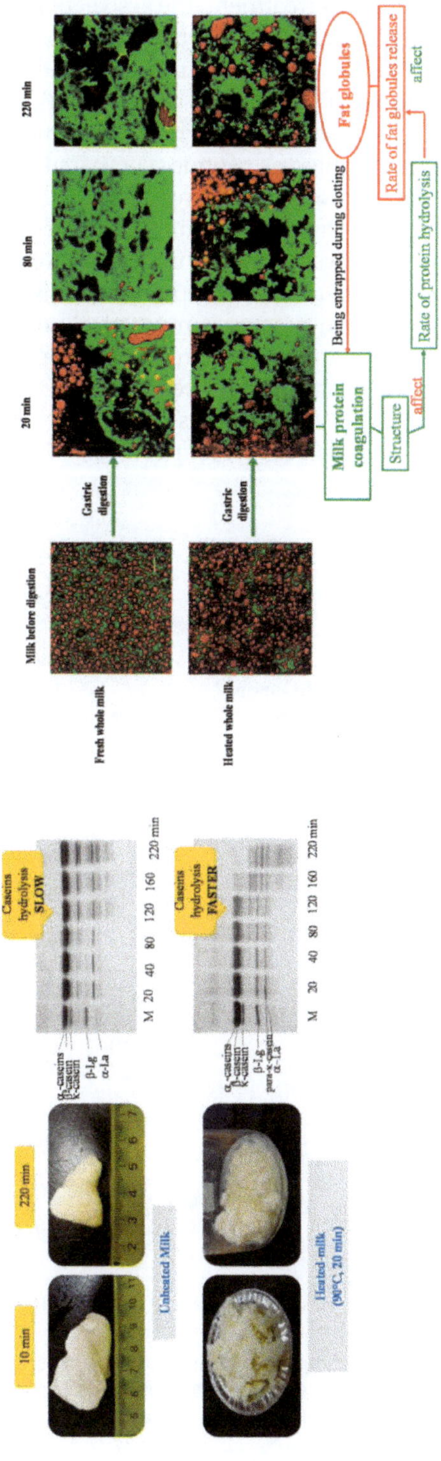

Figure 7.2 Gastric digestion behavior of unheated fresh milk and heated milk. Left: Photographs of clots of unheated and heated milk formed during *in vitro* dynamic gastric digestion in a human gastric simulator (HGS) and sodium dodecyl sulfate polyacrylamide gel electrophoresis (SDS-PAGE) patterns under reducing conditions of proteins in the clots as a function of digestion time. Right: Confocal microscopy images of unheated and heated whole milk at different times (from 0 to 220 min) during gastric digestion. Red indicates fat and green indicates protein. The scale bar in the top left image is 20 µm and applies to all images. Reproduced from ref. 1 with permission from Elsevier, Copyright 2021.

and loose) (Figure 7.2). During heat treatment, the association between casein micelles and denatured whey proteins hinders the formation of a strong clot network because of steric effects and results in the formation of open and loose clot structures in heat-processed milk (Figure 7.3).[6] Further, heat treatment also more rapidly degrades the whey proteins, and they are more rapidly emptied from the gastric phase as peptides compared with native whey proteins. In addition, the fat release is closely related to the clot structure as the milk fat globules are trapped physically within the clot network. Hence the structure of the clot plays a significant role in milk fat release and digestion.[3]

Different processing treatments also modify the gastric digestion behavior of milk, *e.g.* homogenization alone forms comparatively dense curd compared with homogenization + heating, which results in a looser curd structure (Figure 7.4). Further, the degree of processing also impacts the clot structures, *e.g.* homogenized UHT (140 °C/4 s) cow milk forms crumbled unstructured curds and homogenized pasteurized (85 °C/15 s) cow milk

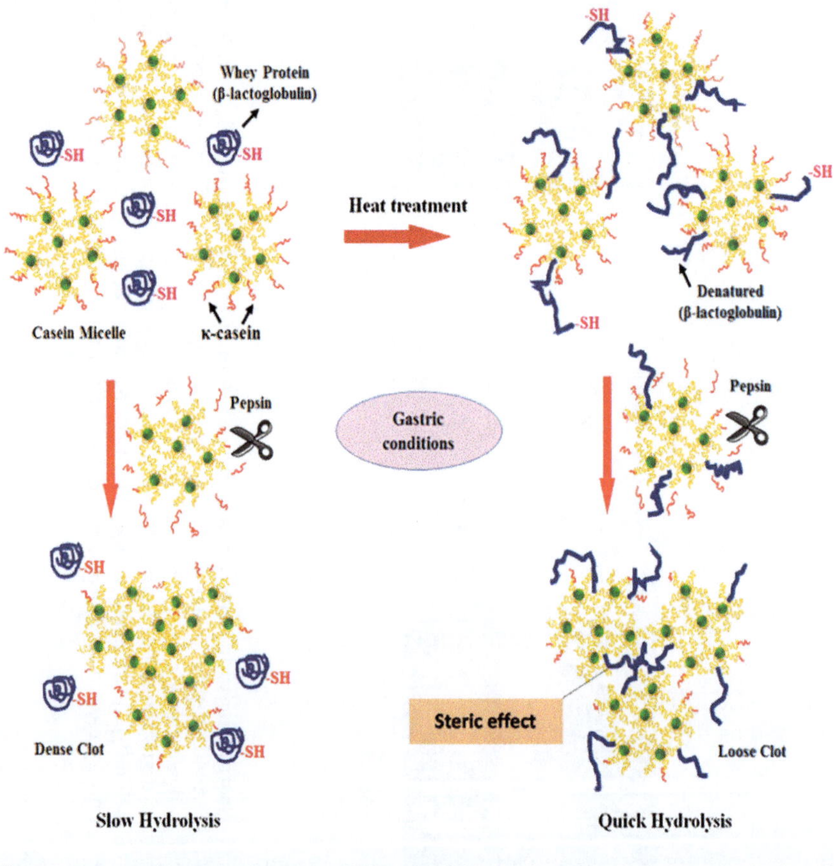

Figure 7.3 Schematic diagram of the possible mechanism of the events during the formation of protein curds from unheated milk and heated milk in a gastric environment. Reproduced from ref. 1 with permission from Elsevier, Copyright 2021.

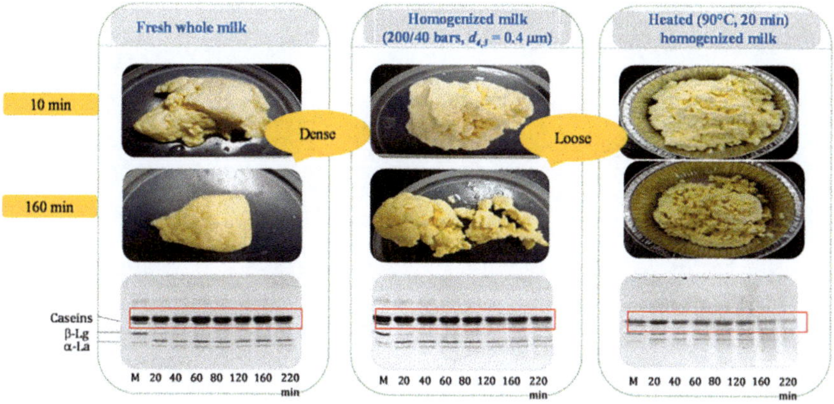

Figure 7.4 Photographs of clots of raw whole milk, homogenized milk, and heated homogenized milk formed during *in vitro* dynamic gastric digestion in an HGS and SDS-PAGE patterns under reducing conditions of proteins in the clots as a function of the digestion time. Reproduced from ref. 1 with permission from Elsevier, Copyright 2021.

forms a soft curd in the rat's stomach, which also supported the previous *in vitro* findings (Figure 7.5).[7] It is noteworthy that the protein hydrolysis and nutrient emptying for raw and processed milk are significantly impacted by this gastric coagulation behavior and clot structural changes.[1] Additionally, milk samples from different animal species vary in their physicochemical properties (such as fat globule size and casein micelle size) and composition (protein content), which was reported to influence their digestion behavior.[4] Regardless of species variations, the gastric digestion mechanism was the same for milk from cows, sheep, and goats, and the rate of fat release from coagulated caseins (curd) was directly related to the protein network breakdown of the clots.[8] However, the relative amount of curds formed during gastric digestion was determined by the distinct casein content in milk of different species.[9]

Interestingly, recent research evidence indicates that the dairy matrix plays an important role in the reduction of adverse nutritional effects that were previously believed for dairy products, such as contributing to high cholesterol levels, a higher risk of hypertension, and saturated fat in the diet.[10,11] The consumption of milk, cheese, and yogurt has significant beneficial effects on health, *e.g.* it aids in reducing the risk of stroke, cardiovascular diseases, and diabetes.[12] In general, the available research on milk suggests that it undergoes substantial structural and physicochemical changes during gastric digestion and those changes were subsequently proven to affect nutrient (protein and fat) gastric emptying rates and absorption in the small intestine. Nevertheless, more *in vivo* human trials are needed to validate the inferences from *in vitro* and animal studies, and also to acquire more information regarding the effects of dairy matrices on potential health benefits.

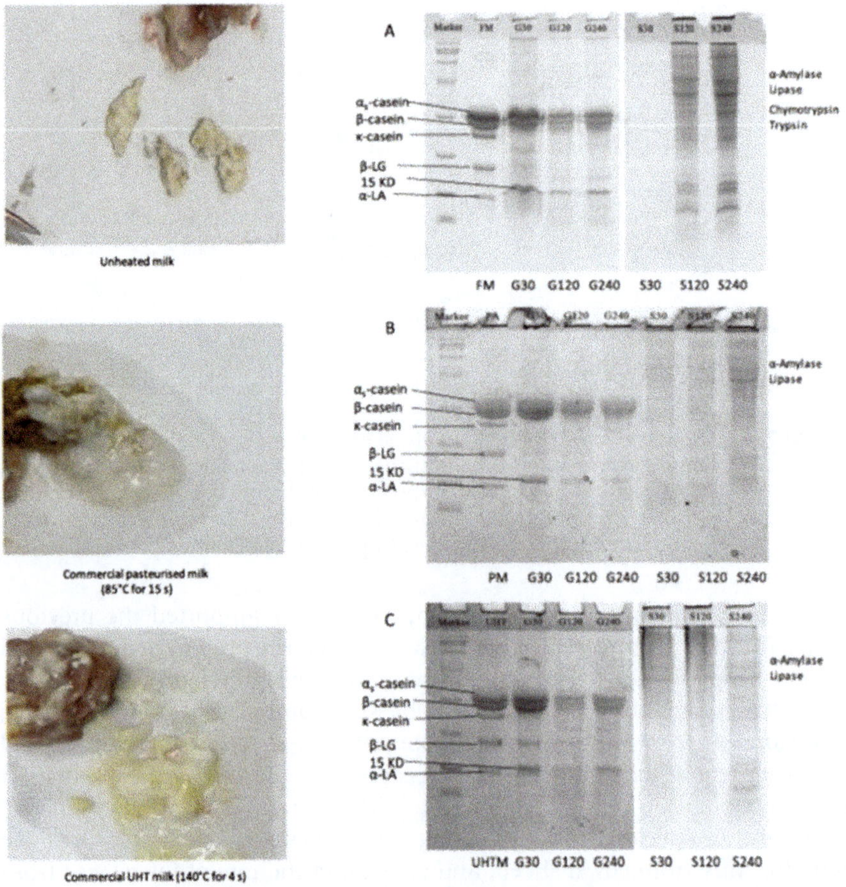

Figure 7.5 Left: Milk clots/curds (photographs) obtained from unheated homogenized milk, homogenized pasteurized milk (85 °C for 15 s), and UHT homogenized milk (140 °C for 4 s) at 30 min in the rat stomach after ingestion. Right: SDS-PAGE patterns under reducing conditions of the curds obtained during the digestion in the rat stomach and small intestine of (A) unheated homogenized milk, (B) homogenized pasteurized milk (85 °C for 15 s), and (C) UHT homogenized milk (140 °C for 4 s) at different times. FM, unheated fresh milk; PM, pasteurized milk; UHTM, UHT milk; G30, 30 min at stomach; G120, 120 min at stomach; G240, 240 min at stomach; S30, 30 min at intestine. S120, 120 min at intestine; S240, 240 min at intestine. Reproduced from ref. 7 with permission from Elsevier, Copyright 2019.

7.2.2 Case Study 2: Almonds

Almonds were chosen to demonstrate the gastrointestinal fate of solid foods owing to their versatility, as they are consumed both raw and processed (heat processing and particle size reduction), and are also incorporated in many food products in various forms and particle sizes such as almond bits, almond powder, and almond paste. Further, many studies have investigated their digestion behavior and nutrient bioavailability such as mastication behavior, gastric protein hydrolysis, physical property changes, lipid digestion and absorption, and nutrient bioaccessibility and bioavailability.[13–16] Almonds are widely consumed because of their superior nutritional benefits and role in disease prevention, such as reducing the risk of heart disease and aiding in the prevention of obesity, weight gain, and type 2 diabetes.[17] They are a good source of proteins, lipids, dietary fiber, vitamins (*e.g.* vitamin E), phenolic compounds, phytosterols, and minerals.[17] The significant reason for the health benefits of consumption of almonds and other hard nuts, despite their high caloric value, is their microstructure (cellular matrix effect). During digestion, their intact cell wall acts as a physical barrier by encapsulating lipids and other nutrients and making them inaccessible for absorption. Further, the size, processing treatment, and microstructure of almond particles have a substantial effect on digestion kinetics and nutrient bioaccessibility and bioavailability, and other physiological processes within the GI tract (such as gut hormone responses).[13]

Various *in vitro* and *in vivo* studies have been carried out with almonds to understand their gastrointestinal fate. In brief, mastication during oral processing has an important role in the physical destruction of the cellular structure of almonds, which consequently affects the accessibility of enzymes during further digestion, bioaccessibility of lipids, postprandial lipidemia, and satiety.[18] The bolus formed at the end of oral processing consists of large multicellular almond tissue particles with a meager amount of free lipid droplets that are readily bioaccessible, and has major quantities of undamaged cells encapsulating intact oil bodies having low lipid bioaccessibility.[19,20] However, roasted almonds are extensively broken down into smaller particles with a greater extent of cell wall degradation during the oral processing stage compared with raw almonds.[21] Further disintegration of the cellular matrix occurs and intracellular nutrients are released during gastric digestion.

In the gastric phase also, raw almonds exhibited a slower breakdown rate than roasted almonds. The faster disintegration rate of roasted almonds was due to the enhanced accessibility of digestive enzymes to the nutrients (proteins and lipids) through the channels and voids (structural changes) developed in the cell wall and intracellular spaces as a result of roasting (heat treatment).[22] However, studies showed that the physical property changes, such as rheological properties, particle size distribution, textural attributes, gastric emptying rate, and subsequent lipid absorption, were similar for both the raw and roasted almonds, implying that almonds disintegrated in the gastric compartment at similar rates regardless of roasting treatment.[14,15] However, possible structural changes that occurred to the almond proteins during roasting could be the reason for the lower gastric emptying rate and protein hydrolysis of roasted almonds compared with raw almonds.[16] It has been demonstrated in *in vitro* studies that a substantial number of plant cells in almonds remain intact until the end of the duodenal stage of the small intestine where most of the lipid hydrolysis occurs.[23] Hence the rate and the extent of lipolysis in almonds are affected by the cell wall and its structural integrity, which acts as a physical barrier to lipase access. Recent studies suggest that there could also be other factors, such as internal constituent properties (lipid and protein bodies) and nut structure, which might influence intestinal lipolysis apart from cell wall intactness.[24]

7.3 Effects of Various Processing Methods on Nutrient Bioaccessibility and Bioavailability

Food processing treatments impact the food microstructure, which thereby has a significant role in the release and absorption of nutrients. The relationship between food processing, resulting food microstructural changes, and their consequence on nutrient release and absorption is a multifaceted subject that is mainly reliant on extrinsic factors (process-induced physical property changes or chemical changes) and intrinsic factors (physical and chemical properties of the food products and interactions among food components). Every processing treatment, such as pretreatment, cooking, boiling, and frying, has both positive and negative impacts on nutrient release and availability, depending on the treatment conditions, treatment time, and the impact on food structure.[25,26] For instance, blanching is a pretreatment that is usually applied to fruits and vegetables to inactivate enzymes responsible for browning. Blanching at low temperatures for a long time can degrade nutrients, lead to loss of soluble nutrients through leaching, and reduce the bioaccessibility of β-carotene due to a higher firmness of texture.[27] Nonetheless, proper processing treatment and conditions can significantly enhance nutrient release, bioaccessibility, and bioavailability.[28,29] Further, fermentation was reported to increase the bioaccessibility and bioavailability of nutrients such as minerals and vitamins by reducing antinutritional factors, such as phytates and tannins, which could interfere with nutrient bioavailability and bioaccessibility.[30,31] For instance, the lactic fermentation of bread dough improved the bioavailability of iron.[32]

Studies have been conducted to investigate the effect of food processing on the bioaccessibility and bioavailability of various functional compounds from plant-based foods, such as carotenoids and polyphenols, due to their functional health benefits.[33] The bioaccessibility of β-carotene from plants such as carrots, sweet potatoes, and spinach is mainly enhanced through the destruction of the cell wall structure by grinding and through thermal treatments.[34,35] Further, the bioaccessibility of lycopene is strongly enhanced by extensive meshing (mechanical treatment), cooking, heat, and enzymatic treatment.[36,37] It is worth noting that the bioavailability of lycopene from tomato sauce paste and purée is higher than that from raw tomato.[38] However, in the case of polyphenols, fruit peeling, dehulling (grains), cooking (leaching), and cellular breakdown due to griding (oxidative degradation) adversely affect polyphenol availability.[39,40]

In addition to processing effects, the bioavailability of nutrients can be affected by various other controlling factors, such as the following:

- Presence of other food compounds. For instance, the bioavailability of carotenoids may decrease in the presence of other unabsorbable fat-soluble compounds and also owing to their interactions with carotenoids.[41] However, in the case of phenolic compounds, the presence of soluble dietary fiber improved their stability and digestibility during duodenal digestion.[42]
- Type of food matrix. For example, carotene absorption is enhanced when it is incorporated in an oil or fat matrix compared with the carotenes in fruits and vegetables.[43]

Processing of foods is a key value additive step which, when applied appropriately, can be exploited to our advantage to increase the bioavailability and bioaccessibility of specific food nutrients, but it is crucial to have an in-depth understanding of how food processing alters the food matrix.

7.4 Conclusion

Food is an essential part of human life, providing all the nutrients required, and it is largely responsible for our health and wellbeing. There is a great demand for the development of nutritious, safe, palatable, and convenient foods owing to the fast-moving lifestyle of present-day humans. Processing of foods plays a crucial role in making food safe, nutritious, and shelf stable, with improved sensory properties. At the same time, food processing also alters the food microstructure, which has a huge impact on the digestion behavior of foods within the GI tract and nutrient release, bioaccessibility, and bioavailability. Food processing can have both positive and negative impacts on food digestion and nutrient absorption, which can be specific to the type of food, processing treatment, processing conditions, and end application. Hence an in-depth understanding of the interrelationships between the effects of processing on food structure at various levels and their impact on food properties and functionality is necessary. Therefore, in the future, more investigations related to the effect of food processing on the structure of foods and thereby on digestion behavior and nutrient absorption are highly desirable to help us identify appropriate processing techniques and conditions to design personalized food structures through which the digestion and absorption behavior of specific nutrients can be exploited to our advantage.

References

1. A. Ye, Gastric colloidal behaviour of milk protein as a tool for manipulating nutrient digestion in dairy products and protein emulsions, *Food Hydrocolloids*, 2021, **115**, 106599.

2. A. Acevedo-Fani, D. Roy, D. T. Do and H. Singh, Understanding food structure modifications during digestion and their implications in nutrient release, *Food Structure Engineering and Design for Improved Nutrition, Health and Well-Being*, Elsevier, 2023, pp. 277–314.
3. A. Ye, J. Cui, D. Dalgleish and H. Singh, The formation and breakdown of structured clots from whole milk during gastric digestion, *Food Funct.*, 2016, **7**(10), 4259–4266.
4. D. Roy, A. Ye, P. J. Moughan and H. Singh, Composition, structure, and digestive dynamics of milk from different species—A review, *Front. Nutr.*, 2020, **7**, 577759.
5. T. Huppertz and L. W. Chia, Milk protein coagulation under gastric conditions: A review, *Int. Dairy J.*, 2021, **113**, 104882.
6. A. Ye, J. Cui, D. Dalgleish and H. Singh, Formation of a structured clot during the gastric digestion of milk: Impact on the rate of protein hydrolysis, *Food Hydrocolloids*, 2016, **52**, 478–486.
7. A. Ye, W. Liu and J. Cui, *et al.*, Coagulation behaviour of milk under gastric digestion: Effect of pasteurization and ultra-high temperature treatment, *Food Chem.*, 2019, **286**, 216–225.
8. D. Roy, A. Ye, P. J. Moughan and H. Singh, Impact of gastric coagulation on the kinetics of release of fat globules from milk of different species, *Food Funct.*, 2021, **12**(4), 1783–1802.
9. D. Roy, A. Ye, P. J. Moughan and H. Singh, Structural changes in cow, goat, and sheep skim milk during dynamic *in vitro* gastric digestion, *J. Dairy Sci.*, 2021, **104**(2), 1394–1411.
10. A. Fardet, Food health potential is primarily due to its matrix structure, then nutrient composition: a new paradigm for food classification according to technological processes applied, *J. Nutr. Health Food Eng.*, 2014, **1**(5), 2.
11. D. I. Givens, Saturated fats, dairy foods and health: a curious paradox?, *Nutr. Bull.*, 2017, **42**(3), 274–282.
12. T. K. Thorning, H. C. Bertram and J.-P. Bonjour, *et al.*, Whole dairy matrix or single nutrients in assessment of health effects: current evidence and knowledge gaps, *Am. J. Clin. Nutr.*, 2017, **105**(5), 1033–1045.
13. M. M. L. Grundy, K. Lapsley and P. R. Ellis, A review of the impact of processing on nutrient bioaccessibility and digestion of almonds, *Int. J. Food Sci. Technol.*, 2016, **51**(9), 1937–1946.
14. G. M. Bornhorst, M. J. Roman, K. C. Dreschler and R. P. Singh, Physical property changes in raw and roasted almonds during gastric digestion *in vivo* and *in vitro*, *Food Biophys.*, 2014, **9**, 39–48.
15. G. M. Bornhorst, M. J. Roman, S. M. Rutherfurd, B. J. Burri, P. J. Moughan and R. P. Singh, Gastric digestion of raw and roasted almonds *in vivo*, *J. Food Sci.*, 2013, **78**(11), H1807–H1813.
16. G. M. Bornhorst, K. C. Drechsler, C. A. Montoya, S. M. Rutherfurd, P. J. Moughan and R. P. Singh, Gastric protein hydrolysis of raw and roasted almonds in the growing pig, *Food Chem.*, 2016, **211**, 502–508.

17. A. Kamil and C.-Y. O. Chen, Health benefits of almonds beyond cholesterol reduction, *J. Agric. Food Chem.*, 2012, **60**(27), 6694–6702.
18. B. A. Cassady, J. H. Hollis, A. D. Fulford, R. V. Considine and R. D. Mattes, Mastication of almonds: effects of lipid bioaccessibility, appetite, and hormone response, *Am. J. Clin. Nutr.*, 2009, **89**(3), 794–800.
19. G. Mandalari, M. M.-L. Grundy and T. Grassby, *et al.*, The effects of processing and mastication on almond lipid bioaccessibility using novel methods of *in vitro* digestion modelling and micro-structural analysis, *Br. J. Nutr.*, 2014, **112**(9), 1521–1529.
20. A. C. Creedon, E. S. Hung, E. Dimidi, T. Grassby, S. E. Berry and K. Whelan, Particle Size Distribution and Predicted Lipid Bioaccessibility of Almonds and the Effect of Almond Processing: A Randomised Mastication Study in Healthy Adults, *Nutrients*, 2023, **15**(3), 489.
21. G. Mandalari, R. M. Faulks and G. T. Rich, *et al.*, Release of protein, lipid, and vitamin E from almond seeds during digestion, *J. Agric. Food Chem.*, 2008, **56**(9), 3409–3416.
22. F. Kong and R. P. Singh, Digestion of raw and roasted almonds in simulated gastric environment, *Food Biophys.*, 2009, **4**, 365–377.
23. M. M. Grundy, F. Carrière, A. R. Mackie, D. A. Gray, P. J. Butterworth and P. R. Ellis, The role of plant cell wall encapsulation and porosity in regulating lipolysis during the digestion of almond seeds, *Food Funct.*, 2016, 7(1), 69–78.
24. B. McArthur and R. Mattes, Energy extraction from nuts: walnuts, almonds and pistachios, *Br. J. Nutr.*, 2020, **123**(4), 361–371.
25. X. Jin and X. D. Chen, Microstructure and its relationship with release behavior of different vehicles, *Food Microstructure and Its Relationship with Quality and Stability*, Elsevier, 2018, pp. 83–96.
26. S. Van Buggenhout, M. Alminger and L. Lemmens, *et al.*, In vitro approaches to estimate the effect of food processing on carotenoid bioavailability need thorough understanding of process induced microstructural changes, *Trends Food Sci. Technol.*, 2010, **21**(12), 607–618.
27. L. Lemmens, S. Van Buggenhout, I. Oey, A. Van Loey and M. Hendrickx, Towards a better understanding of the relationship between the β-carotene *in vitro* bio-accessibility and pectin structural changes: a case study on carrots, *Food Res. Int.*, 2009, **42**(9), 1323–1330.
28. P. Y. Nugrahedi, R. Verkerk, B. Widianarko and M. Dekker, A mechanistic perspective on process-induced changes in glucosinolate content in Brassica vegetables: A review, *Crit. Rev. Food Sci. Nutr.*, 2015, **55**(6), 823–838.
29. B. Hiranvarachat, S. Devahastin and N. Chiewchan, In vitro bioaccessibility of β-carotene in dried carrots pretreated by different methods, *Int. J. Food Sci. Technol.*, 2012, **47**(3), 535–541.
30. J. A. Adebo, P. B. Njobeh and S. Gbashi, *et al.*, Fermentation of cereals and legumes: Impact on nutritional constituents and nutrient bioavailability, *Fermentation*, 2022, **8**(2), 63.

31. M. Samtiya, R. E. Aluko, A. K. Puniya and T. Dhewa, Enhancing micronutrients bioavailability through fermentation of plant-based foods: A concise review, *Fermentation*, 2021, **7**(2), 63.
32. Z. Khodaii, M. N. Zadeh, J. Kamali and M. M. Natanzi, Enhanced iron absorption from lactic acid fermented bread (an *in vivo/ex vivo* study), *Gene Rep.*, 2019, **15**, 100389.
33. T. Bohn, G. J. McDougall and A. Alegría, *et al.*, Mind the gap—deficits in our knowledge of aspects impacting the bioavailability of phytochemicals and their metabolites—a position paper focusing on carotenoids and polyphenols, *Mol. Nutr. Food Res.*, 2015, **59**(7), 1307–1323.
34. C. L. Rock, J. L. Lovalvo, C. Emenhiser, M. T. Ruffin, S. W. Flatt and S. J. Schwartz, Bioavailability of β-carotene is lower in raw than in processed carrots and spinach in women, *J. Nutr.*, 1998, **128**(5), 913–916.
35. A. Bengtsson, C. Brackmann, A. Enejder, M. L. Alminger and U. Svanberg, Effects of thermal processing on the *in vitro* bioaccessibility and microstructure of β-carotene in orange-fleshed sweet potato, *J. Agric. Food Chem.*, 2010, **58**(20), 11090–11096.
36. E. A. Tibäck, C. A. Svelander and I. J. Colle, *et al.*, Mechanical and thermal pretreatments of crushed tomatoes: effects on consistency and *in vitro* accessibility of lycopene, *J. Food Sci.*, 2009, **74**(7), E386–E395.
37. L. Yonekura and A. Nagao, Intestinal absorption of dietary carotenoids, *Mol. Nutr. Food Res.*, 2007, **51**(1), 107–115.
38. I. Sensoy, A review on the relationship between food structure, processing, and bioavailability, *Crit. Rev. Food Sci. Nutr.*, 2014, **54**(7), 902–909.
39. J. Parada and J. Aguilera, Food microstructure affects the bioavailability of several nutrients, *J. Food Sci.*, 2007, **72**(2), R21–R32.
40. M. Van Boekel, V. Fogliano and N. Pellegrini, *et al.*, A review on the beneficial aspects of food processing, *Mol. Nutr. Food Res.*, 2010, **54**(9), 1215–1247.
41. K. H. van het Hof, C. E. West, J. A. Weststrate and J. G. Hautvast, Dietary factors that affect the bioavailability of carotenoids, *J. Nutr.*, 2000, **130**(3), 503–506.
42. N. Ortega, A. Macià, M.-P. Romero, J. Reguant and M.-J. Motilva, Matrix composition effect on the digestibility of carob flour phenols by an *in-vitro* digestion model, *Food Chem.*, 2011, **124**(1), 65–71.
43. D. Hornero-Méndez and M. I. Mínguez-Mosquera, Bioaccessibility of carotenes from carrots: Effect of cooking and addition of oil, *Innovative Food Sci. Emerging Technol.*, 2007, **8**(3), 407–412.
44. M. L. Rojas, M. T. Kubo, M. E. Caetano-Silva, G. R. Carvalho and P. E. Augusto, How food structure influences the physical, sensorial, and nutritional quality of food products, *Food Structure Engineering and Design for Improved Nutrition, Health and Well-Being*, Elsevier, 2023, pp. 113–138.

45. A. Ye, J. Cui, D. Dalgleish and H. Singh, Effect of homogenization and heat treatment on the behavior of protein and fat globules during gastric digestion of milk, *J. Dairy Sci.*, 2017, **100**(1), 36–47.
46. M. Oghbaei and J. Prakash, Effect of primary processing of cereals and legumes on its nutritional quality: A comprehensive review, *Cogent Food Agric.*, 2016, **2**(1), 1136015.
47. J. L. Slavin, D. Jacobs and L. Marquart, Grain processing and nutrition, *Crit. Rev. Food Sci. Nutr.*, 2000, **40**(4), 309–326.
48. G. Maiani, M. J. Periago Castón and G. Catasta, *et al.*, Carotenoids: actual knowledge on food sources, intakes, stability and bioavailability and their protective role in humans, *Mol. Nutr. Food Res.*, 2009, **53**(S2), S194–S218.
49. D. M. Barrett, J. C. Beaulieu and R. Shewfelt, Color, flavor, texture, and nutritional quality of fresh-cut fruits and vegetables: desirable levels, instrumental and sensory measurement, and the effects of processing, *Crit. Rev. Food Sci. Nutr.*, 2010, **50**(5), 369–389.
50. A. G. A. Sá, Y. M. F. Moreno and B. A. M. Carciofi, Food processing for the improvement of plant proteins digestibility, *Crit. Rev. Food Sci. Nutr.*, 2020, **60**(20), 3367–3386.
51. Z. F. Bhat, J. D. Morton, A. E.-D. A. Bekhit, S. Kumar and H. F. Bhat, Non-thermal processing has an impact on the digestibility of the muscle proteins, *Crit. Rev. Food Sci. Nutr.*, 2022, **62**(28), 7773–7800.
52. M. A. Dandago, Changes in nutrients during storage and processing of foods; a review, *Technol. Sci. Afr. J.*, 2009, **3**, 24–27.
53. Y. Zhu, X. Chen, D. J. McClements, L. Zou and W. Liu, pH-, ion-and temperature-dependent emulsion gels: Fabricated by addition of whey protein to gliadin-nanoparticle coated lipid droplets, *Food Hydrocolloids*, 2018, **77**, 870–878.
54. D. J. McClements, E. A. Decker and D. J. Weiss, Emulsion-based delivery systems for lipophilic bioactive components, *J. Food Sci.*, 2007, **72**(8), R109–R124.
55. L. de Souza Simões, D. A. Madalena, A. C. Pinheiro, J. A. Teixeira, A. A. Vicente and Ó. L. Ramos, Micro-and nano bio-based delivery systems for food applications: In vitro behavior, *Adv. Colloid Interface Sci.*, 2017, **243**, 23–45.
56. S. Mishra and J. Monro, Wholeness and primary and secondary food structure effects on *in vitro* digestion patterns determine nutritionally distinct carbohydrate fractions in cereal foods, *Food Chem.*, 2012, **135**(3), 1968–1974.
57. M. E. Dalmau, G. M. Bornhorst, V. Eim, C. Rosselló and S. Simal, Effects of freezing, freeze drying and convective drying on *in vitro* gastric digestion of apples, *Food Chem.*, 2017, **215**, 7–16.
58. G. Mandalari, Z. Merali and P. Ryden, *et al.*, Durum wheat particle size affects starch and protein digestion *in vitro*, *Eur. J. Nutr.*, 2018, **57**, 319–325.

59. L. Kaur, T. Astruc and A. Vénien, *et al.*, High pressure processing of meat: Effects on ultrastructure and protein digestibility, *Food Funct.*, 2016, **7**(5), 2389–2397.
60. Z. Bhat, J. D. Morton, S. L. Mason and A. E.-D. A. Bekhit, Pulsed electric field improved protein digestion of beef during *in-vitro* gastrointestinal simulation, *LWT*, 2019, **102**, 45–51.
61. J. M. Guevara-Zambrano, S. Verkempinck and L. Hernandez-Ruiz, *et al.*, Digestion kinetics of lipids and proteins in plant-based shakes: Impact of processing conditions and resulting structural properties, *Food Chemistry*, 2022, **382**, 132306.

CHAPTER 8

Designing Foods with New Food Ingredients

8.1 Introduction

In recent years, rational food design has become crucial in the field of food science and technology to cater for the specific needs and requirements of consumers.[1] Consumers have become increasingly aware of their food choices and their implications for their health and the environment. Further, a paradigm shift in consumer behavior to consume healthier food products is due to their awareness that unhealthy foods and diets contribute significantly to the incidence of various non-communicable diseases and health conditions such as cardiovascular diseases, type 2 diabetes, obesity, high blood pressure, *etc.*[2] Governments and health organizations together have developed and framed policies and regulations to regulate food manufacturers to produce healthier food products through modification of their composition by reducing some ingredients of concern (*e.g.* salt, sugar, and fat) and addition of ingredients with proven health benefits (*e.g.* fiber, proteins, micronutrients, and bioactives).[3,4]

In addition, the food fraternity is also in another critical situation, requiring it to ensure food security (both macronutrients and micronutrients) for the rising global population, which is forecast to reach 9.1 billion by 2050, and also to minimize the impact of food systems on the environment and climate change.[5] Many researchers have cautioned that we need to shift immediately to more sustainable yet nutritious alternative food sources to alleviate the environmental damage and resource depletion caused by various food production and processing activities.[6–8] In addition, many food researchers have also predicted that it will be

Food Chemistry, Function and Analysis No. 42
Food Digestion and Absorption: Its Role in Food Product Development
By C. Anandharamakrishnan, Jeyan Arthur Moses and S. Priyanka

Published by the Royal Society of Chemistry, www.rsc.org

strategies that can be employed with regard to food reformulation include the following:

- educating, informing, and creating awareness among consumers;
- implementing nutrition profiling for fast foods;
- influencing food preferences of children through awareness and promotions;
- "silent" reformulation approaches for retail food products;
- stringent policies with respect to nutrition labels and use of claims;
- incentives and tax exemptions for healthy food retailing, manufacturing, and services; and
- restricting food advertising and commercial promotion.[17,20,21]

Various countries around the globe have implemented one or more of these strategies effectively and the details can be found elsewhere in greater depth.[3,4,22–24]

8.3 Reduction of Sugars in Food Products

Sugars are naturally present in various foods such as honey, milk, fruits, and vegetables and they provide a characteristic sweet taste and energy upon consumption. Naturally sweet and sugary foods (honey and fruits) have been part of the human diet for centuries. The "sugar" that we are referring to and concerned about reducing here is the refined versions of sugar (such as sucrose, glucose, and fructose) that are added to processed food products in various forms, such as crystals, fine powder, and liquids. In recent times, excessive and frequent consumption of refined sugar and processed food products containing refined sugar has led to many health issues, *e.g.* weight gain, obesity, type 2 diabetes, and cancers.[25,26] Mechanisms underlying key health problems developed as a result of the consumption of refined sugar are illustrated in Figure 8.1. The World Health Organization (WHO) has recommended limiting the intake of free sugars to less than 10%, and if possible less than 5%, of the total energy intake for a healthy lifestyle.[27]

Sugars are not only added to food products just to confer a sweet taste, they also play a significant technological role as a preservative (reducing water activity), bulking agent, dispersant, texturizer, fermentation substrate, and moisturizer, and further they increase viscosity, improve shelf-life, act as stabilizers and flavor carriers, inhibit protein coagulation, and act as browning and decorative agents.[26,28] The reduction/removal of sugar from a food formulation demands the addition of other preservatives to maintain the food's shelf-life and safety and the inclusion of other food additives to compensate for the sensorial attributes and functionality offered by sugar. However, the addition of preservatives and additives may sometimes do more harm than good, hence it is wise to weigh up the pros and cons of sugar reduction in a selected product and the level of reduction necessary. For instance, sugar usually acts as a bulking agent in various

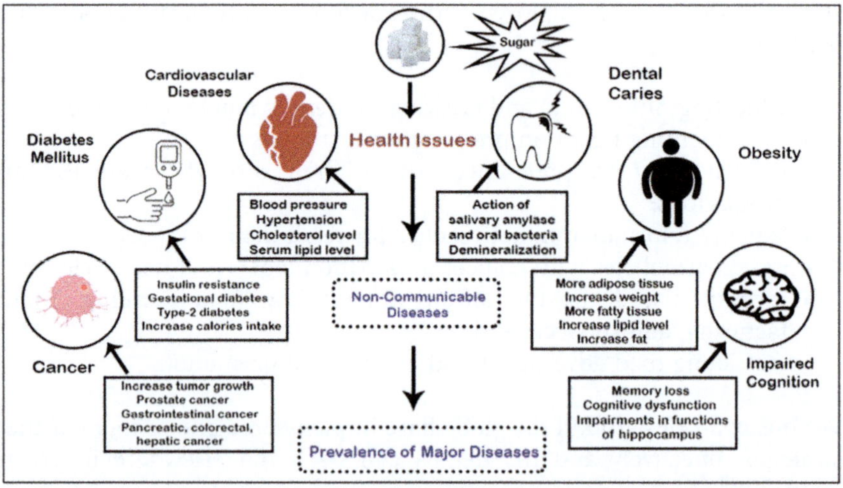

Figure 8.1 Mechanisms underlying key health problems developed due to the consumption of refined sugar. Reproduced from ref. 26, https://doi.org/10.1016/j.heliyon.2022.e10711, under the terms of the CC BY 4.0 license, http://creativecommons.org/licenses/by/4.0/.

foods, but then it is usually replaced with maltodextrin, which also provides the same calories as sugar, thus negating the purpose of replacing the sugar.[29]

Common strategies used to reduce/replace sugars in food products are as follows:

(i) non-calorific artificial sweeteners (sucralose, aspartame, and saccharin);[25,30]
(ii) non-calorific natural sweeteners (glycyrrhizin, stevioside, and thaumatin);[31–34]
(iii) low-calorie natural sugar alcohols/polyols (sorbitol, xylitol, mannitol, erythritol, maltitol, isomalt, and lactitol);[35,36]
(iv) functional alternative sugar sources such as inulin (prebiotic) and arabinoxylan oligosaccharides (dietary fiber);[37–39]
(v) novel and clean-label sweetening ingredients (such as whey permeate, polydextrose, apple pomace, and oligofructose);[40]
(vi) other alternative sugar sources (such as date palm, jaggery, honey, beetroot, and coconut);[26,41]
(vii) size reduction of sugar particles;[42,43]
(viii) stealth approach: gradual and progressive reduction of sugar to shift consumer preference to low sugar levels;
(ix) enhancing sweetness perception by microstructural modifications of foods such as texture modification through serum release and change of breakdown properties,[44,45] and inhomogeneous spatial distribution of sugar;[46–48] and

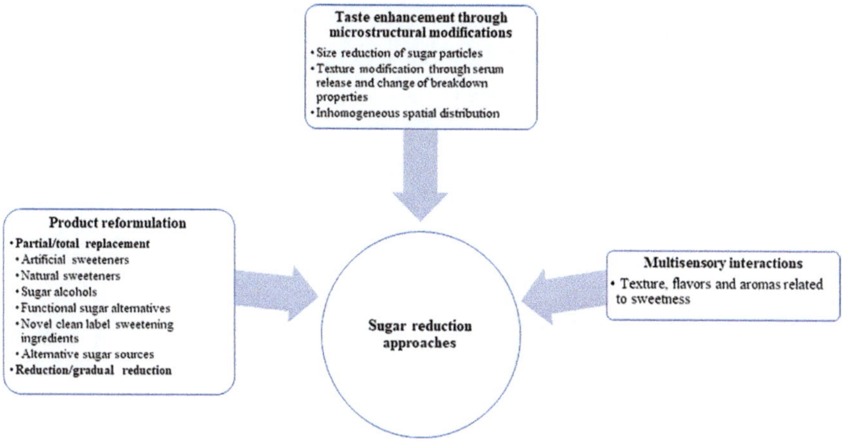

Figure 8.2 Various sugar reduction approaches. Reproduced with modifications from ref. 111 with permission from Elsevier, Copyright 2018.

(x) the use of multisensory interactions such as texture, flavors, and aromas related to sweetness enhances the sweet intensity due to cross-modal interactions.[49–51]

These strategies are illustrated in Figure 8.2.

8.4 Reduction of Sodium/Salt in Foods

Salt (NaCl) in food is the major contributor (90%) of sodium to the human body.[52] Excessive salt/sodium consumption plays a significant role in increasing blood pressure and urinary secretion of calcium, blood vessel damage, altering hormonal and inflammatory pathways, and affecting the immune system and metabolism. All of these physiological risk factors are associated with the development of various disease conditions such as cardiovascular disease, chronic kidney disease, brain damage (stroke and dementia), gastrointestinal diseases (stomach cancer and gut dysbiosis), bone disorders, and metabolic disorders.[21] Excess dietary salt intake leading to chronic diseases and organ damage is illustrated in Figure 8.3. The WHO has recommended a maximum consumption limit for salt (NaCl) of 5 g per day, consisting of no more than 2300 mg of sodium.[53] Further, education and awareness of growing health concerns due to excessive sodium/salt intake have compelled the public and food manufacturers to take progressive steps to lessen salt consumption and salt content in food formulations, respectively.[12] Also, enforcement of stringent salt-reduction policies and programs from the government side on food manufacturers will play a vital role in accomplishing the target goal of a 30% reduction in salt intake.[54]

Salt plays a multifunctional role in a food system. Predominantly, it acts as a tastant, providing a characteristic salt taste. In addition, it also plays various functional roles in food processing (meat, bakery, and dairy products),

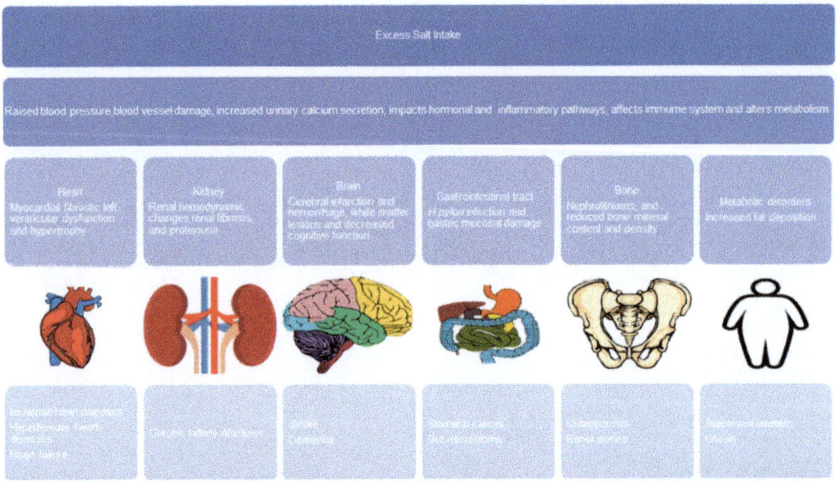

Figure 8.3 Excess dietary salt intake leading to chronic diseases and organ damage. Reproduced from ref. 21 with permission from Elsevier, Copyright 2022.

preservation, and sensory aspects of foods and it is crucial in several physiological roles and for the proper functioning of the human body.[21,55–57] The role of salt in food and health is illustrated in Figure 8.4. Various technological approaches have been investigated to reduce salt levels in foods: (i) salt substitutes (such as potassium chloride, magnesium chloride, magnesium sulfate, calcium chloride, and ammonium chloride); (ii) taste potentiators (such as monosodium glutamate, acids, amino acids, yeast extracts, edible seaweed, whey peptides and plant-based compounds such as herbs and spices); (iii) food structural design approaches (such as particle size reduction, optimization of salt shape, and controlling the spatial distribution and encapsulation of salt) and the driving forces for enhanced salt perception and salt reduction as illustrated in Figure 8.5; (iv) non-thermal techniques [such as high-pressure processing (HPP), pulsed electric field (PEF), and ultrasonication (US)]; (v) addition of filler materials (such as emulsions, air bubbles, and solid particles); (vi) enhancing salt taste through multisensory perception; and (vii) shifting consumer preferences. Table 8.1 summarizes the various mechanisms of action and methods of implementation of these salt reduction approaches.[21] Further, it is noteworthy that the methods of implementation of these salt reduction strategies vary depending on the product type, processing methods, and presence of other ingredients.

8.5 Reduction of Fats in Food Products

Fats and oils are an essential part of our daily diet. Additionally, they also are widely used in food formulations owing to their functionality, nutritional value, and the sensory attributes that they can offer such as taste, flavor, and

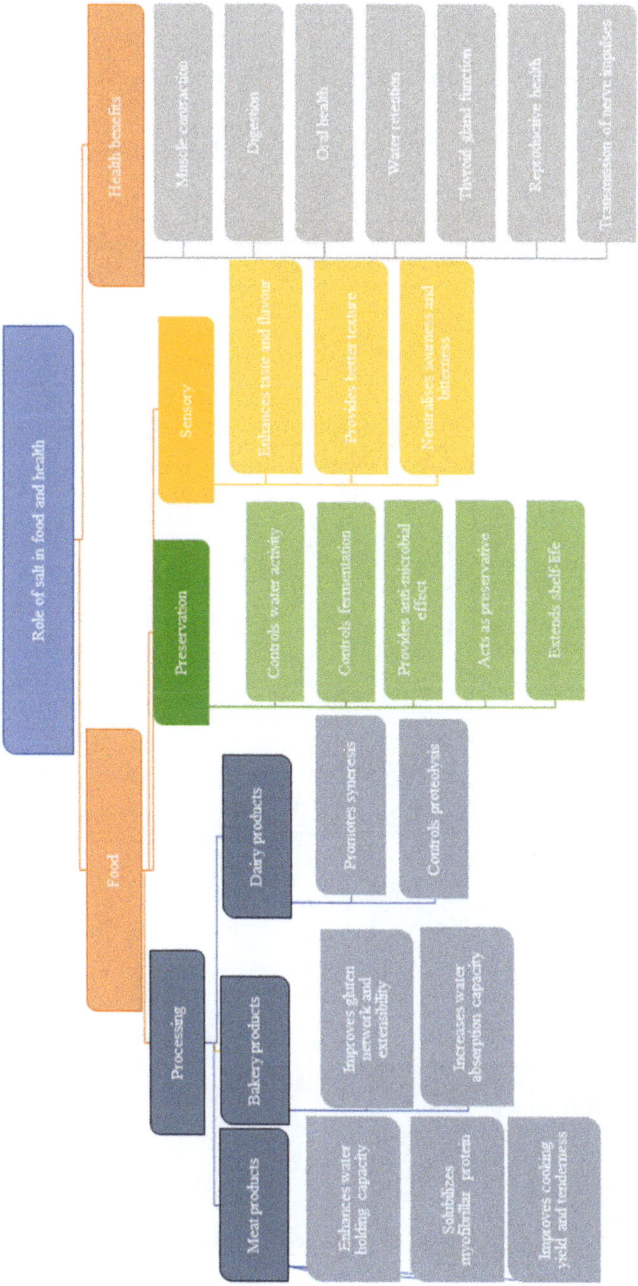

Figure 8.4 Role of salt in food and health. Reproduced from ref. 21 with permission from Elsevier, Copyright 2022.

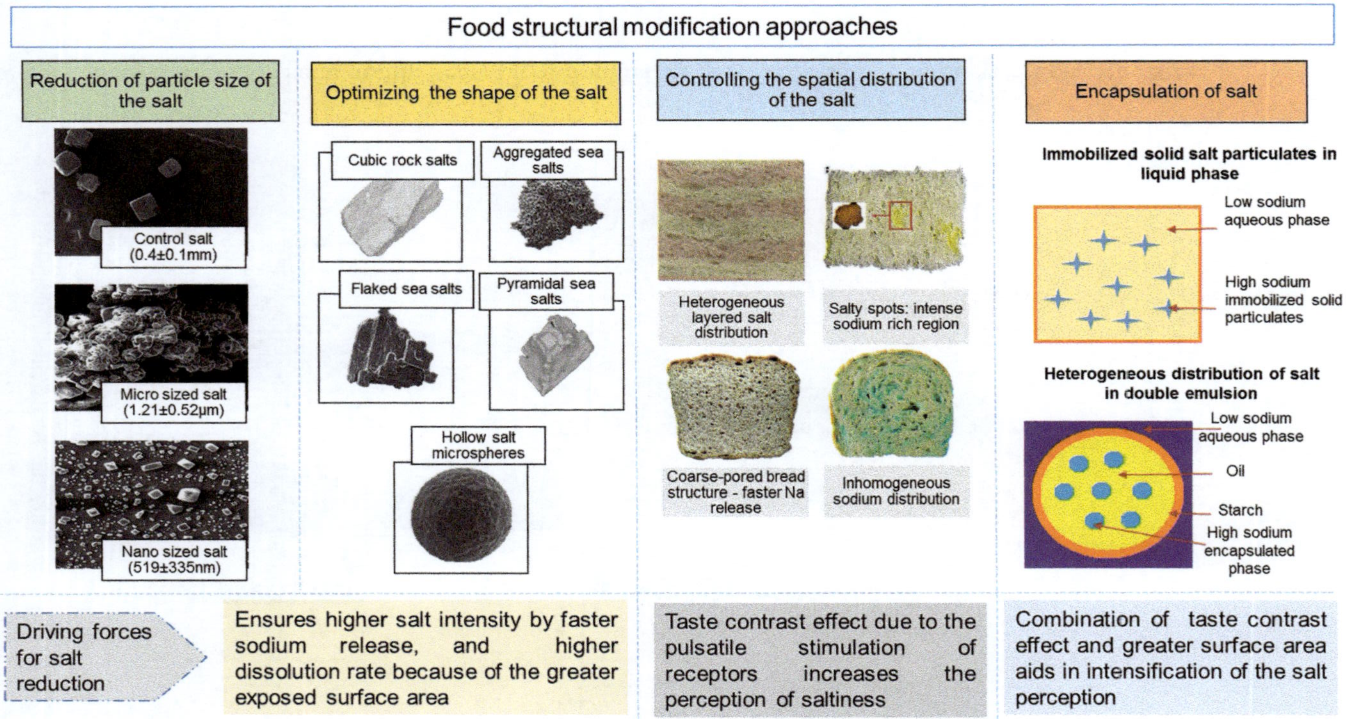

Figure 8.5 Food structural design approaches as driving forces for enhanced salt perception and salt reduction. Reproduced from ref. 21 with permission from Elsevier, Copyright 2022.

Table 8.1 Approaches to reduce the salt content in foods: mechanism of action and methods to implement. Reproduced from ref. 21 with permission from Elsevier, Copyright 2022.

Salt reduction approach	Mechanism of action	Methods to implement	Ref.
Food structural design	Delivery of Na^+ ions from the food matrix to the taste receptor is substantially altered by modification and optimization of product structure	Altering the size, shape, and morphology of the salt crystal accordingly	112–114
		Heterogeneous spatial distribution	115 and 116
		Encapsulation of salt crystals	117 and 118
Effect of filler materials	Diffusion of tastants for perception is aided by the addition of a variety of filler materials, which maximizes the diffusion and transport of Na^+ ions to taste receptors	Adding fillers such as emulsions, air bubbles, and solid particles in combination with salt	119–121
Salt alternatives	Through biochemical physiology, substituting other salty-tasting substances for salt improves the perception of salt taste	By substituting different salts such as KCl, $MgCl_2$, NH_4Cl, $CaCl_2$, *etc.* for NaCl	122 and 123
Taste potentiators	Taste potentiators are added to foods to improve the magnitude of the salt taste perception by stimulating taste receptors related to salt and umami taste.	Through the use of flavoring agents such as monosodium glutamate, hydrolyzed vegetable protein, 5′-adenosine monophosphate, edible seaweed	124 and 125
	They also help to suppress and cover up other unpleasant tastes, such as bitterness	Through the use of herbs, spices, and by-products from winemaking	126 and 127
		Through the use of acidic ingredients such as organic acids, amino acids, and dibasic acids	128 and 129
Emerging technologies	The perception of saltiness is enhanced through increasing diffusion, gain, distribution, and release rate of salt in the mouth	Through applying non-thermal techniques such as HPP, PEF, and US	130–132
		Through merging various other approaches with non-thermal techniques	133 and 134
Shifting consumer preferences	Educating and creating awareness among consumers about the ill effects of excess salt consumption will result in reduced salt consumption and a shift in consumers' preference towards low-salt products	Through improving public knowledge about the negative effects of a high-salt diet on health	2 and 135
		Employing a stealth reduction approach *via* product reformulation (with lower salt content)	136 and 137

Table 8.1 (*Continued*)

Salt reduction approach	Mechanism of action	Methods to implement	Ref.
Salt perception in food systems	The utilization of multisensory perception of taste–taste and taste–aroma interactions enhances the salty taste and compensates for the salt reduction in products	By incorporating various flavoring and aromatic substances	138–140
		Adding flavoring agents with salts to mask the metallic aftertaste and unpleasant bitter taste	127 and 141

textural effects. Saturated fats are present in large amounts in foods of animal origin such as meat and meat-based products and dairy products such as cheese, butter, and ice-cream. Further, plant-based foods such as nuts, seeds and vegetable oils (sunflower, corn, olive, and canola oil), and fatty fish are rich in monounsaturated and polyunsaturated fats. Certain food groups, such as meat products, spreads, bakery products, dairy products, cheese, and convenience foods, are considered priority food categories for reformulation to reduce saturated fatty acids and *trans*-fatty acids.[12]

In the case of lipids and fats, high intakes of *trans*-fatty acids (TFAs) and saturated fatty acids (SFAs) from food products have been well correlated with increased risk of cardiovascular diseases and stroke.[58] The consumption of TFAs and SFAs was reported to have negative effects on blood lipoprotein profiles such as increasing the plasma concentration of low-density lipoprotein (LDL) cholesterol and lipoprotein (a), reducing the concentration of high-density lipoprotein (HDL) cholesterol, and promoting endothelial dysfunction, inflammation, and insulin resistance.[59,60] Additionally, another major source of oil in the diet is deep-fat fried foods, and it is desirable to avoid these foods. Various strategies have been studied to reduce the oil content in deep-fat fried foods, such as coating with bread and batter to reduce oil absorption, modification of frying oil's viscosity and temperature (oil absorption is a minimum at low viscosity and high temperature), and altering the frying technique, such as pre-frying, vacuum frying, microwave heating, de-oiling after frying, and air frying.

The WHO has recommended reducing the intake of TFAs to less than 1% and SFAs to less than 10% of the total energy intake.[61] Based on WHO recommendations and due to national policies, the TFAs that are produced in foods during industrial processing have been effectively eliminated by the replacement of partially hydrogenated vegetable oil, which is the main source of TFAs in processed foods, with alternative fats.[62] Further, the WHO has given guidelines to reduce the SFAs in foods using polyunsaturated fatty acids, monounsaturated fatty acids, polysaccharides, and proteins, with a caution not to carry out any replacement with other ingredients of public health concern (such as free sugars).[61] However, there are many technical issues with the replacement of saturated fats with unsaturated fats, such as poor textural characteristics (spreadability, tenderness, mouthfeel, bite, juiciness), flavor and rancidity, palatability, appearance, and hence reduced shelf-life and product quality. Therefore, the approach used for fat replacement and reduction should be robust in terms of nutritional value and should provide relevant structural and sensorial attributes.

In recent times, several fat reduction/replacement approaches have been extensively investigated and can be broadly classified into three categories:

(i) Replacing animal fats with vegetable oils and fish oils to improve the ratio of unsaturated to saturated FAs.[63]
(ii) Fat substitutes. These have chemical structures that closely resemble those of fats, exhibiting similar physicochemical properties to fats but usually low in calories or indigestible, such as olestra, caprenin, and salatrim.[64,65]
(iii) Fat mimetics. These have distinct chemical structures compared with fats. They can simulate certain sensory attributes of fats such as appearance, mouthfeel, and viscosity. Fat mimetics are usually hydrocolloids such as a variety of proteins and polysaccharides with different structuring mechanisms,[66,67] namely:
    (a) Indirect oil structuring *via* colloidal templates (emulsion and foam templates) and solvent exchange (protein hydrogels and carrageenan aerogels).
    (b) Oleogelation. Structuring oil using non-lipidic gelators (*e.g.* ethylcellulose and chitin) and lipidic gelators such as monocomponent gels (*e.g.* partial glycerides, natural waxes, fatty alcohols, and hydroxylated fatty acids) and mixed gels (*e.g.* fatty acid + a fatty alcohol, sitosterol + oryzanol, lecithin + tocopherols, and lecithin + sorbitan tristearate).
    (c) Structured biphasic systems (emulsion gels, gelled emulsions, and structured emulsions).

In addition, based on the composition, fat mimetics can be classified as carbohydrate based (mostly polysaccharides) and protein based (usually animal or plant proteins). Further, the effect of fat reduction and replacement has been investigated for various food products such as cakes, cookies and biscuits, muffins, laminated pastries, cheese, butter and spreads, ice-creams, yogurt, milk and flavored milk, meat products, and salad dressings and sauces using one or more of the above-mentioned fat reduction/replacement strategies, which are discussed in detail elsewhere.[68–72]

8.6 Development of Low Glycemic Index Foods

There is a need to develop low glycemic index (GI) foods due to the influence of the glycemic response of foods on diabetes. Diabetes is a chronic disease that is characterized by persistent elevated blood glucose levels. Diabetes treatment recommends controlling the diet through portion control and monitoring the nutritional content of meals consumed to maintain blood sugar levels in the normal range.[73] It is also reported that a high glycemic load leads to insulin resistance, glucose intolerance, and diabetes. Further, consumption of low-GI foods has been shown to provide various health benefits, such as reducing blood glucose levels and type 2 diabetes, reducing the risk of various cancers, helping with managing obesity, lowering the risk of heart diseases, managing cholesterol, and reducing age-related macular degeneration and polycystic ovarian syndrome.[74]

Amongst macronutrients, carbohydrates affect the blood glucose level the most, as they are broken down to glucose, which is then rapidly absorbed, increasing the blood glucose level.[75] Further, not all carbohydrates are equal when it comes to digestion, as some are rapidly digested and absorbed (simple sugars) compared with complex carbohydrates. Based on the *in vitro* digestion time, starches can be classified into three main types, namely (i) rapidly digestible starch (RDS) (if it is digested within 0–20 min), (ii) slowly digestible starch (SDS) (if it is digested within 20–120 min), and (iii) resistant starch (RS) (if it remains undigested even after 120 min).[76] The utilization of starch-based ingredients with higher proportions of SDS, RS, and fiber can reduce the GI of food products and provide other health benefits such as gut health by facilitating the growth of beneficial microorganisms in the colon.[77] Nevertheless, RDS in foods will increase blood glucose levels if ingested in large quantities.

The glycemic response (change in blood glucose concentration) upon consumption of different food products with the same amount of carbohydrates varies. The concept of the GI of food was established to classify different carbohydrate-based food sources depending on their impact on postprandial glycemia. The GI of food is estimated by comparing the postprandial blood glucose level of food against reference foods such as pure glucose (GI = 100).[78] Based on the glycemic response, foods are classified into low glycemic foods (GI ≤ 55), medium glycemic foods (GI = 56–69), and high glycemic foods (GI ≥ 70).[79] Knowledge of the GIs of different carbohydrate sources can be useful when choosing ingredients for food formulations as the GI provides a measure of carbohydrate quality. However, the total glycemic response of a food product is dependent on the quality of carbohydrates present in the food and the quantity of food consumed. To address this, the concept of glycemic load (GL) was introduced, which considers the GI of the food and the amount of available carbohydrates in the quantity of food consumed. Both GI and GL play significant roles in determining the glycemic response of foods. Based on GL, foods can be classified as low (GL ≤ 10), medium (GL > 10 to <20), and high (GL ≥ 20).[80]

Various factors influence the GI and glycemic response of a food, such as the nature and complexity of carbohydrates, food processing treatments, food structure, presence of other nutrients (proteins, fats, polyphenols, and bioactives), presence of fiber and monosaccharides, starch composition, maturity levels in fruits and vegetables, the acidity of foods, enzymes, food digestion stages (oral, gastric, and small intestine), chyme viscosity and gastric emptying rate, and rate of digestion and absorption of carbohydrates.[74,81] Figures 8.6 and 8.7 illustrate the influence of various factors on the GI of food products and the impact of food digestion stages on glycemic response, respectively.

Various strategies have been investigated and used in the development of low-GI foods, as follows:

(i) Using non-digestible carbohydrates:
   (a) Non-starch polysaccharides such as whole grains with bran, higher fiber content,[82,83] β-glucan,[84,85] guar gum,[86] and fructooligosaccharides.[87]
   (b) Using ingredients with higher quantities of resistant starch.[88,89]
(ii) Using low-GI pseudo-cereal ingredients such as amaranth, buckwheat, and quinoa,[90] and legumes such as chickpeas, lentils, and peas.[91]
(iii) Increasing the protein, lipid, and polyphenol content of foods.[92]
(iv) Using artificial sweeteners.[93,94]
(v) Addition of ingredients containing bioactives with antidiabetic properties and/or enzyme-inhibiting capacities such as fenugreek, pomelo, seaweed, mangiferin, and cinnamon.[74,95]
(vi) Modification of starch using physical, chemical, and enzymatic methods to obtain high RS and SDS fractions.[73,75]
(vii) Certain processing techniques play a significant role in the reduction of the GI of food products, such as fermentation with lactic acid bacteria (increased RS and production of exopolysaccharides),[96,97] microwave cooking of basmati rice,[98] and autoclaving, boiling, parboiling, and extrusion cooking (increase RS).[73,99]
(viii) Modification of the processing conditions also reduces the GI of foods through various mechanisms. For instance, in the case of bread, a longer baking time at low temperatures hinders starch gelatinization and facilitates the formation of RS, and steaming of bread instead of baking forms a compact bread structure that inhibits the accessibility of amylase.[100]

8.7 Redesigning Foods with Novel and Alternative Food Sources and Ingredients

In recent years, great efforts in terms of research, policy, and regulations have been made to redesign foods with sustainable, novel, and alternative ingredients. The key aim was to make future food systems more robust and sustainable, and various novel and alternative ingredients have been widely explored recently. Especially plant-based ingredients have been studied in detail and recommended as alternatives to replace animal-based products (such as milk, eggs, meat, and fish) and those products which require animal-based ingredients (cheese, dressings, sauces, spreads, and yogurts). Plant-based ingredients can be broadly classified into plant-based protein

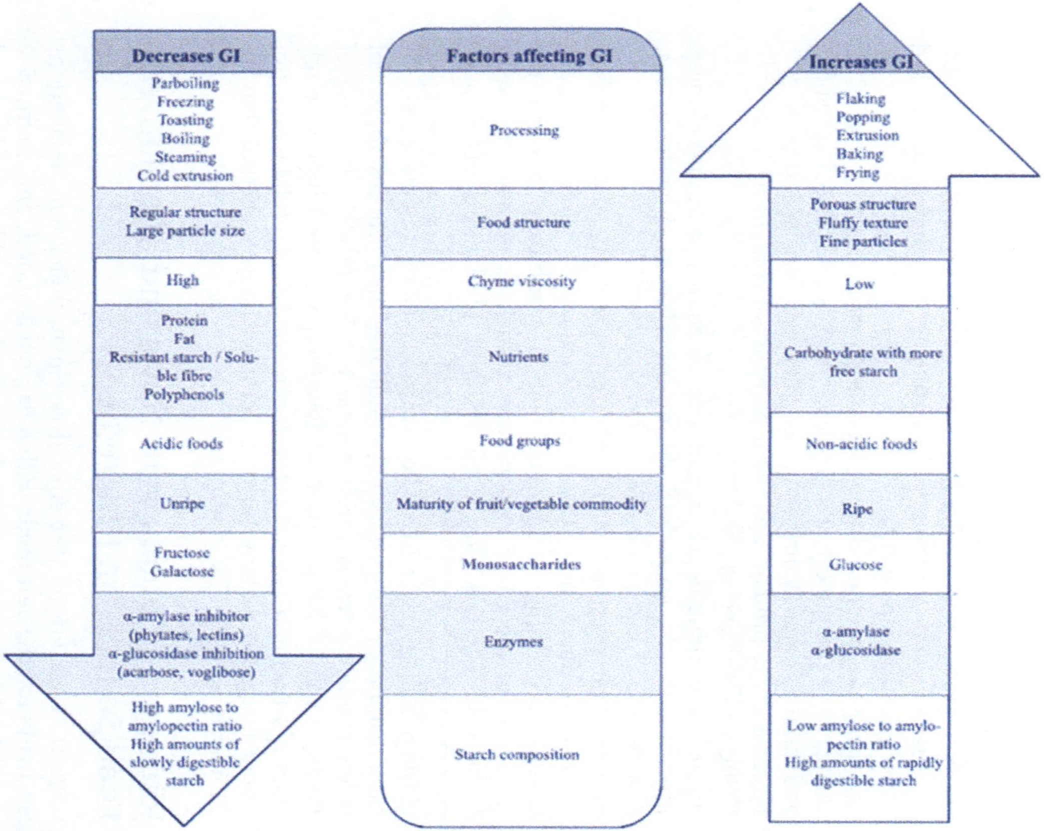

Figure 8.6 Factors that influence the GI of food products. Reproduced from ref. 81 with permission from Cambridge University Press, Copyright 2022.

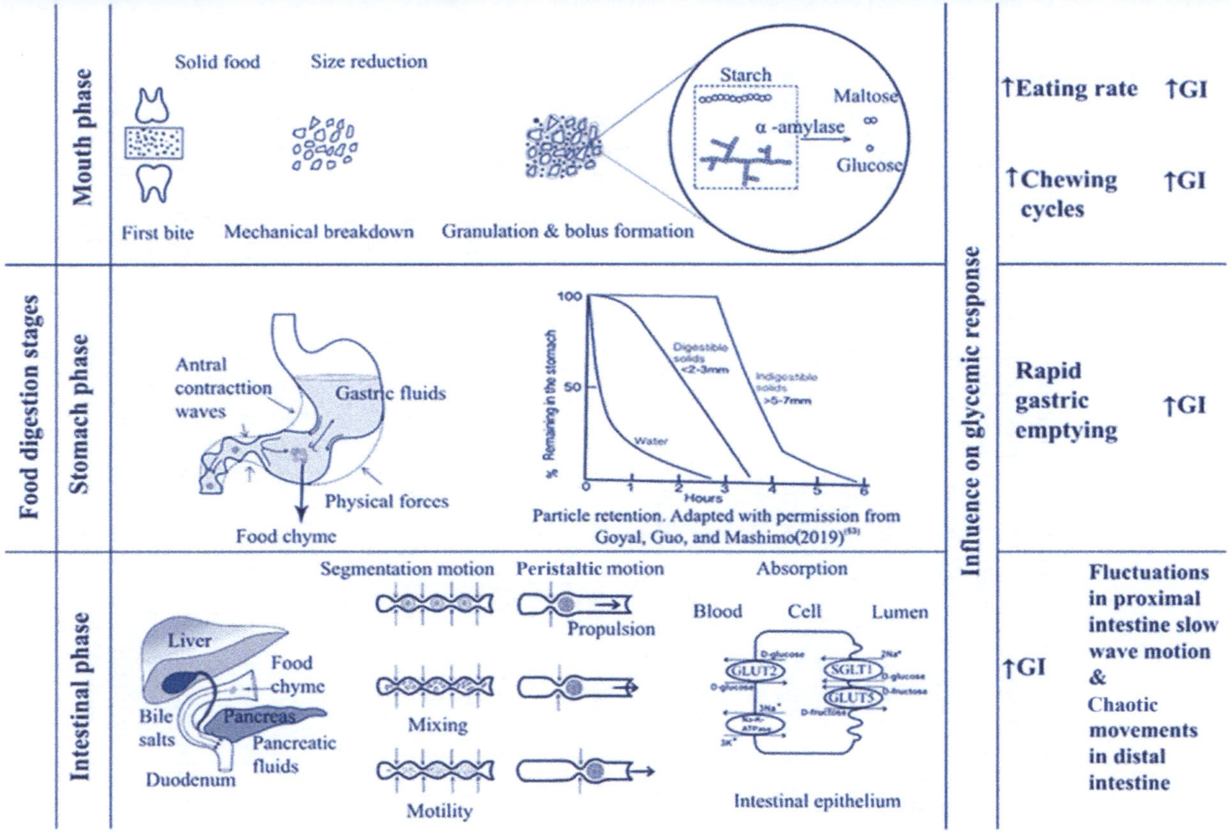

Figure 8.7 Impact of food digestion stages on the glycemic response. The reference cited near the center relates to the original publication. Reproduced from ref. 81 with permission from Cambridge University Press, Copyright 2022.

(*e.g.* derived from soybeans, faba beans, lentils, mung beans, and peas), plant-based carbohydrates (such as sugars, oligosaccharides, polysaccharides, and fibers), plant-based lipids (*e.g.* derived from soybeans, sunflowers, coconuts, flaxseeds, canola, corn, olives, palm, safflower, and cocoa), and other additives such as preservatives, buffers, pigments, flavors, and cross-linking agents.[9]

Further, other novel sustainable ingredients that have been extensively investigated include the following:

- By-products of food manufacturing industries such as dairy processing by-products (whey, whey protein, milk permeate, buttermilk, *etc.*),[101] by-products from cereals, pulses, nuts, fruits, and vegetables,[102] and by-products from the egg, meat, poultry, fish, and animal processing industries.[103]
- Edible insects such as grasshoppers, locusts, crickets, beetles, cockroaches, bees, wasps, ants, termites, flies, and caterpillars.[104,105]
- Algae such as macroalgae (seaweeds) and microalgae. They are very good sources of polysaccharides, polyunsaturated fatty acids, proteins, dietary fiber, amino acids, pigments, antioxidants, vitamins, and minerals.[106–108]

Certain critical issues need to be addressed when redesigning foods with novel alternative ingredients, such as the following:

- Consumer-based hurdles: Consumer attitudes towards novel ingredients such as plant-based meats, insect proteins, and seaweed must be confronted. There is a need to create awareness and educate consumers to choose sustainable products and we also require more knowledge and tools to understand consumer behavior towards novel food sources and ingredients.[109]
- Technological difficulties: Structuring complex foods (whole meat muscles, fish, yogurt, and cheese) is still challenging, there is a lack of high-quality novel alternative ingredients in sufficient quantities, and a deep understanding of the properties of alternative ingredients and their structure–function relationships is required for the reformulation and redesign of foods with novel alternative ingredients.[110]
- Commercial-based hurdles: There is a lack of knowledge about novel alternative ingredients with respect to processing, manufacturing, relative advantages and disadvantages among various novel ingredients, supply chain issues, safety concerns, and regulations in different countries, which hinders the successful commercialization of foods made with novel ingredients.[108]

8.8 Conclusion

Consumer demand for ethical and sustainable foods, animal cruelty-free and clean-label products with pleasant sensorial attributes (such as texture and taste), and improved health and nutrition potential is increasing exponentially. In order to satisfy the ever-expanding consumer needs and global environmental demands (sustainable, smaller carbon footprint, less water, and land requirements), food scientists and manufacturers must exploit the versatile multidisciplinary approach of rational food design (*i.e.* logical and science-based planning for product conceptualization) to deliver the target psychophysical, nutritional, and health functionalities utilizing green technologies and sustainable approaches. In this chapter, we have summarized the scientific and technological aspects employed in food reformulation and redesign and various strategies employed in the reduction of ingredients of concern (salt, sugar, and fat) and the development of low-GI foods, and also briefly considered the utilization of novel alternative ingredients from plants, insects, algae, and by-products from food manufacturing to design and develop sustainable food systems.

References

1. J. M. Aguilera, Rational food design and food microstructure, *Trends Food Sci. Technol.*, 2022, **122**, 256–264.
2. N. Belc, I. Smeu, A. Macri, D. Vallauri and K. Flynn, Reformulating foods to meet current scientific knowledge about salt, sugar and fats, *Trends Food Sci. Technol.*, 2019, **84**, 25–28.
3. M. Gressier, F. Sassi and G. Frost, Healthy foods and healthy diets, How government policies can steer food reformulation, *Nutrients*, 2020, **12**(7), 1992.
4. M. Gressier, B. Swinburn, G. Frost, A. B. Segal and F. Sassi, What is the impact of food reformulation on individuals' behaviour, nutrient intakes and health status? A systematic review of empirical evidence, *Obes. Rev.*, 2021, **22**(2), e13139.
5. S. Naheed, An overview of the influence of climate change on food security and human health, *Life*, 2023, **3**, 15.
6. D. Knorr and M. Augustin, From value chains to food webs: The quest for lasting food systems, *Trends Food Sci. Technol.*, 2021, **110**, 812–821.
7. E. M. Berry, Sustainable food systems and the Mediterranean diet, *Nutrients*, 2019, **11**(9), 2229.
8. R. Gormley, Shaping the Production of Sustainable, Healthy Foods for the Future, *Trends Food Sci. Technol.*, 2023, **134**, 77–79.
9. D. J. McClements and L. Grossmann, A brief review of the science behind the design of healthy and sustainable plant-based foods, *npj Sci. Food*, 2021, **5**(1), 17.
10. A. M. Salter and C. Lopez-Viso, Role of novel protein sources in sustainably meeting future global requirements, *Proc. Nutr. Soc.*, 2021, **80**(2), 186–194.
11. R. Tso, A. J. Lim and C. G. Forde, A critical appraisal of the evidence supporting consumer motivations for alternative proteins, *Foods*, 2020, **10**(1), 24.

12. World Health Organization, *Reformulation of food and beverage products for healthier diets: policy brief*, World Health Organization, 2022.
13. P. Strazzullo, L. D'Elia, N.-B. Kandala and F. P. Cappuccio, Salt intake, stroke, and cardiovascular disease: meta-analysis of prospective studies, *BMJ*, 2009, **339**, b4567.
14. R. P. Mensink and W. H. Organization, *Effects of saturated fatty acids on serum lipids and lipoproteins: a systematic review and regression analysis*, 2016.
15. L. Te Morenga, S. Mallard and J. Mann, Dietary sugars and body weight: systematic review and meta-analyses of randomised controlled trials and cohort studies, *BMJ*, 2013, **346**, e7492.
16. B. Lauby-Secretan, C. Scoccianti, D. Loomis, Y. Grosse, F. Bianchini and K. Straif, Body fatness and cancer—viewpoint of the IARC Working Group, *N. Engl. J. Med.*, 2016, **375**(8), 794–798.
17. V. Raikos and V. Ranawana, *Reformulation as a Strategy for Developing Healthier Food Products*, Springer, 2019.
18. J. Brambila-Macias, B. Shankar and S. Capacci, *et al.*, Policy interventions to promote healthy eating: a review of what works, what does not, and what is promising, *Food Nutr. Bull.*, 2011, **32**(4), 365–375.
19. C. A. Roberto and N. Khandpur, Improving the design of nutrition labels to promote healthier food choices and reasonable portion sizes, *Int. J. Obes.*, 2014, **38**(1), S25–S33.
20. H. K. De Zoysa and V. Y. Waisundara, The Importance of Food Reformulation in Developing Countries, *Reformulation as a Strategy for Developing Healthier Food Products: Challenges, Recent Developments and Future Prospects*, 2019, pp. 127–149.
21. K. Vinitha, P. Sethupathy, J. Moses and C. Anandharamakrishnan, Conventional and emerging approaches for reducing dietary intake of salt, *Food Res. Int.*, 2022, 110933.
22. K. Trieu, B. Neal and C. Hawkes, *et al.*, Salt reduction initiatives around the world–a systematic review of progress towards the global target, *PLoS One*, 2015, **10**(7), e0130247.
23. F. He, H. Brinsden and G. MacGregor, Salt reduction in the United Kingdom: a successful experiment in public health, *J. Hum. Hypertens.*, 2014, **28**(6), 345–352.
24. A. Al-Jawaldeh, M. Taktouk and S. Naalbandian, *et al.*, Sugar Reduction Initiatives in the Eastern Mediterranean Region: A Systematic Review, *Nutrients*, 2023, **15**(1), 55.
25. C. H. Edwards, M. Rossi, C. P. Corpe, P. J. Butterworth and P. R. Ellis, The role of sugars and sweeteners in food, diet and health: Alternatives for the future, *Trends Food Sci. Technol.*, 2016, **56**, 158–166.
26. S. Arshad, T. Rehman and S. Saif, *et al.*, Replacement of refined sugar by natural sweeteners: focus on potential health benefits, *Heliyon*, 2022, e10711.
27. World Health Organization, *Reformulation of food and beverage products for healthier diets: policy brief*, World Health Organization, 2022.

28. S. Erickson and J. Carr, The technological challenges of reducing the sugar content of foods, *Nutr. Bull.*, 2020, **45**(3), 309–314.
29. M. O'Sullivan, Nutritional optimization-reduced-sugar products and challenges, *Salt, Fat and Sugar Reduction: Sensory Approaches for Nutritional Reformulation of Foods and Beverages*, 2020, pp. 29–61.
30. S. Chattopadhyay, U. Raychaudhuri and R. Chakraborty, Artificial sweeteners–a review, *J. Food Sci. Technol.*, 2014, **51**, 611–621.
31. A. El-Lahot, A. El-Razek, M. Amal, M. I. Massoud and E. Gomaa, Utilization of glycyrrhizin and licorice extract as natural sweetener in some food products and biological impacts, *J. Food Dairy Sci.*, 2017, **8**(3), 127–136.
32. J. Wang, H. Zhao and Y. Wang, *et al.*, A review of stevia as a potential healthcare product: Up-to-date functional characteristics, administrative standards and engineering techniques, *Trends Food Sci. Technol.*, 2020, **103**, 264–281.
33. M. A. A. Gasmalla, R. Yang and X. Hua, Stevia rebaudiana Bertoni: An alternative sugar replacer and its application in food industry, *Food Eng. Rev.*, 2014, **6**, 150–162.
34. J. A. Joseph, S. Akkermans, P. Nimmegeers and J. F. Van Impe, Bioproduction of the recombinant sweet protein thaumatin: Current state of the art and perspectives, *Front. Microbiol.*, 2019, **10**, 695.
35. S. Martínez-Cervera, A. Salvador and T. Sanz, Comparison of different polyols as total sucrose replacers in muffins: Thermal, rheological, texture and acceptability properties, *Food Hydrocolloids*, 2014, **35**, 1–8.
36. S. Ghosh and M. Sudha, A review on polyols: new frontiers for health-based bakery products, *Int. J. Food Sci. Nutr.*, 2012, **63**(3), 372–379.
37. K. Tsatsaragkou, L. Methven, A. Chatzifragkou and J. Rodriguez-Garcia, The functionality of inulin as a sugar replacer in cakes and biscuits; highlighting the influence of differences in degree of polymerisation on the properties of cake batter and product, *Foods*, 2021, **10**(5), 951.
38. M. Shoaib, A. Shehzad and M. Omar, *et al.*, Inulin: Properties, health benefits and food applications, *Carbohydr. Polym.*, 2016, **147**, 444–454.
39. B. Pareyt, M. Goovaerts, W. F. Broekaert and J. A. Delcour, Arabinoxylan oligosaccharides (AXOS) as a potential sucrose replacer in sugar-snap cookies, *LWT – Food Sci. Technol.*, 2011, **44**(3), 725–728.
40. L. Milner, J. P. Kerry, M. G. O'Sullivan and E. Gallagher, Physical, textural and sensory characteristics of reduced sucrose cakes, incorporated with clean-label sugar-replacing alternative ingredients, *Innovative Food Sci. Emerging Technol.*, 2020, **59**, 102235.
41. P. Singh, Y. Ban, L. Kashyap, A. Siraree and J. Singh, Sugar and sugar substitutes: recent developments and future prospects, *Sugar and Sugar Derivatives: Changing Consumer Preferences*, 2020, pp. 39–75.
42. A. M. Richardson, A. A. Tyuftin, K. N. Kilcawley, E. Gallagher, M. G. O'Sullivan and J. P. Kerry, The impact of sugar particle size manipulation on the physical and sensory properties of chocolate brownies, *LWT*, 2018, **95**, 51–57.

43. R. Caporizzi, C. Severini and A. Derossi, Study of different technological strategies for sugar reduction in muffin addressed for children, *NFS J.*, 2021, **23**, 44–51.
44. G. Sala, M. Stieger and F. van de Velde, Serum release boosts sweetness intensity in gels, *Food Hydrocolloids*, 2010, **24**(5), 494–501.
45. M. Stieger and F. van de Velde, Microstructure, texture and oral processing: New ways to reduce sugar and salt in foods, *Curr. Opin. Colloid Interface Sci.*, 2013, **18**(4), 334–348.
46. A. C. Mosca, F. van de Velde, J. H. Bult, M. A. van Boekel and M. Stieger, Effect of gel texture and sucrose spatial distribution on sweetness perception, *LWT – Food Sci. Technol.*, 2012, **46**(1), 183–188.
47. A. C. Mosca, F. van de Velde, J. H. Bult, M. A. van Boekel and M. Stieger, Enhancement of sweetness intensity in gels by inhomogeneous distribution of sucrose, *Food Qual. Prefer.*, 2010, **21**(7), 837–842.
48. K. Holm, K. Wendin and A.-M. Hermansson, Sweetness and texture perceptions in structured gelatin gels with embedded sugar rich domains, *Food Hydrocolloids*, 2009, **23**(8), 2388–2393.
49. A. S. Bertelsen, Y. Zeng, L. A. Mielby, Y.-X. Sun, D. V. Byrne and U. Kidmose, Cross-modal effect of vanilla aroma on sweetness of different sweeteners among Chinese and Danish consumers, *Food Qual. Prefer.*, 2021, **87**, 104036.
50. A. S. Bertelsen, L. A. Mielby, D. V. Byrne and U. Kidmose, Ternary cross-modal interactions between sweetness, aroma, and viscosity in different beverage matrices, *Foods*, 2020, **9**(4), 395.
51. F. Alcaire, L. Antúnez, L. Vidal, A. Giménez and G. Ares, Aroma-related cross-modal interactions for sugar reduction in milk desserts: Influence on consumer perception, *Food Res. Int.*, 2017, **97**, 45–50.
52. F. J. He, M. Tan, Y. Ma and G. A. MacGregor, Salt reduction to prevent hypertension and cardiovascular disease: JACC state-of-the-art review, *J. Am. Coll. Cardiol.*, 2020, **75**(6), 632–647.
53. World Health Organization, *The SHAKE technical package for salt reduction*, 2016.
54. G. S. Aljuraiban, A. P. Jose, P. Gupta, K. Shridhar and D. Prabhakaran, Sodium intake, health implications, and the role of population-level strategies, *Nutr. Rev.*, 2021, **79**(3), 351–359.
55. E. S. Inguglia, Z. Zhang, B. K. Tiwari, J. P. Kerry and C. M. Burgess, Salt reduction strategies in processed meat products–A review, *Trends Food Sci. Technol.*, 2017, **59**, 70–78.
56. C. Silow, C. Axel, E. Zannini and E. K. Arendt, Current status of salt reduction in bread and bakery products–a review, *J. Cereal Sci.*, 2016, **72**, 135–145.
57. C. Sun, X. Zhou, Z. Hu, W. Lu, Y. Zhao and Y. Fang, Food and salt structure design for salt reducing, *Innovative Food Sci. Emerging Technol.*, 2021, **67**, 102570.
58. R. Uauy, A. Aro and R. Clarke, *et al.*, WHO Scientific Update on *trans* fatty acids: summary and conclusions, *Eur. J. Clin. Nutr.*, 2009, **63**(2), S68–S75.

59. D. Mozaffarian, M. B. Katan, A. Ascherio, M. J. Stampfer and W. C. Willett, Trans fatty acids and cardiovascular disease, *N. Engl. J. Med.*, 2006, **354**(15), 1601–1613.
60. D. Mozaffarian, A. Aro and W. C. Willett, Health effects of trans-fatty acids: experimental and observational evidence, *Eur. J. Clin. Nutr.*, 2009, **63**(2), S5–S21.
61. World Health Organization, *Draft guidelines on saturated fatty acid and trans-fatty acid intake for adults and children*, World Health Organization, Geneva, Switzerland, 2018, p. 103.
62. D. Mozaffarian, M. F. Jacobson and J. S. Greenstein, Food reformulations to reduce *trans* fatty acids, *N. Engl. J. Med.*, 2010, **362**(21), 2037–2039.
63. T. L. S. Lima, G. F. da Costa, R. D. N. Alves, C. D. Leal de Araújo, G. F. G. da Silva, N. L. Ribeiro, C. F. V. de Figueiredo and R. O. de Andrade, *et al.*, Vegetable oils in emulsified meat products: a new strategy to replace animal fat, *Food Sci. Technol.*, 2022, **42**, e103621.
64. S. Shaheen, M. Kamal, C. Zhao and M. A. Farag, Fat substitutes and low-calorie fats: A compile of their chemical, nutritional, metabolic and functional properties, *Food Rev. Int.*, 2022, 1–27.
65. V. Syan, J. Kaur and K. Sharma, *et al.*, An overview on the types, applications and health implications of fat replacers, *J. Food Sci. Technol.*, 2022, 1–12.
66. A. G. Marangoni, J. P. Van Duynhoven, N. C. Acevedo, R. A. Nicholson and A. R. Patel, Advances in our understanding of the structure and functionality of edible fats and fat mimetics, *Soft Matter*, 2020, **16**(2),289–306.
67. E. M. Vélez-Erazo, P. K. Okuro, A. Gallegos-Soto, R. L. da Cunha and M. D. Hubinger, Protein-based strategies for fat replacement: Approaching different protein colloidal types, structured systems and food applications, *Food Res. Int.*, 2022, 111346.
68. M. O'Sullivan, *Salt, Fat and Sugar Reduction: Sensory Approaches for Nutritional Reformulation of Foods and Beverages*, Woodhead Publishing, 2020.
69. X. Peng and Y. Yao, Carbohydrates as fat replacers, *Annu. Rev. Food Sci. Technol.*, 2017, **8**, 331–351.
70. M. Akbari, M. H. Eskandari and Z. Davoudi, Application and functions of fat replacers in low-fat ice cream: A review, *Trends Food Sci. Technol.*, 2019, **86**, 34–40.
71. A. R. Patel, R. A. Nicholson and A. G. Marangoni, Applications of fat mimetics for the replacement of saturated and hydrogenated fat in food products, *Curr. Opin. Food Sci.*, 2020, **33**, 61–68.
72. M. Yashini, C. K. Sunil, S. Sahana, S. D. Hemanth, D. V. Chidanand and A. Rawson, Protein-based fat replacers–a review of recent advances, *Food Rev. Int.*, 2021, **37**(2), 197–223.
73. A. Skendi and M. Papageorgiou, Low glycemic index ingredients and modified starches in food products, *The Role of Alternative and Innovative Food Ingredients and Products in Consumer Wellness*, Elsevier, 2019, pp. 167–195.

74. J. Kaur, K. Kaur, B. Singh, A. Singh and S. Sharma, Insights into the latest advances in low glycemic foods, their mechanism of action and health benefits, *J. Food Meas. Charact.*, 2021, 1–14.
75. S. B. Kumar and P. Prabhasankar, Low glycemic index ingredients and modified starches in wheat based food processing: A review, *Trends Food Sci. Technol.*, 2014, **35**(1), 32–41.
76. H. N. Englyst, J. Veenstra and G. J. Hudson, Measurement of rapidly available glucose (RAG) in plant foods: a potential *in vitro* predictor of the glycaemic response, *Br. J. Nutr.*, 1996, **75**(3), 327–337.
77. D. Belobrajdic, I. Brownlee, G. Hendrie, M. Rebuli and T. Bird, *Gut health and weight loss: An overview of the scientific evidence of the benefits of dietary fibre during weight loss*, 2018.
78. F. Brouns, I. Bjorck and K. Frayn, *et al.*, Glycaemic index methodology, *Nutr. Res. Rev.*, 2005, **18**(1), 145–171.
79. L. S. Augustin, C. W. Kendall and D. J. Jenkins, *et al.*, Glycemic index, glycemic load and glycemic response: an International Scientific Consensus Summit from the International Carbohydrate Quality Consortium (ICQC), *Nutr., Metab. Cardiovasc. Dis.*, 2015, **25**(9), 795–815.
80. B. Venn and T. Green, Glycemic index and glycemic load: measurement issues and their effect on diet–disease relationships, *Eur. J. Clin. Nutr.*, 2007, **61**(1), S122–S131.
81. S. Priyadarshini, J. Moses and C. Anandharamakrishnan, Determining the glycaemic responses of foods: conventional and emerging approaches, *Nutr. Res. Rev.*, 2022, **35**(1), 1–27.
82. R. Nagaraju, P. P. Sobhana, D. Thappatla, S. Epparapalli, B. Kandlakunta and D. Korrapati, Glycemic index and sensory evaluation of whole grain based multigrain Indian breads (Rotis), *Prev. Nutr. Food Sci.*, 2020, **25**(2), 194.
83. C. Breen, M. Ryan, M. J. Gibney, M. Corrigan and D. O'Shea, Glycemic, insulinemic, and appetite responses of patients with type 2 diabetes to commonly consumed breads, *Diabetes Educ.*, 2013, **39**(3), 376–386.
84. H. J. Kim and P. J. White, Impact of the molecular weight, viscosity, and solubility of β-glucan on *in vitro* oat starch digestibility, *J. Agric. Food Chem.*, 2013, **61**(13), 3270–3277.
85. M. C. Casiraghi, M. Garsetti, G. Testolin and F. Brighenti, Post-prandial responses to cereal products enriched with barley β-glucan, *J. Am. Coll. Nutr.*, 2006, **25**(4), 313–320.
86. S. Giri, A. Banerji, S. Lele and L. Ananthanarayan, Effect of addition of enzymatically modified guar gum on glycemic index of selected Indian traditional foods (idli, chapatti), *Bioact. Carbohydr. Diet. Fibre*, 2017, **11**, 1–8.
87. P. Sethupathy, P. Suriyamoorthy, J. A. Moses and A. Chinnaswamy, Physical, sensory, in-vitro starch digestibility and glycaemic index of granola bars prepared using sucrose alternatives, *Int. J. Food Sci. Technol.*, 2020, **55**(1), 348–356.

88. E. Fuentes-Zaragoza, M. Riquelme-Navarrete, E. Sánchez-Zapata and J. Pérez-Álvarez, Resistant starch as functional ingredient: A review, *Food Res. Int.*, 2010, **43**(4), 931–942.
89. K. Kahraman, E. Aktas-Akyildiz, S. Ozturk and H. Koksel, Effect of different resistant starch sources and wheat bran on dietary fibre content and *in vitro* glycaemic index values of cookies, *J. Cereal Sci.*, 2019, **90**, 102851.
90. S. Punia Bangar, N. Sharma, A. Singh, Y. Phimolsiripol and C. S. Brennan, Glycaemic response of pseudocereal-based gluten-free food products: a review, *Int. J. Food Sci. Technol.*, 2022, **57**(8), 4936–4944.
91. M. Di Cairano, N. Condelli, M. C. Caruso, A. Marti, N. Cela and F. Galgano, Functional properties and predicted glycemic index of gluten free cereal, pseudocereal and legume flours, *LWT*, 2020, **133**, 109860.
92. M. S. Wee and C. J. Henry, Reducing the glycemic impact of carbohydrates on foods and meals: Strategies for the food industry and consumers with special focus on Asia, *Compr. Rev. Food Sci. Food Saf.*, 2020, **19**(2), 670–702.
93. S. D. Anton, C. K. Martin and H. Han, *et al.*, Effects of stevia, aspartame, and sucrose on food intake, satiety, and postprandial glucose and insulin levels, *Appetite*, 2010, **55**(1), 37–43.
94. P. Shankar, S. Ahuja and K. Sriram, Non-nutritive sweeteners: review and update, *Nutrition*, 2013, **29**(11–12), 1293–1299.
95. S. Banerji and S. Banerjee, A formulation of grape seed, Indian gooseberry, turmeric and fenugreek helps controlling type 2 diabetes mellitus in advanced-stage patients, *Eur. J. Integr. Med.*, 2016, **8**(5), 645–653.
96. M. De Angelis, N. Damiano, C. G. Rizzello, A. Cassone, R. Di Cagno and M. Gobbetti, Sourdough fermentation as a tool for the manufacture of low-glycemic index white wheat bread enriched in dietary fibre, *Eur. Food Res. Technol.*, 2009, **229**, 593–601.
97. X. Sun, S. Wu and W. Li, *et al.*, The effects of cooperative fermentation by yeast and lactic acid bacteria on the dough rheology, retention and stabilization of gas cells in a whole wheat flour dough system–A review, *Food Hydrocolloids*, 2022, 108212.
98. M. Gunathilaka and S. Ekanayake, Effect of different cooking methods on glycaemic index of Indian and Pakistani basmati rice varieties, *Ceylon Med. J.*, 2015, **60**, 57–61.
99. S. Sivakamasundari, S. Priyanga, J. Moses and C. Anandharamakrishnan, Impact of processing techniques on the glycemic index of rice, *Crit. Rev. Food Sci. Nutr.*, 2022, **62**(12), 3323–3344.
100. E. Lau, Y. Y. Soong, W. Zhou and J. Henry, Can bread processing conditions alter glycaemic response?, *Food Chem.*, 2015, **173**, 250–256.
101. D. Oliveira, P. Fox and J. A. O'Mahony, Byproducts from dairy processing, *Byproducts from Agriculture and Fisheries: Adding Value for Food, Feed, Pharma, and Fuels*, 2019, pp. 57–106.
102. T. O. Akanbi, K. O. Dare and A. N. Aryee, High-Value Products from Cereal, Nuts, Fruits, and Vegetables Wastes, *Byproducts from*

agriculture and fisheries: adding value for food, feed, pharma, and fuels, 2019, pp. 347–368.

103. B. K. Simpson, A. N. Aryee and F. Toldrá, *Byproducts from agriculture and fisheries: Adding value for food, feed, pharma and fuels*, Wiley Online Library, 2019.
104. B. A. Rumpold and O. K. Schlüter, Nutritional composition and safety aspects of edible insects, *Mol. Nutr. Food Res.*, 2013, **57**(5), 802–823.
105. G. Melgar-Lalanne, A. J. Hernández-Álvarez and A. Salinas-Castro, Edible insects processing: Traditional and innovative technologies, *Compr. Rev. Food Sci. Food Saf.*, 2019, **18**(4), 1166–1191.
106. S. Aumeerun, J. Soulange-Govinden, M. F. Driver, R. A. Ranga, G. A. Ravishankar and N. Hudaa, Macroalgae and microalgae: novel sources of functional food and feed, *Handbook of algal technologies and phytochemicals*, CRC Press, 2019, pp. 207–219.
107. A. P. Batista, A. Niccolai and P. Fradinho, *et al.*, Microalgae biomass as an alternative ingredient in cookies: Sensory, physical and chemical properties, antioxidant activity and *in vitro* digestibility, *Algal Res.*, 2017, **26**, 161–171.
108. N. Hosseinkhani, J. I. McCauley and P. J. Ralph, Key challenges for the commercial expansion of ingredients from algae into human food products, *Algal Res.*, 2022, **64**, 102696.
109. M. C. Onwezen, E. P. Bouwman, M. J. Reinders and H. Dagevos, A systematic review on consumer acceptance of alternative proteins: Pulses, algae, insects, plant-based meat alternatives, and cultured meat, *Appetite*, 2021, **159**, 105058.
110. S. M. Loveday, Plant protein ingredients with food functionality potential, *Nutr. Bull.*, 2020, **45**(3), 321–327.
111. R. Di Monaco, N. A. Miele, E. K. Cabisidan and S. Cavella, Strategies to reduce sugars in food, *Curr. Opin. Food Sci.*, 2018, **19**, 92–97.
112. D. M. Rodrigues, V. R. de Souza, J. F. Mendes, C. A. Nunes and A. C. M. Pinheiro, Microparticulated salts mix: An alternative to reducing sodium in shoestring potatoes, *LWT – Food Sci. Technol.*, 2016, **69**, 390–399.
113. N. Chindapan, C. Niamnuy and S. Devahastin, Physical properties, morphology and saltiness of salt particles as affected by spray drying conditions and potassium chloride substitution, *Powder Technol.*, 2018, **326**, 265–271.
114. M. Quilaqueo, L. Duizer and J. M. Aguilera, The morphology of salt crystals affects the perception of saltiness, *Food Res. Int.*, 2015, **76**, 675–681.
115. M. Emorine, C. Septier, T. Thomas-Danguin and C. Salles, Heterogeneous salt distribution in hot snacks enhances saltiness without loss of acceptability, *Food Res. Int.*, 2013, **51**(2), 641–647.
116. A. C. Mosca, J. H. Bult and M. Stieger, Effect of spatial distribution of tastants on taste intensity, fluctuation of taste intensity and consumer preference of (semi-) solid food products, *Food Qual. Prefer.*, 2013, **28**(1), 182–187.

117. N. Chiu, A. Tarrega, C. Parmenter, L. Hewson, B. Wolf and I. D. Fisk, Optimisation of octinyl succinic anhydride starch stablised w1/o/w2 emulsions for oral destablisation of encapsulated salt and enhanced saltiness, *Food Hydrocolloids*, 2017, **69**, 450–458.
118. H. I. Buyukkestelli and S. N. El, Preparation and characterization of double emulsions for saltiness enhancement by inhomogeneous spatial distribution of sodium chloride, *LWT*, 2019, **101**, 229–235.
119. N. Chiu, L. Hewson, N. Yang, R. Linforth and I. Fisk, Controlling salt and aroma perception through the inclusion of air fillers, *LWT – Food Sci. Technol.*, 2015, **63**(1), 65–70.
120. H.-Y. Cho, B. Kim, J.-Y. Chun and M.-J. Choi, Effect of spray-drying process on physical properties of sodium chloride/maltodextrin complexes, *Powder Technol.*, 2015, **277**, 141–146.
121. W. A. Cerrato Rodriguez, D. D. Torrico, L. F. Osorio, J. Cardona and W. Prinyawiwatkul, Taste perception and purchase intent of oil-in-water spreads: effects of oil types and salt (NaCl or KCl) concentrations, *Int. J. Food Sci. Technol.*, 2017, **52**(10), 2138–2147.
122. A.-M. Reißner, J. Wendt, S. Zahn and H. Rohm, Sodium-chloride reduction by substitution with potassium, calcium and magnesium salts in wheat bread, *LWT*, 2019, **108**, 153–159.
123. C. Beeren, K. Groves and P. M. Titoria, *Reducing salt in foods*, Woodhead Publishing, 2019.
124. P. Gullón, G. Astray, B. Gullón, D. Franco, P. C. B. Campagnol and J. M. Lorenzo, Inclusion of seaweeds as healthy approach to formulate new low-salt meat products, *Curr. Opin. Food Sci.*, 2021, **40**, 20–25.
125. Y. Khetra, S. K. Kanawjia and R. Puri, Selection and optimization of salt replacer, flavour enhancer and bitter blocker for manufacturing low sodium Cheddar cheese using response surface methodology, *LWT – Food Sci. Technol.*, 2016, **72**, 99–106.
126. M. Bampi, N. Domschke, F. Schmidt and J. Laurindo, Influence of vacuum application, acid addition and partial replacement of NaCl by KCl on the mass transfer during salting of beef cuts, *LWT*, 2016, **74**, 26–33.
127. V. A. Vidal, J. B. Santana and C. S. Paglarini, *et al.*, Adding lysine and yeast extract improves sensory properties of low sodium salted meat, *Meat Sci.*, 2020, **159**, 107911.
128. G. Nugrahani and D. Afifah, Effect of garlic as a spice on salt preferences of hypertensive individuals, *Pak. J. Nutr.*, 2016, **15**(7), 633–638.
129. D. Taladrid, L. Laguna, B. Bartolomé and M. V. Moreno-Arribas, Plant-derived seasonings as sodium salt replacers in food, *Trends Food Sci. Technol.*, 2020, **99**, 194–202.
130. K. Li, Z.-L. Kang, Y.-F. Zou, X.-L. Xu and G.-H. Zhou, Effect of ultrasound treatment on functional properties of reduced-salt chicken breast meat batter, *J. Food Sci. Technol.*, 2015, **52**, 2622–2633.
131. Z. F. Bhat, J. D. Morton, S. L. Mason and A. E.-D. A. Bekhit, The application of pulsed electric field as a sodium reducing strategy for meat products, *Food Chem.*, 2020, **306**, 125622.

132. A. Tamm, T. Bolumar, B. Bajovic and S. Toepfl, Salt (NaCl) reduction in cooked ham by a combined approach of high pressure treatment and the salt replacer KCl, *Innovative Food Sci. Emerging Technol.*, 2016, **36**, 294–302.
133. Y. Wang, Y. Zhou, P.-J. Li, X.-X. Wang, K.-Z. Cai and C.-G. Chen, Combined effect of $CaCl_2$ and high pressure processing on the solubility of chicken breast myofibrillar proteins under sodium-reduced conditions, *Food Chem.*, 2018, **269**, 236–243.
134. C. M. O'Neill, M. C. Cruz-Romero, G. Duffy and J. P. Kerry, The application of response surface methodology for development of sensory acceptable, low-salt, shelf-stable frankfurters using high-pressure processing and a mix of organic acids, *Eur. Food Res. Technol.*, 2019, **245**, 1277–1291.
135. T. J. van Rompay and S. Groothedde, The taste of touch: Enhancing saltiness impressions through surface texture design, *Food Qual. Prefer.*, 2019, **73**, 248–254.
136. H. Fosberg and H. S. Joyner, The impact of salt reduction on cottage cheese cream dressing rheological behavior and consumer acceptance, *Int. Dairy J.*, 2018, **79**, 62–72.
137. B. D. Tobin, M. G. O'Sullivan, R. M. Hamill and J. P. Kerry, The impact of salt and fat level variation on the physiochemical properties and sensory quality of pork breakfast sausages, *Meat Sci.*, 2013, **93**(2), 145–152.
138. S. K. Ghawi, I. Rowland and L. Methven, Enhancing consumer liking of low salt tomato soup over repeated exposure by herb and spice seasonings, *Appetite*, 2014, **81**, 20–29.
139. S. Fellendorf, J. P. Kerry, R. M. Hamill and M. G. O'Sullivan, Impact on the physicochemical and sensory properties of salt reduced corned beef formulated with and without the use of salt replacers, *LWT*, 2018, **92**, 584–592.
140. J. F. Rodrigues, G. Junqueira, C. S. Goncalves, J. D. Carneiro, A. C. M. Pinheiro and C. A. Nunes, Elaboration of garlic and salt spice with reduced sodium intake, *An. Acad. Bras. Cienc.*, 2014, **86**, 2065–2075.
141. I. K. Carmi and O. Benjamin, Reduction in sodium content of fresh, semihard Tzfat cheese using salt replacer mixtures: taste, texture and shelf life evaluation, *Int. J. Dairy Technol.*, 2017, **70**(3), 354–364.

CHAPTER 9

Structuring Foods and the Case of Encapsulation

9.1 Introduction

In recent years, foods have been designed and developed with the aim of tailored functionality and targeted delivery of active and functional ingredients owing to growing consumer awareness of the health benefits and effects of functional foods, nutraceuticals, and bioactives. In addition, consumers are also demanding foods that are superior in their sensory attributes and clean-label (*i.e.* healthy foods free from artificial additives). Thus, there is always a necessity to add certain active ingredients such as colors, flavors, nutraceuticals, bioactive compounds, microorganisms, antimicrobials, antioxidants, and preservatives to incorporate specific functional attributes into foods. However, several challenges arise with the addition of such active ingredients into food matrices owing to unfavorable functional properties such as low solubility, poor sensory attributes (off-flavor and bitter taste), incompatibility with product matrices, and instability under prevailing environmental and digestion conditions.[1] Further, the addition of these bioactives and functional ingredients to food matrices and the processing conditions should not adversely affect their stability, functionality, and bioavailability.[2] All of these issues with the incorporation of bioactive and functional components into foods can be efficiently resolved through the use of various delivery systems (*e.g.* micelles, micro- and nanoemulsions, hydrogels, solid lipid nanoparticles, oleosomes, *etc.*) that encapsulate the active ingredients before they can be incorporated into the final product.[3]

Various food structuring and size modification processes, such as shearing, homogenization, grinding, injection, spraying, dehydration/desolvation, electrostatic interactions, and self-assembly, have a key role in the

Food Chemistry, Function and Analysis No. 42
Food Digestion and Absorption: Its Role in Food Product Development
By C. Anandharamakrishnan, Jeyan Arthur Moses and S. Priyanka

Published by the Royal Society of Chemistry, www.rsc.org

design and fabrication of delivery systems for bioactives and functional ingredients.[4] Further, the size reduction of encapsulated functional ingredients to micro- and nano-levels increases their surface area, thereby improving their solubility, adsorption, stability, and bioavailability.[5,6] Most of the engineered delivery systems investigated for the delivery of bioactives have several concerns that impact consumer behavior and commercialization, such as the use of synthetic surfactants, expensive and high energy-consuming equipment, laborious preparation methods, and toxicity in the case of nano-scale delivery systems.[7,8] Therefore, in recent years, researchers have been keen to use sustainable and clean-label natural food ingredients (such as plant- and animal-derived ingredients) and nature-inspired delivery systems (*e.g.* casein micelles and oil bodies) for the encapsulation and delivery of bioactive compounds.[9,10] In this chapter, we discuss the need for food structuring and size reduction for the delivery of bioactives and other functional ingredients, various delivery systems, fabrication methods, specific food applications, and their gastrointestinal fate.

9.2 Active Ingredients: Need for Encapsulation and Its Significance

Numerous food-related active ingredients, such as flavors, colors, essential oils, lipids, acidulants, antioxidants, vitamins and minerals, preservatives, colorants, enzymes, peptides, polyphenols, and probiotics, are usually encapsulated for various reasons, as follows:

- Low solubility.
- Inappropriate physical state.
- Poor sensory profile: the bitterness or astringency of most bioactive compounds restricts their inclusion in foods and beverages.
- Poor biochemical stability: they undergo physical or chemical changes upon contact with different gastrointestinal conditions.
- Poor physicochemical stability: they may aggregate or sediment during storage, undergo oxidation, and change in color and flavor, usually under accelerated conditions such as exposure to light, in the presence of pro-oxidants (*e.g.* transition metals), elevated temperatures, and high oxygen levels.
- Poor handling characteristics: they can be difficult to transport, store, or handle.

Further, the encapsulation of active ingredients into various delivery systems has several significant advantages and food-related applications, such as the following:

- Aids in the incorporation of bioactive ingredients and functional compounds in food without seriously impacting food product quality features such as flavor, appearance, taste, texture, and stability.

- Protects against degradation due to physical, chemical, and biological factors (*e.g.* oxygen, light, temperature, enzymes, and oral and gastrointestinal conditions).
- Has the potential for targeted delivery of active ingredients to specific regions and thereby increases their bioavailability and efficacy.
- Aids in controlling the release mechanisms and release rates of functional ingredients during digestion.
- Enhances product storage, handling, and shelf-life.

Table 9.1 summarizes the need for the encapsulation of various active ingredients and their food-related applications.

9.3 Delivery Systems and Fabrication Methods

Delivery systems are predominantly classified into five categories based on the main ingredients used in their development, namely (i) surfactant-based, (ii) emulsion-based, (iii) biopolymer-based, (iv) hybrid systems, and (v) nature-inspired systems.[11] Figure 9.1 illustrates various structured delivery systems. Further, the delivery systems should possess certain desirable functional characteristics (such as high loading capacity, encapsulation efficiency, and delivery efficiency, a suitable delivery mechanism, provide protection against physicochemical and biochemical degradation, and enhance bioavailability) and technical characteristics (compatible with the food matrix, food process, and economic production).[12] In addition, the ingredients used in the development of delivery systems should be of food grade, easy to use, readily available, and relatively inexpensive.[13] Common food-grade materials used for the fabrication of delivery systems are lipids, surfactants, and biopolymers.

The fabrication methods utilized to develop delivery systems are classified into top-down (mechanical methods), bottom-up (physicochemical methods), and a combination of methods based on the physicochemical processes involved.[14–16] Top-down methods utilize various mechanical approaches such as homogenization, grinding, injection, or spraying to break down bulk materials/suspensions comprising large particles into smaller particles (size reduction),[17] for instance the preparation of emulsion droplets through homogenization[18] and nano- or microparticles through spray drying.[19] On the other hand, in bottom-up approaches, delivery systems are fabricated by assembling molecules or colloidal particles into larger particles through either spontaneous self-assembly (micelle and microemulsion formation)[20] or as directed in the case of electrostatic deposition.[21]

Top-down and bottom-up approaches can be used in combination to develop several complex structured emulsions such as multilayer emulsions, where homogenization (top-down) is used to develop an emulsion and electrostatic deposition (bottom-up) can be utilized to coat the oil droplets with biopolymer.[22] Further, the fabrication methods employed in designing delivery systems should be robust, economical, suitable for large-scale production, reproducible and reliable.

Table 9.1 Benefits of food structuring and size reduction (encapsulation) for various active ingredients.

Active ingredients	Significance of encapsulation	Ref.
Probiotics: *Lactobacillus* *Bifdobacterium*	• Protects from deterioration during processing, storage, and digestion (oral, gastric, and intestinal conditions) • Maintains cell viability and functionality • Targeted delivery to the active site (gut)	50 and 51
Bioactive lipids: Omega-3 fatty acids Conjugated linoleic acid	• Permits their incorporation into an aqueous medium • Prevents chemical degradation such as oxidation • Aids in controlled delivery within the gastrointestinal tract • Improves ease of utilization	52 and 53
Vitamins: Vitamins A, D, E, and C	• Permits their incorporation into an aqueous medium • Prevents chemical deterioration • Avoids interactions with other food ingredients • Improves ease of utilization	54 and 55
Minerals: Iron Calcium	• Prevents unwanted oxidative reactions • Prevents sedimentation and precipitation • Increases bioavailability	56 and 57
Antioxidants: Tocopherols Carotenoids Flavonoids Phenolics	• Permits their incorporation into an aqueous medium • Eases their utilization and storage • Prevents chemical degradation • Improves their efficiency • Masks off-flavors	58 and 59
Bioactive peptides: Peptides from milk, meat, and plants	• Hinders deterioration in the stomach • Decreases bitterness and astringency • Regulates release profile and improves bioactivity	60 and 61
Flavors: Natural extracts and citrus oils	• Permits their addition into an aqueous medium • Simplifies their storage and utilization • Aids in controlling their release profile • Inhibits adverse interactions with other components • Prevents chemical degradation	62 and 63
Antimicrobials: Essential oils Nisin	• Enhances matrix compatibility • Increases potency and high antimicrobial activity • Simplifies storage and utilization • Masks undesirable off-flavors • Retards chemical degradation	64–66
Colors/pigments: Anthocyanins	• Prevents chemical degradation • Prevents interactions with other substances	67 and 68

Table 9.1 (*Continued*)

Active ingredients	Significance of encapsulation	Ref.
Carmine Azo dyes	• Improves solubility and facilitates incorporation into food matrices	
Enzymes: β-Galactosidase Flavourzyme Lipase	• Immobilizes enzymes • Retains biological activity • Protects against environmental stress and processing conditions • Improves stability and ease of handling	69 and 70

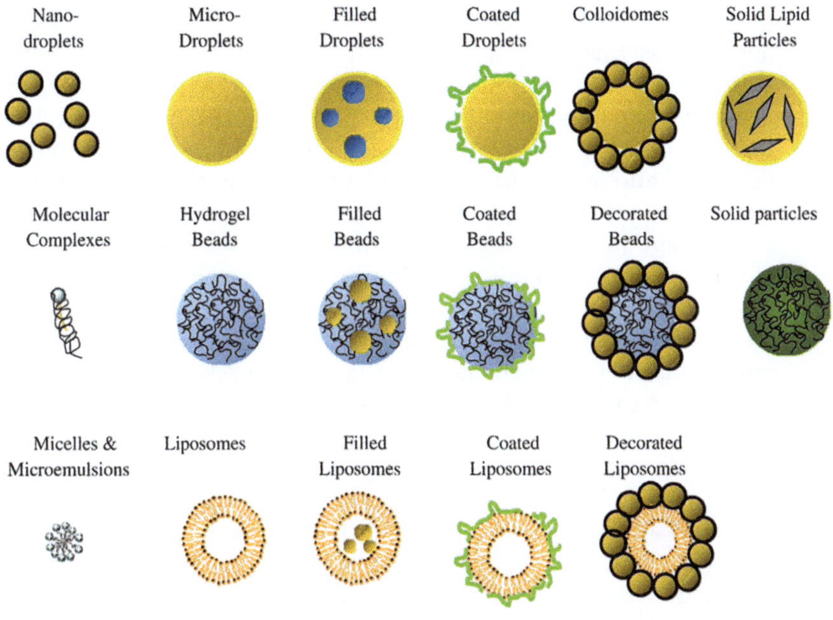

Figure 9.1 Illustrations of various structured delivery systems. Reproduced from ref. 12 with permission from Elsevier, Copyright 2010.

(i) *Surfactant-based delivery systems*: the phospholipid and surfactant molecules are key building blocks of numerous colloidal delivery systems such as micelles, microemulsions, and liposomes.[23] The surfactant molecules have a polar head group and a non-polar tail group and they vary based on the nature of their head (dimensions, polarity, charge, and chemical reactivity) and tail (number, length, unsaturation, and branching) groups.[11] Some of the common surfactants are lecithins, polyol esters of fatty acids (Span, polysorbate, and Tween), fatty acid salts (sodium stearoyl lactylate and calcium stearoyl lactylate), and mono- and diglycerides of fatty acids (glyceryl monostearate and glyceryl distearate).[24,25] It is self-assembly that mainly promotes the formation of surfactant-based delivery systems. In addition, mechanical fabrication methods such as homogenization are employed for mixing (surfactant, oil, and water) to facilitate initial formation, modify structure and reduce size.[11]

(ii) *Emulsion-based delivery systems*: these are formed by mixing/blending emulsifiers (milk proteins, egg proteins, plant proteins, polysaccharides, and protein–polysaccharide complexes), oil, and water components. The emulsifier-stabilized oil droplets are the key structural feature, and they can vary in size, composition, interfacial characteristics, physical state, and structural organization.[12,26–28] Solid lipid nanoparticles (SLNs), conventional emulsions, nanoemulsions, and Pickering emulsions are the primary emulsion-based delivery systems.[29] Further, a combination of primary emulsion-based delivery systems can be utilized to develop complex structures, namely colloidosomes, microclusters, filled hydrogel particles, and multilayer emulsions.[11] Various high-energy approaches, such as the use of microfluidizers, colloidal mills, high-pressure homogenizers, and ultrasonicators, are utilized for the fabrication of emulsion-based systems. Further, certain low-energy approaches, such as spontaneous emulsification, phase inversion temperature methods, and emulsion inversion point methods, can be used to develop emulsion-based systems.[30]

(iii) *Biopolymer-based delivery systems*: biopolymers of food origin, namely polysaccharides (carrageenan, xanthan gum, pectin, gum arabic, inulin, chitosan, and alginate) and proteins (β-lactoglobulin, caseins, lactoferrin, ovalbumin, lysozyme, gelatin, soy glycinin, and zein), are key building blocks of biopolymer-based delivery systems.[31,32] Further, these delivery systems are developed utilizing physiochemical and mechanical processing methods (extrusion, microfluidics, spray drying, antisolvent precipitation, emulsion templating, application of shearing forces, and molding) that use a bottom-up and a top-down approach, respectively, based on the type of biopolymers used and the functional performance required.[33] Molecular complexes and hydrogel particles are the most common biopolymer-based delivery systems and they can be used to develop other complex structures, such as filled hydrogel particles.[34]

(iv) *Hybrid systems*: more complicated structures can be fabricated using basic building blocks of surfactant-, emulsion-, and biopolymer-based delivery systems for novel or improved functional performance.[35] Various approaches for fabricating complex hybrid delivery systems can be classified into three main groups:

(a) *Coating* – liposomes, emulsions, or hydrogel particles can be coated by electrostatically depositing biopolymer layers around their charged particles.[36]

(b) *Clustering* – colloidosomes and microclusters are created by clustering nanoemulsions, emulsions, and multiple emulsions.[37]

(c) *Embedding* – trapping microemulsions, emulsions, and nanoemulsions within hydrogel particles.[38]

(v) *Nature-inspired systems*: various natural structures from nature could be useful for the delivery and encapsulation of active substances. Recently, various nature-assembled structures, such as casein micelles, milk fat globules, oleosomes (plant oil bodies), yeast cells, and spores isolated from specific varieties of plants, mosses, fungi, and algae, have been explored for entrapping and targeted delivery of bioactive ingredients.[39–41] The use of these natural structures as delivery systems will be a clean, green, and sustainable option compared with engineered delivery systems fabricated from food-grade ingredients. Nevertheless, there might be difficulties owing to their complex nature.[9]

Table 9.2 summarizes recent studies on selected delivery systems for specific functionality.

9.4 Gastrointestinal Fate of Delivery Systems

The main objective of the encapsulation of bioactive components and fabrication of a food-grade delivery system is to protect and preserve the active ingredient from various environmental conditions and gastrointestinal tract (GIT) conditions, control their digestion behavior, and target their release within the GIT to achieve maximum bioavailability. The delivery systems are usually susceptible to a range of physiological and physicochemical changes within the various regions of the GIT (mouth, stomach, small intestine, and colon) owing to varied gastrointestinal conditions (mechanical forces, dilution, action of enzymes and microorganisms, and changes in temperature, pH, and ionic strength) before the release of active ingredients.[7] The key changes that happen to delivery systems in GIT and gastrointestinal conditions are summarized in Table 9.3.

The active ingredient from delivery systems can be released by a variety of release mechanisms, such as diffusion, dissolution (encapsulation–dissolution and matrix dissolution), erosion (surface erosion and bulk erosion), swelling, osmosis, degradation (homogeneous and heterogeneous), and fragmentation.[7,42] Further, various factors impact the active ingredient's release from the delivery system, such as the following:

(a) Gastrointestinal conditions:
- Residence time, pH, enzymes, salts, biopolymers, mechanical shear, bile salts, phospholipids, and gut microbiome.

(b) Environmental conditions:
- Temperature, pH changes, salts, additives, light, ultrasound, oxidation/reduction potential, moisture, enzymes, and processing conditions.

(c) Active ingredient-related factors:
- Size, charge, stability, solubility, diffusivity, initial amounts, type of ingredients, means of entrapment, concentration gradient, and localization.

Table 9.2 Summary of recent studies on delivery systems for specific functionality.

Type of delivery system	Fabrication method	Encapsulated active ingredient	Specific functionality	Ref.
Nanoparticles	Electrospraying	Resveratrol	Improved bioavailability. Better stability and sustained release profiles	71
Nanoliposomes	Solvent evaporation followed by sonication	Chia oil (α-linolenic acid)	Enhanced thermal and gastrointestinal stability. Sustained intestinal release of chia oil	72
Nanoencapsulates incorporated in 3D printed hydrogel	Electrospraying and 3D printing	Curcumin and resveratrol	Enhanced bioaccessibility and intestinal permeability. Effective protection from harsh acidic conditions of the stomach. Increased release under the intestinal conditions	73
Nanoemulsion	Microfluidization	Docosahexaenoic acid	Higher physicochemical stability during storage. Enhanced lipid bioavailability	74
Emulsion–alginate beads in yogurt	Homogenization	Safflower oil	Delayed release of encapsulated lipids suppresses appetite sensations and reduces food intake	75
Pickering emulsion gel	Homogenization followed by 3D printing	*Bifidobacterium lactis*	Improved cell viability against heat treatment and simulated gastrointestinal environment. A printing temperature of 65 °C reduced viable probiotics	76
Microcapsules	Complex coacervation–spray drying	Vanilla oleoresin	Higher retention of vanillin. Increased protection of volatile compounds. Higher stability to environmental conditions	77

Table 9.2 (*Continued*)

Type of delivery system	Fabrication method	Encapsulated active ingredient	Specific functionality	Ref.
Microencapsulated into oil bodies from *Brassica napu*	Passive diffusion	Astaxanthin	Improved oxidative stability to light and air	78
β-Cyclodextrin nanocomplexes	Homogenization	Fish oil	Improved stability against oxidation. Masked the off-flavor of fish oil	79
Alginate–chitosan nanoparticles	Ionotropic gelation and polyelectrolyte complexation	Rennet	Improved enzyme activity. Improved stability to processing conditions pH and temperature	80

(d) Carrier/encapsulant-related factors:
- Composition and other properties of biopolymers; initial size, shape, and structure; physical state and rheology of phases; stability and flocculated state in the case of emulsion droplets; and particle charges, porosity, pore formation, and closure.

The release profile is defined as *the events of releasing encapsulated compounds from carrier showing a specific concentration–time curve at the target site.*[42] Subsequently, the release profiles of the delivery systems are influenced by the type of release mechanism. Further, the release profiles are classified into several types based on the increase in the concentration of the active ingredient at the action site as a function of time, such as the following:

(i) Burst release (rapid release in a short period).
(ii) Delayed release (active ingredient is released after a lag time after consumption and is site specific).
(iii) Sustained release (constant prolonged release at a fixed rate).
(iv) Triggered release (burst release as a reply to a surrounding stimulus such as enzymes, microbes, pH, and temperature in specific regions of the GIT).
(v) Targeted release (designed to be released near the designated site of the physiological activity).

It is important to design a delivery system with suitable release mechanisms and release profiles depending on the specific application. Figure 9.2 illustrates different release mechanisms and release profiles of delivery

Table 9.3 Summary of characteristic features of the GIT and key changes that happen to delivery systems in the GIT.

Region in the GIT	Characteristic features	Key changes in delivery systems	Ref.
Mouth	Residence time: 5 s–5 min pH: 5–7 Mastication: structural breakdown, high shear Change in temperature Dilution by saliva Involves salivary proteins (mucins and proline-rich proteins) salts, and biopolymers Enzymes: amylase, lysozymes, and lipase	• Change in structure, composition, properties, pH, ionic strength, and temperature • Solids (such as gels) are broken down into smaller fragments due to chewing • Hydrolysis by digestive enzymes based on composition (starch hydrolyzed by amylase) causes a decrease in viscosity • Increase in particle size of micro- and nanoemulsions because of aggregation (depletion or bridging flocculation) favored by mucins • Environmental conditions (amylase, temperature, or pH) in the mouth can activate the breakdown of the matrix surrounding active compounds and promote their release • Commonly oral phase-targeted delivery systems are designed for burst release and prolong the retention times of desirable flavors, to provide a pleasant sensory experience, and to deliver antimicrobials for oral health and certain oil-soluble drugs	81–84
Stomach	Residence time: 30 min–4 h pH: 1–3 Enzymes: pepsin and gastric lipase Involves salts, hormones, acid (HCl), and gastric juice Mechanical agitation: peristaltic mixing and propulsion	• Bolus may further change its structure and composition • High acidity may degrade some components of the delivery system • Ionic strength and pH may affect electrical characteristics and interactions, and thereby directly have an impact on the permeability, aggregation, and integrity of the colloidal delivery systems • Some components of delivery systems can be hydrolyzed by enzymes (gastric lipase and proteases initiate lipid and protein digestion, respectively) • Gastric motility and gastric emptying can promote the	16 and 85–87

Table 9.3 (*Continued*)

Region in the GIT	Characteristic features	Key changes in delivery systems	Ref.
		breakdown of any left-out large fragments and transfer digested materials to the small intestine, respectively • Surface-active compounds (*e.g.* proteins and phospholipids) present in gastric secretions may adsorb on the surfaces of the delivery systems and alter their surface characteristics • The gastric behavior of delivery systems is highly influenced by their composition (*e.g.* protein-stabilized emulsions are prone to coalescence and flocculation, resulting in creaming, whereas surfactant-stabilized emulsions are stable) • The presence of polysaccharides (type and concentration) can influence the aggregation stability of protein-coated lipids • Most of the delivery systems containing bioactives, nutraceuticals, vitamins, and minerals are mainly designed to withstand harsh gastric conditions, but acid-stable nutrients can be released in the stomach	
Small intestine	Residence time: 1–2 h pH: 6–7.5 Enzymes: pancreatic lipase, protease, and amylase Involves bile, salts, bicarbonate, phospholipids, and biopolymers Low shear	• Delivery systems with bioactive compounds (such as polyphenols, peptides, curcumin, and carotenoids), vitamins, minerals, and antioxidants are targeted predominantly to the small intestine as it is the main site for digestion and absorption • Acidic chyme is converted to neutral when mixed with alkaline small intestinal secretions to facilitate pancreatic enzyme activity • Surface-active compounds in the small intestine such as bile salts and phospholipids adsorb competitively on the lipid droplets and displace the surface-active substance present initially, thereby altering the aggregation state of the droplets and interfacial	88–90

Table 9.3 (*Continued*)

Region in the GIT	Characteristic features	Key changes in delivery systems	Ref.
		composition and resulting in changes in particle size • Enzymes digest various components of the delivery system into absorbable forms • Mixed micelle and vesicle formation plays a crucial role in the solubilization, transport, and absorption of lipophilic bioactive components • The intestinal digestion behavior of delivery systems is dependent on the composition of the delivery system and properties of intestinal fluids • The charge of the emulsifier used in the formulation of an emulsion-based delivery system has a significant impact on size, charge, and microstructure during digestion, thereby influencing the release and bioavailability of active compounds	
Colon	Residence time: 12–24 h pH: 5–7 Involves enzymes and gut microbiome	• Only an indigestible fraction of food reaches the colon after most of the ingested food has been digested and absorbed in the upper GIT • In the case of delivery systems, those containing indigestible components such as indigestible oil or dietary fibers reach the colon • The gut microbiome also plays a key role in enhancing health benefits by improving the bioavailability of phenolic compounds, producing short-chain fatty acids, and metabolizing protein-based nutraceuticals • Delivery systems with prebiotics, probiotics, and nutraceuticals are mainly targeted at the colon for gut health and treatment of colon cancer and inflammatory bowel disease	51, 91 and 92

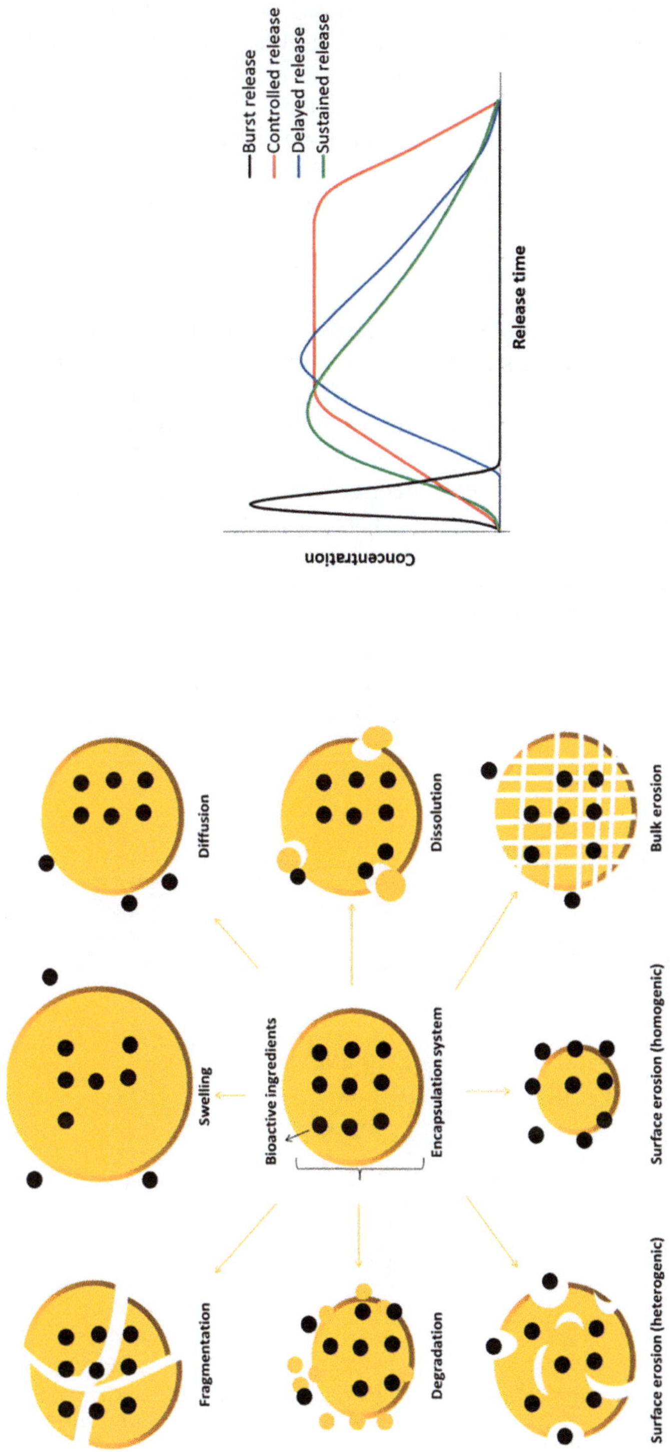

Figure 9.2 Examples of release mechanisms and release profiles. Reproduced from ref. 42 with permission from Elsevier, Copyright 2021.

systems. For instance, if a delivery system is designed for the rapid release of flavor in the mouth, then the carrier particle chosen should quickly dissolve/erode/fragment in response to oral triggers such as pH, temperature, enzymes, saliva, and mastication. On the other hand, if the application is to deliver probiotic microbes to the gut then one should select carrier particles that will be undamaged by gastrointestinal conditions in the mouth, stomach, and small intestine and only release the active ingredient within the large intestine because of the activity of colonic bacteria. Figure 9.3 illustrates the gastrointestinal fate of a hydrogel-based delivery system for targeted delivery.

After their release from the carrier matrix, the particles of active ingredient usually become absorbed in the small intestine or at other appropriate sites/release locations in the GIT. Most of the bioactive food ingredients are targeted for small intestinal delivery and their absorption occurs at the surface of the GIT through two main types of cells, namely epithelial cells and M cells, through different mechanisms such as paracellular, transcellular, and perception routes.[43] The particle properties of active ingredients, such as surface chemistry, shape, charge, and size, influence the absorption route and rate and extent of particle absorption.[44] For instance, cationic nanoparticles are more readily absorbed than anionic counterparts,[45] hydrophobic particles are more easily absorbed than hydrophilic particles,[46] and smaller particles have higher absorption than larger particles; however, there might be an optimal size.[47] Nevertheless, there is only

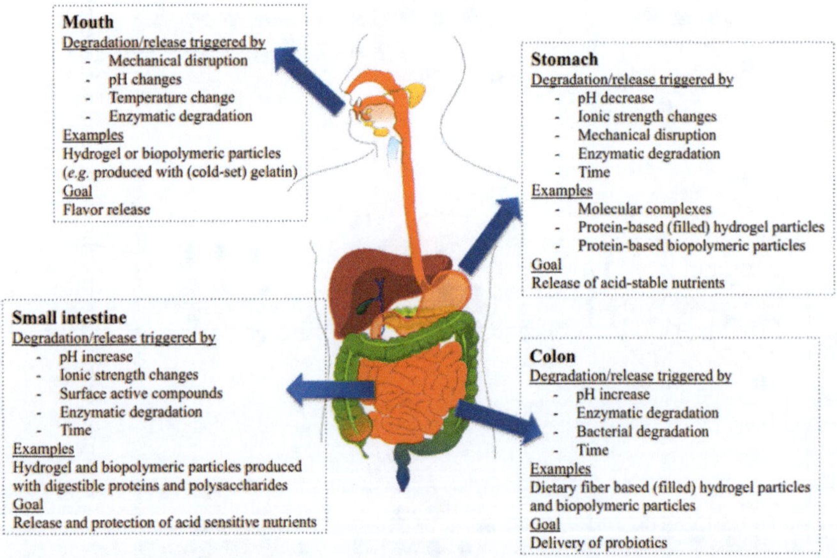

Figure 9.3 Gastrointestinal fate of a hydrogel-based delivery system for targeted delivery. Reproduced from ref. 33 with permission from Elsevier, Copyright 2014.

a poor understanding of the relationship between particle properties and their absorption characteristics (such as route and extent of absorption).

The bioavailability of the bioactive components is based on different fractions, namely the fraction of components liberated from the food matrix (delivery system) into the lumen of the GIT, the fraction of released bioactive components that were absorbed by epithelial cells, the fraction of absorbed component that reached the action site in the body after distribution, the fraction of bioactive component that was metabolically active at the site of action, and the fraction of bioactive component that remained within the body without excretion, and factors that affect them such as the nature of the bioactive component (molecular weight, solubility, dissolution, permeability, polarity, and physical state).[48] Further, certain metabolic reactions that occur within the GIT can also modify the structure and functional groups of the bioactive component, thereby changing its biological fate and activity. Several digestion and absorption studies investigated the gastrointestinal fate of delivery systems using *in vitro* and *in vivo* models to gain a clearer understanding of the gastrointestinal behavior of these delivery systems, such as their release mechanisms, release profiles, absorption, and bioavailability.[49]

9.5 Conclusion

In this chapter, we have discussed food structuring and size modification with specific emphasis on the encapsulation of active ingredients for specific functionalities. Encapsulation of active ingredients such as vitamins, minerals, flavors, preservatives, antioxidants, probiotics, bioactive compounds, and enzymes improves their stability against adverse environmental conditions, harsh processing, and gastrointestinal conditions, modifies their functional properties such as solubility, which aids in their easy incorporation into food matrices, and helps in masking their unfavorable sensory attributes (off-flavors and bitter taste). Further, size reduction to micro- and nano-levels during the encapsulation of active ingredients improves their absorption and bioavailability. Various structured delivery systems have been designed for the targeted delivery of active ingredients and controlling digestion by modifying the release behavior of lipids. Hence it is crucial to design delivery systems with appropriate release mechanisms and release profiles for specific functionality. The release mechanisms and release profiles can be controlled by varying the components of the delivery system (surfactant, emulsifier, lipid, biopolymer), fabrication method, size distribution, interfacial layer properties, particle charge, and porosity. In recent times, delivery systems have been designed using food-grade ingredients and nature-inspired structures as consumers increasingly demand clean-label and sustainable foods. Rapid progress is being made to understand the gastrointestinal fate of the delivery systems using various *in vitro*, *in vivo*, and *ex vivo* approaches. However, more investigations using standardized and authenticated methods such as human clinical trials are needed in the case of bioavailability studies.

References

1. V. Nedovic, A. Kalusevic, V. Manojlovic, S. Levic and B. Bugarski, An overview of encapsulation technologies for food applications, *Procedia. food sci.*, 2011, **1**, 1806–1815.
2. J. Ubbink and J. Krüger, Physical approaches for the delivery of active ingredients in foods, *Trends Food Sci. Technol.*, 2006, **17**(5), 244–254.
3. D. J. McClements, Recent developments in encapsulation and release of functional food ingredients: delivery by design, *Curr. Opin. Food Sci.*, 2018, **23**, 80–84.
4. E. Assadpour and S. M. Jafari, An overview of biopolymer nanostructures for encapsulation of food ingredients, in *Biopolymer Nanostructures for Food Encapsulation Purposes*, 2019, pp. 1–35.
5. S. M. Jafari and D. J. McClements, Nanotechnology approaches for increasing nutrient bioavailability, in *Advances in Food and Nutrition Research*, 2017, vol. 81, pp. 1–30.
6. S. Gorantla, G. Wadhwa and S. Jain, *et al.*, Recent advances in nanocarriers for nutrient delivery, *Drug Delivery Transl. Res.*, 2021, 1–26.
7. R. F. Gonçalves, J. T. Martins, C. M. Duarte, A. A. Vicente and A. C. Pinheiro, Advances in nutraceutical delivery systems: From formulation design for bioavailability enhancement to efficacy and safety evaluation, *Trends Food Sci. Technol.*, 2018, **78**, 270–291.
8. D. J. McClements and H. Xiao, Is nano safe in foods? Establishing the factors impacting the gastrointestinal fate and toxicity of organic and inorganic food-grade nanoparticles, *npj Sci. Food*, 2017, **1**(1), 6.
9. A. Acevedo-Fani, A. Dave and H. Singh, Nature-assembled structures for delivery of bioactive compounds and their potential in functional foods, *Front. Chem.*, 2020, **8**, 564021.
10. D. J. McClements, L. Bai and C. Chung, Recent advances in the utilization of natural emulsifiers to form and stabilize emulsions, *Annu. Rev. Food Sci. Technol.*, 2017, **8**, 205–236.
11. D. J. McClements, *Nanoparticle-and Microparticle-based Delivery Systems: Encapsulation, Protection and Release of Active Compounds*, CRC Press, 2014.
12. D. J. McClements and Y. Li, Structured emulsion-based delivery systems: Controlling the digestion and release of lipophilic food components, *Adv. Colloid Interface Sci.*, 2010, **159**(2), 213–228.
13. P. de Vos, M. M. Faas, M. Spasojevic and J. Sikkema, Encapsulation for preservation of functionality and targeted delivery of bioactive food components, *Int. Dairy J.*, 2010, **20**(4), 292–302.
14. U. Lesmes and D. J. McClements, Structure–function relationships to guide rational design and fabrication of particulate food delivery systems, *Trends Food Sci. Technol.*, 2009, **20**(10), 448–457.
15. O. Torres, B. Murray and A. Sarkar, Emulsion microgel particles: Novel encapsulation strategy for lipophilic molecules, *Trends Food Sci. Technol.*, 2016, **55**, 98–108.

16. P. Ezhilarasi, P. Karthik, N. Chhanwal and C. Anandharamakrishnan, Nanoencapsulation techniques for food bioactive components: a review, *Food Bioprocess Technol.*, 2013, **6**, 628–647.
17. D. McClements, Requirements for food ingredient and nutraceutical delivery systems, in *Encapsulation technologies and delivery systems for food ingredients and nutraceuticals*, Elsevier, 2012, pp. 3–18.
18. K. Ahmed, Y. Li, D. J. McClements and H. Xiao, Nanoemulsion-and emulsion-based delivery systems for curcumin: Encapsulation and release properties, *Food Chem.*, 2012, **132**(2), 799–807.
19. L. Mahalakshmi, P. Choudhary, J. Moses and C. Anandharamakrishnan, Emulsion electrospraying and spray drying of whey protein nano and microparticles with curcumin, *Food Hydrocolloids Health*, 2023, **3**, 100122.
20. N. Garti and A. Aserin, Micelles and microemulsions as food ingredient and nutraceutical delivery systems, in *Encapsulation Technologies and Delivery Systems for Food Ingredients and Nutraceuticals*, Elsevier, 2012, pp. 211–251.
21. R. Indiarto, L. P. A. Indriana, R. Andoyo, E. Subroto and B. Nurhadi, Bottom–up nanoparticle synthesis: a review of techniques, polyphenol-based core materials, and their properties, *Eur. Food Res. Technol.*, 2022, **248**(1), 1–24.
22. D. Guzey and D. J. McClements, Formation, stability and properties of multilayer emulsions for application in the food industry, *Adv. Colloid Interface Sci.*, 2006, **128**, 227–248.
23. I. J. Joye, G. Davidov-Pardo and D. J. McClements, Nanotechnology for increased micronutrient bioavailability, *Trends Food Sci. Technol.*, 2014, **40**(2), 168–182.
24. N. A. F. M. Anuar, F. Kormin and N. A. Z. Abidin, A Review on Natural-based Active Compounds Delivery System and its Potential in Food Preservative Application, *IOP Conf. Ser.: Earth Environ. Sci.*, 2019, **269**(1), 012010.
25. N. Gutiérrez-Méndez, D. R. Chavez-Garay and M. Y. Leal-Ramos, Lecithins: A comprehensive review of their properties and their use in formulating microemulsions, *J. Food Biochem.*, 2022, **46**(7), e14157.
26. C. E. Gumus, E. A. Decker and D. J. McClements, Formation and stability of ω-3 oil emulsion-based delivery systems using plant proteins as emulsifiers: Lentil, pea, and faba bean proteins, *Food Biophys.*, 2017, **12**, 186–197.
27. D. J. McClements, Crystals and crystallization in oil-in-water emulsions: Implications for emulsion-based delivery systems, *Adv. Colloid Interface Sci.*, 2012, **174**, 1–30.
28. R. Zhang, T. Belwal, L. Li, X. Lin, Y. Xu and Z. Luo, Recent advances in polysaccharides stabilized emulsions for encapsulation and delivery of bioactive food ingredients: A review, *Carbohydr. Polym.*, 2020, **242**, 116388.
29. C. Tan and D. J. McClements, Application of advanced emulsion technology in the food industry: A review and critical evaluation, *Foods*, 2021, **10**(4), 812.

30. J. S. Komaiko and D. J. McClements, Formation of food-grade nanoemulsions using low-energy preparation methods: A review of available methods, *Compr. Rev. Food Sci. Food Saf.*, 2016, **15**(2), 331–352.
31. I. J. Joye and D. J. McClements, Biopolymer-based delivery systems: challenges and opportunities, *Curr. Top. Med. Chem.*, 2016, **16**(9), 1026–1039.
32. A. Matalanis, O. G. Jones and D. J. McClements, Structured biopolymer-based delivery systems for encapsulation, protection, and release of lipophilic compounds, *Food Hydrocolloids*, 2011, **25**(8), 1865–1880.
33. I. J. Joye and D. J. McClements, Biopolymer-based nanoparticles and microparticles: Fabrication, characterization, and application, *Curr. Opin. Colloid Interface Sci.*, 2014, **19**(5), 417–427.
34. Z. Zhang, E. A. Decker and D. J. McClements, Encapsulation, protection, and release of polyunsaturated lipids using biopolymer-based hydrogel particles, *Food Res. Int.*, 2014, **64**, 520–526.
35. D. J. McClements, Advances in fabrication of emulsions with enhanced functionality using structural design principles, *Curr. Opin. Colloid Interface Sci.*, 2012, **17**(5), 235–245.
36. Y. Zhu, X. Chen, D. J. McClements, L. Zou and W. Liu, pH-, ion-and temperature-dependent emulsion gels: Fabricated by addition of whey protein to gliadin-nanoparticle coated lipid droplets, *Food Hydrocolloids*, 2018, **77**, 870–878.
37. Y. Mao and D. J. McClements, Fabrication of functional micro-clusters by heteroaggregation of oppositely charged protein-coated lipid droplets, *Food Hydrocolloids*, 2012, **27**(1), 80–90.
38. A. Kwan and G. Davidov-Pardo, Controlled release of flavor oil nanoemulsions encapsulated in filled soluble hydrogels, *Food Chem.*, 2018, **250**, 46–53.
39. U. Sadiq, H. Gill and J. Chandrapala, Casein micelles as an emerging delivery system for bioactive food components, *Foods*, 2021, **10**(8), 1965.
40. A. Jash, A. Ubeyitogullari and S. S. Rizvi, Synthesis of multivitamin-loaded heat stable liposomes from milk fat globule membrane phospholipids by using a supercritical-CO_2 based system, *Green Chem.*, 2020, **22**(16), 5345–5356.
41. Y. Sun, M. Zhong, Y. Liao, M. Kang, B. Qi and Y. Li, Pickering emulsions stabilized by reassembled oleosome protein nanoparticles for co-encapsulating hydrophobic nutrients, *Food Hydrocolloids*, 2023, **138**, 108445.
42. S. Boostani and S. M. Jafari, A comprehensive review on the controlled release of encapsulated food ingredients; fundamental concepts to design and applications, *Trends Food Sci. Technol.*, 2021, **109**, 303–321.
43. C. Dima, E. Assadpour, S. Dima and S. M. Jafari, Bioavailability of nutraceuticals: Role of the food matrix, processing conditions, the gastrointestinal tract, and nanodelivery systems, *Compr. Rev. Food Sci. Food Saf.*, 2020, **19**(3), 954–994.

44. D. J. McClements, Edible lipid nanoparticles: Digestion, absorption, and potential toxicity, *Prog. Lipid Res.*, 2013, **52**(4), 409–423.
45. W. I. Hagens, A. G. Oomen, W. H. de Jong, F. R. Cassee and A. J. Sips, What do we (need to) know about the kinetic properties of nanoparticles in the body?, *Regul. Toxicol. Pharmacol.*, 2007, **49**(3), 217–229.
46. L. M. Ensign, R. Cone and J. Hanes, Oral drug delivery with polymeric nanoparticles: the gastrointestinal mucus barriers, *Adv. Drug Delivery Rev.*, 2012, **64**(6), 557–570.
47. L. Hu, Z. Mao and C. Gao, Colloidal particles for cellular uptake and delivery, *J. Mater. Chem.*, 2009, **19**(20), 3108–3115.
48. D. J. McClements, Advances in edible nanoemulsions: Digestion, bioavailability, and potential toxicity, *Prog. Lipid Res.*, 2021, **81**, 101081.
49. P. Wu and X. D. Chen, Validation of *in vitro* bioaccessibility assays—A key aspect in the rational design of functional foods towards tailored bioavailability, *Curr. Opin. Food Sci.*, 2021, **39**, 160–170.
50. M. S. Abbas, F. Saeed and M. Afzaal, *et al.*, Recent trends in encapsulation of probiotics in dairy and beverage: A review, *J. Food Process. Preserv.*, 2022, **46**(7), e16689.
51. K. Yoha, S. Nida, S. Dutta, J. Moses and C. Anandharamakrishnan, Targeted delivery of probiotics: perspectives on research and commercialization, *Probiotics Antimicrob. Proteins*, 2021, 1–34.
52. J. A. Vellido-Perez, J. M. Ochando-Pulido, E. Brito-de la Fuente and A. Martinez-Ferez, Novel emulsions–based technological approaches for the protection of omega–3 polyunsaturated fatty acids against oxidation processes–A comprehensive review, *Food Struct.*, 2021, **27**, 100175.
53. D. J. McClements and B. Öztürk, Utilization of nanotechnology to improve the handling, storage and biocompatibility of bioactive lipids in food applications, *Foods*, 2021, **10**(2), 365.
54. J. Mujica-Álvarez, O. Gil-Castell and P. A. Barra, *et al.*, Encapsulation of vitamins A and E as spray-dried additives for the feed industry, *Molecules*, 2020, **25**(6), 1357.
55. T. Comunian, A. Babazadeh and A. Rehman, *et al.*, Protection and controlled release of vitamin C by different micro/nanocarriers, *Crit. Rev. Food Sci. Nutr.*, 2022, **62**(12), 3301–3322.
56. C. Liyanage and S. Zlotkin, Bioavailability of iron from microencapsulated iron sprinkle supplement, *Food Nutr. Bull.*, 2002, **23**(3_suppl1), 133–137.
57. R. Arshad, L. Gulshad and I. U. Haq, *et al.*, Nanotechnology: A novel tool to enhance the bioavailability of micronutrients, *Food Sci. Nutr.*, 2021, **9**(6), 3354–3361.
58. S. Sharma, S-F Cheng, B. Bhattacharya and S. Chakkaravarthi, Efficacy of free and encapsulated natural antioxidants in oxidative stability of edible oil: Special emphasis on nanoemulsion-based encapsulation, *Trends Food Sci. Technol.*, 2019, **91**, 305–318.
59. I. Ferreira, S. Rocha and M. Coelho, Encapsulation of antioxidants by spray-drying, *Chem. Eng. Trans.*, 2007, **11**(9), 713–717.

60. J. Aguilar-Toalá, D. Quintanar-Guerrero, A. Liceaga and M. Zambrano-Zaragoza, Encapsulation of bioactive peptides: a strategy to improve the stability, protect the nutraceutical bioactivity and support their food applications, *RSC Adv.*, 2022, **12**(11), 6449–6458.
61. K. Sarabandi, P. Gharehbeglou and S. M. Jafari, Spray-drying encapsulation of protein hydrolysates and bioactive peptides: Opportunities and challenges, *Drying Technol.*, 2020, **38**(5–6), 577–595.
62. Y. Premjit, S. Pandhi, A. Kumar, D. C. Rai, R. K. Duary and D. K. Mahato, Current trends in flavor encapsulation: A comprehensive review of emerging encapsulation techniques, flavour release, and mathematical modelling, *Food Res. Int.*, 2022, **151**, 110879.
63. S. M. Jafari, E. Assadpoor, Y. He and B. Bhandari, Encapsulation efficiency of food flavours and oils during spray drying, *Drying Technol.*, 2008, **26**(7), 816–835.
64. A. Bahrami, R. Delshadi, S. M. Jafari and L. Williams, Nanoencapsulated nisin: An engineered natural antimicrobial system for the food industry, *Trends Food Sci. Technol.*, 2019, **94**, 20–31.
65. F. Donsì, M. Annunziata, M. Sessa and G. Ferrari, Nanoencapsulation of essential oils to enhance their antimicrobial activity in foods, *LWT–Food Sci. Technol.*, 2011, **44**(9), 1908–1914.
66. M. Sherry, C. Charcosset, H. Fessi and H. Greige-Gerges, Essential oils encapsulated in liposomes: a review, *J. Liposome Res.*, 2013, **23**(4), 268–275.
67. S. A. Mahdavi, S. M. Jafari, E. Assadpour and M. Ghorbani, Storage stability of encapsulated barberry's anthocyanin and its application in jelly formulation, *J. Food Eng.*, 2016, **181**, 59–66.
68. J. Gómez-Estaca, R. Gavara and P. Hernández-Muñoz, Encapsulation of curcumin in electrosprayed gelatin microspheres enhances its bioaccessibility and widens its uses in food applications, *Innovative Food Sci. Emerging Technol.*, 2015, **29**, 302–307.
69. A. Mohammadi, S. M. Jafari, A. S. Mahoonak and M. Ghorbani, Liposomal/nanoliposomal encapsulation of food-relevant enzymes and their application in the food industry, *Food Bioprocess Technol.*, 2021, **14**, 23–38.
70. H. Jiang, Y. Sheng and T. Ngai, Pickering emulsions: Versatility of colloidal particles and recent applications, *Curr. Opin. Colloid Interface Sci.*, 2020, **49**, 1–15.
71. H. Jayan, M. M. Leena, S. S. Sundari, J. Moses and C. Anandharamakrishnan, Improvement of bioavailability for resveratrol through encapsulation in zein using electrospraying technique, *J. Funct. Foods*, 2019, **57**, 417–424.
72. P. Choudhary, S. Dutta, J. A. Moses and C. Anandharamakrishnan, Nanoliposomal encapsulation of chia oil for sustained delivery of α-linolenic acid, *Int. J. Food Sci. Technol.*, 2021, **56**(9), 4206–4214.
73. M. M. Leena, T. Anukiruthika, J. Moses and C. Anandharamakrishnan, Co-delivery of curcumin and resveratrol through electrosprayed core–shell nanoparticles in 3D printed hydrogel, *Food Hydrocolloids*, 2022, **124**, 107200.

74. P. Karthik and C. Anandharamakrishnan, Enhancing omega-3 fatty acids nanoemulsion stability and *in vitro* digestibility through emulsifiers, *J. Food Eng.*, 2016, **187**, 92–105.
75. M. N. Corstens, F. J. Troost and A. M. Alleleyn, *et al*, Encapsulation of lipids as emulsion-alginate beads reduces food intake: A randomized placebo-controlled cross-over human trial in overweight adults, *Nutr. Res.*, 2019, **63**, 86–94.
76. D. Xu, Z. Liu and Z. An, *et al.*, Incorporation of probiotics into 3D printed Pickering emulsion gel stabilized by tea protein/xanthan gum, *Food Chem.*, 2023, **409**, 135289.
77. M. Á. Hernández-Fernández, S. García-Pinilla and O. I. Ocampo-Salinas, *et al.*, Microencapsulation of vanilla oleoresin (V. Planifolia andrews) by complex coacervation and spray drying: Physicochemical and microstructural characterization, *Foods*, 2020, **9**(10), 1375.
78. F. Acevedo, M. Rubilar and I. Jofré, *et al.*, Oil bodies as a potential microencapsulation carrier for astaxanthin stabilisation and safe delivery, *J. Microencapsulation*, 2014, **31**(5), 488–500.
79. T. Ghorbanzade, S. Akhavan-Mahdavi, M. S. Kharazmi, S. A. Ibrahim and S. M. Jafari, Loading of fish oil into β-cyclodextrin nanocomplexes for the production of a functional yogurt, *Food Chem.: X*, 2022, **15**, 100406.
80. S. Hosseini and M. Varidi, Optimization of microbial rennet encapsulation in alginate–chitosan nanoparticles, *Food Chem.*, 2021, **352**, 129325.
81. Z. Zhang, R. Zhang, E. A. Decker and D. J. McClements, Development of food-grade filled hydrogels for oral delivery of lipophilic active ingredients: pH-triggered release, *Food Hydrocolloids*, 2015, **44**, 345–352.
82. R. Zhang and D. J. McClements, Characterization of gastrointestinal fate of nanoemulsions, in *Nanoemulsions*, Elsevier, 2018, pp. 577–612.
83. C. W. Pouton and C. J. Porter, Formulation of lipid-based delivery systems for oral administration: materials, methods and strategies, *Adv. Drug Delivery Rev.*, 2008, **60**(6), 625–637.
84. C. B. Huang, B. George and J. L. Ebersole, Antimicrobial activity of n-6, n-7 and n-9 fatty acids and their esters for oral microorganisms, *Arch. Oral Biol.*, 2010, **55**(8), 555–560.
85. J. Chai, P. Jiang and P. Wang, *et al*, The intelligent delivery systems for bioactive compounds in foods: Physicochemical and physiological conditions, absorption mechanisms, obstacles and responsive strategies, *Trends Food Sci. Technol.*, 2018, **78**, 144–154.
86. D. J. McClements and H. Xiao, Potential biological fate of ingested nanoemulsions: influence of particle characteristics, *Food Funct.*, 2012, **3**(3), 202–220.
87. H. Singh and A. Sarkar, Behaviour of protein-stabilised emulsions under various physiological conditions, *Adv. Colloid Interface Sci.*, 2011, **165**(1), 47–57.
88. H. Singh, A. Ye and D. Horne, Structuring food emulsions in the gastrointestinal tract to modify lipid digestion, *Prog. Lipid Res.*, 2009, **48**(2), 92–100.

89. D. J. McClements and Y. Li, Review of *in vitro* digestion models for rapid screening of emulsion-based systems, *Food Funct.*, 2010, **1**(1), 32–59.
90. A. C. Pinheiro, M. Lad, H. D. Silva, M. A. Coimbra, M. Boland and A. A. Vicente, Unravelling the behaviour of curcumin nanoemulsions during *in vitro* digestion: effect of the surface charge, *Soft Matter*, 2013, **9**(11), 3147–3154.
91. K. Feng, Y.-S. Wei, T.-G. Hu, R. J. Linhardt, M.-H. Zong and H. Wu, Colon-targeted delivery systems for nutraceuticals: A review of current vehicles, evaluation methods and future prospects, *Trends Food Sci. Technol.*, 2020, **102**, 203–222.
92. H.-Y. Tang, Z. Fang and K. Ng, Dietary fiber-based colon-targeted delivery systems for polyphenols, *Trends Food Sci. Technol.*, 2020, **100**, 333–348.

CHAPTER 10

Developing Novel Personalized Foods

10.1 Introduction

In general, humans are complex living organisms, and their food preferences and dietary habits are very distinct on various levels, namely individual (based on age, gender, preferences, food allergies, and health), family (income, locality, and dietary habits), specific profession and field for enhanced performance (sports person, bodybuilders, boxers, martial artists, gymnasts, *etc.*), health problems and disease conditions, and population level (geography and availability).[1,2] The idea of personalization of food and nutrition is age-old and exists across various cultures and populations. For instance, the use of herbs, spices, and roots with health and functional benefits to treat certain ailments and diseases has been practiced since ancient times.[3] In recent times, the concept of personalized foods, diet, and nutrition has been growing exponentiality owing to personal choice and preferences, health consciousness, necessity, affordability, and availability.[4] For example, several personalized food products targeted towards specific populations are commercially available, such as lactose-free and plant-based milk for lactose-intolerant people, zero-sugar and low-sugar varieties for diabetic patients, low-salt products for people with high blood pressure and cardiovascular disease (CVD), products fortified/enriched with iron and calcium targeting women, and instant energy drinks for sports persons and youngsters, *etc.*

On a larger scale, public health recommendations suggested by health organizations and governments are generally focused on preventing chronic health/disease conditions of the population and are mainly based on age and gender factors. However, various other factors also have key roles in

Food Chemistry, Function and Analysis No. 42
Food Digestion and Absorption: Its Role in Food Product Development
By C. Anandharamakrishnan, Jeyan Arthur Moses and S. Priyanka

Published by the Royal Society of Chemistry, www.rsc.org

public health, such as genotype, behavior, personality, phenotype, and socio-psychological environment, and need to be taken into consideration when formulating a personalized diet plan to reap maximum benefits.

Figure 10.1 illustrates the personalized food perspective.

Personalized nutrition has been defined as the use of individual-specific information, founded in evidence-based science, to promote dietary behavior change that may result in measurable health benefits.[5] In a recent study, a four-step cycle of personalized nutrition encompassing (i) collection of individual-specific information, (ii) translation of individual data into evidence-based dietary recommendations, (iii) promotion of dietary behavior change, and (iv) measurement of the success of the advice and behavior change support was suggested for improving and sustaining health[6] (Figure 10.2).

In this chapter, we consider the need for and significance of personalized foods for different age groups, genders, and people with specific health conditions, needs, and preferences. Further, the chapter also covers various functional food ingredients used in the development of personalized foods for the above-mentioned population groups. Additionally, the last two sections summarize clinical evidence concerning functional ingredients used in the treatment of various diseases and tailor-made foods for specific conditions and health benefits.

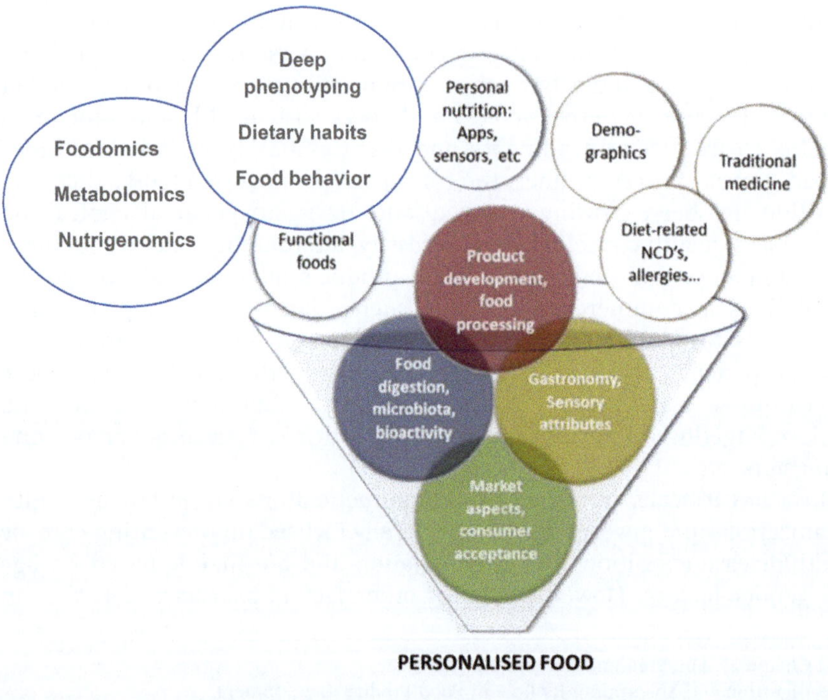

Figure 10.1 The personalized food perspective. Reproduced from ref. 4, https://doi.org/10.1016/j.tifs.2020.05.021, under the terms of the CC BY 4.0 license, http://creativecommons.org/licenses/by/4.0/.

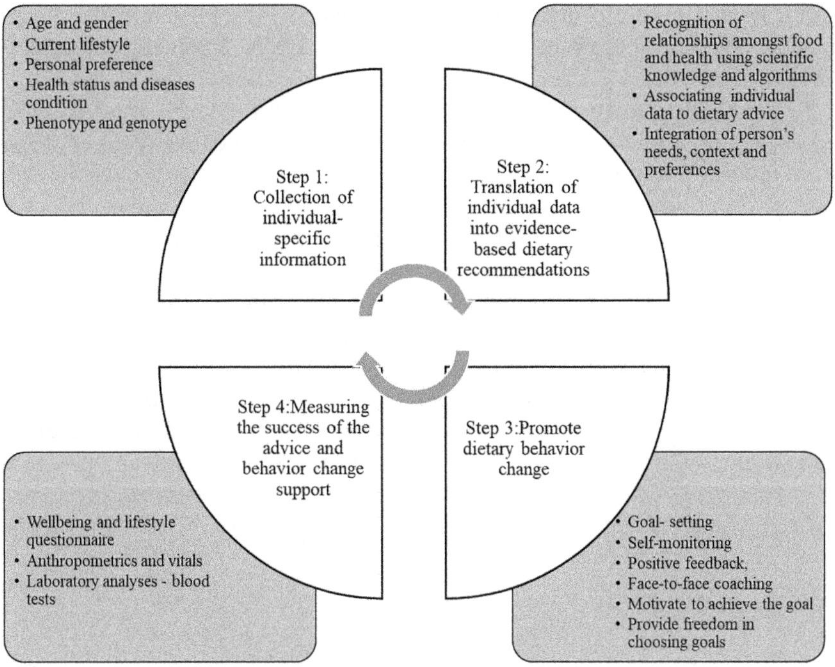

Figure 10.2 A four-step cycle of personalized nutrition. Reproduced from ref. 6, https://doi.org/10.3390/nu13061763, under the terms of the CC BY 4.0 license, http://creativecommons.org/licenses/by/4.0/.

10.2 Personalized Foods for Different Age Groups

First, the key main objective behind the personalization of foods is to improve the quality of an individual's life by providing all the nourishment to their body through food. The nutritional requirements change for an individual throughout their lifetime from birth to death. Based on age, the human population can be broadly divided into six major groups: infants (0–1 years), toddlers (1–3 years), children (5–12 years), adolescents and youths (13–19 years), adults (20–65 years), and the elderly/geriatric population (≥65 years).[7]

10.2.1 Infants

An infant's nutrition and food choices are mostly breast milk and commercial infant formulas.[8] The mother's breast milk contains various protective substances, including antibodies, white cells, bifidus factor, and lactoferrin, which protect infants from infections, inactivate harmful bacteria, aid in the growth of healthy bacteria such as *Lactobacillus bifidus*, and act as antibacterial agents and also help in iron binding and absorption.[9,10] In addition, breast milk contains various vital nutrients such as lactose (easily digestible and absorbable form), α-lactalbumin (protein), vitamins and minerals except vitamin D, and linolenic acid (an essential omega-6 fatty acid). Concerning minerals, breast milk is rich in magnesium, calcium, phosphorus and highly absorbable iron and zinc, but is low in sodium.[11]

For newborns until the age of 1 year, breast milk is an ideal food that provides efficient nourishment and protection. In addition to breast milk, solid foods such as single-grain, iron-fortified cereals, such as rice, barley, or oats, can be included in infants' diets after 4–6 months of age to meet their energy needs.[12] However, after 1 year of age or due to some health reason or personal preference, infants are fed with commercial or home-made infant formulas. The infant formulas and replacements for breast milk must be selected carefully. Supplementation of some essential nutrients, such as essential fatty acids (linoleic acid, α-linolenic acid, arachidonic acid, docosahexaenoic acid, and eicosapentaenoic acid), vitamins (A, D, E, K, and B_{12}) and minerals (such as iron, zinc, copper, chromium, manganese, selenium, and fluoride), must be introduced as they are crucial for the baby's development.[13] In addition to these, other macronutrients such as carbohydrates, proteins, and fat and an adequate amount of fluids need to be fed.

10.2.2 Toddlers

This is the second stage in the human lifecycle at ages between 1 and 3 years where the child learns to walk and move, *i.e.* "toddle". During this stage, the child will have lost almost all of their baby fat and will have a higher body mass, denser muscle tissues, and greater bone density than infants. During this stage, the toddler must be fed a diet consisting of a wide range of foods and beverages from various food groups to ensure good health and nutrition.[7]

10.2.3 Children

Based on age, children can be broadly classified into two age groups, namely (i) 4–8 years old and (ii) 9–12 years old. The US Department of Agriculture (USDA) MyPlate plan recommends 1200 calories (girls) and 1400 calories (boys) daily in the case of children between 4 and 8 years old.[14] In the case of children aged 9–12 years, girls require 1400–2200 calories daily and boys require 1600–2600 calories daily.[7]

10.2.4 Adolescents (Youths)

Adolescence or the teenage period, which ranges between 13 and 19 years, demands good nutrition as these are the years of rapid growth and development. The calorie requirements per day vary based on gender and age (13-year-old girls 1600 calories, 14–18-year-old girls 1800 calories, 13-year-old boys 2000 calories, and 14–18-year-old boys 2000–2400 calories) to meet their growth, enhanced energy and muscular needs.[7] In recent years, being overweight and obese have become two major nutritional problems faced by teenagers owing to changes in their eating habits such as the consumption of fast foods (*e.g.* burgers, fries, tacos, hot dogs, pasta, pizza, and macaroni and cheese) and foods high in sugar and fat, together with a reduction in the consumption of healthy foods such as fruits and vegetables.[15]

10.2.5 Adults

The age between 20 and 65 years is classified as adulthood, and this is a period of stability and not growth.[16] Hence factors such as food, weight, activity, and nutrition intake should be normal. Further, nutritionists advise people in adulthood to practice and follow the "four Ms", namely modification (old eating and activity habits), moderation (health behaviors such as food intake, sleep, exercise, and alcohol), movement (exercise for a minimum of 30 min or walk 10 000 steps), and management (disease or health conditions as they arise).[7] Figure 10.3 summarizes dietary recommendations provided by health professionals for "normal healthy" adults.

10.2.6 Elderly/Geriatric Population

People aged ≥65 years are classified as elderly, and they are further subclassified into early elderly (aged 65–74 years) and late elderly (aged ≥75 years).[17] However, biologically, aging starts after a person reaches their full growth potential. In the case of women, the growth stops in their late teenage years and for men in their early twenties. Further, the rate at which each person gets physically older is governed by various factors such as genetics, diet, exercise, and healthy/unhealthy lifestyle choices.[7] A range of factors must be taken into consideration when formulating a personalized nutrition or diet, apart from the nutrition needs of the geriatric population such as biological determinants, economic elements, structural determinants, social characteristics, and psychological determinants (Figure 10.4).

Various experimental studies have reported that because of aging there is a decline in food intake (senile anorexia) due to changes in the regulation of peripheral feedback signals, modified gastrointestinal control, altered taste threshold, and taste dysfunction (due to certain diseases, medications, bad oral hygiene, use of dental prostheses, malnutrition, surgical interventions, and environmental exposure), smell disorders, alcoholism; central nervous system (CNS) diseases (such as Alzheimer's disease and Parkinson's disease), dysphagia (chewing and swallowing difficulties), hypermetabolism and hypercatabolism, chronic/acute clinical conditions and medications, and environmental and psychological causes.[18] All of these aspects must be taken into consideration when designing personalized food for the geriatric population. Further, there is also a heightened risk of foodborne illness in the case of the aging population. Therefore, proper care and attention must be applied during food selection, storage, and preparation, and carers must be educated about food safety practices and be made aware of the consequences.[19]

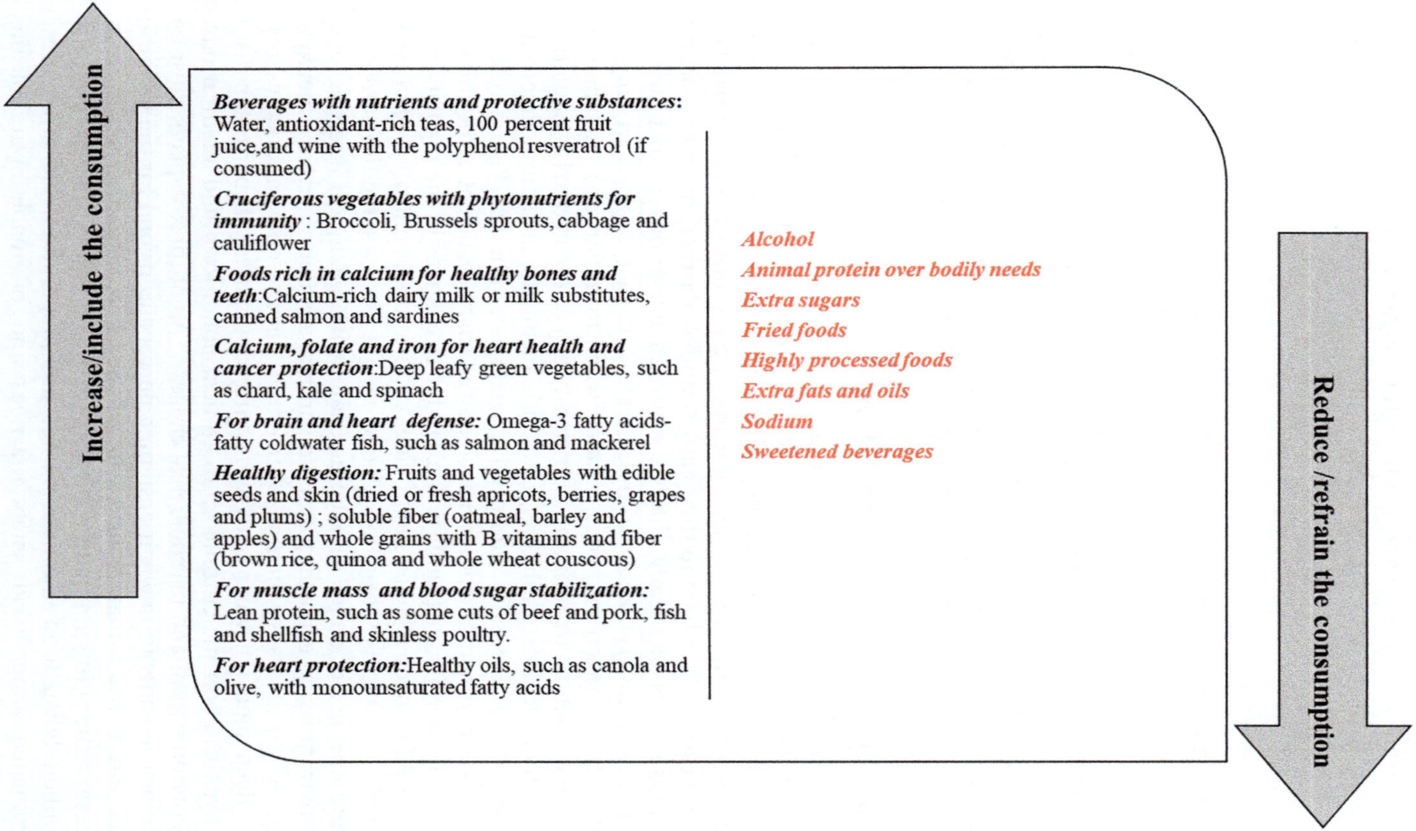

Figure 10.3 Dietary recommendations by health professionals to "normal healthy" adults.

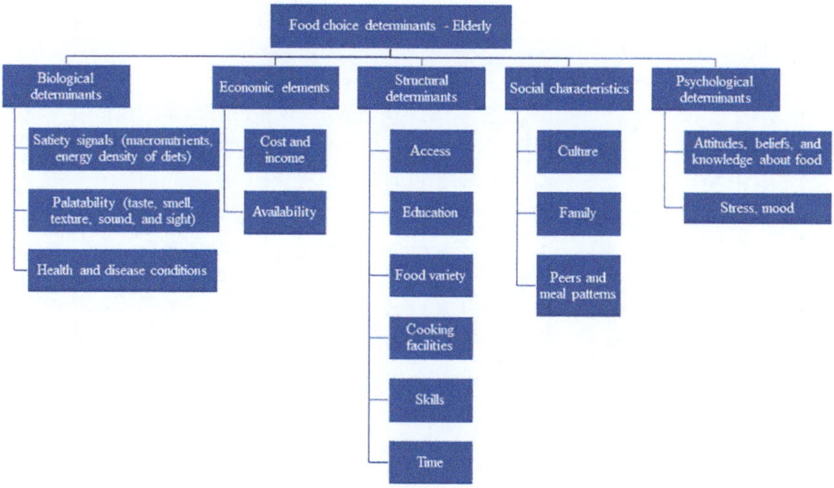

Figure 10.4 Factors determining food choice in the elderly population. Reproduced from ref. 18 with permission from Elsevier, Copyright 2017.

10.3 Gender-specific Novel and Personalized Foods

The relationship between food and gender is complicated.[20] The nutritional and energy needs of men and women differ significantly. In recent years, the double burden of malnutrition, which is the coexistence of over-nutrition (overweight and obesity) and under-nutrition, along with various non-communicable diseases within the same community and households, has complicated decision-making in terms of public health measures and food choices and diets.[21] Therefore, developing personalized nutrition guidelines that are unique for men and women based on their specific requirements is necessary. Table 10.1 summarizes recent studies on gender-specific clinical nutritional trials to improve the health of women and men. Various food manufacturers are designing and developing personalized foods with targeted health benefits based on gender-specific needs and requirements, although they can be consumed by all. Also, food manufacturers and researchers need to give more attention and importance to designing personalized foods for women rather than men owing to their unique and changing health and nutrition needs across their lifetime. Commercially available gender-targeted personalized foods are listed in Table 10.2.

In the case of women, personalized food and nutrition are much needed to satisfy their specific nutritional needs during their preconception, pregnancy, and maternity. It has been reported that during pregnancy (third trimester) there is a sharp increase in need (between 6 and 50%) for both macronutrients (such as energy and proteins) and micronutrients (such as iron, iodine, zinc, magnesium, selenium, folate, vitamin B_6, niacin, riboflavin, thiamine, pantothenic acid, vitamin C, A, and B_{12}, and choline) to meet the elevated demands of the growing fetus and maternal tissue deposition.[22]

Table 10.1 Summary of recent studies on gender-specific clinical nutritional trials to improve women's and men's health.

Targeted group	Key objective	Functional ingredient	Significant findings	Ref.
Women with type 2 diabetes	To evaluate the efficacy of a resistant dextrin on immune-mediated inflammation and the hypothalamic–pituitary–adrenal axis	10 g per day of maltodextrin or Nutriose® 06 (a resistant dextrin) for 8 weeks	Nutriose® 06 supplementation may improve immune system function, biomarkers of inflammation/anti-inflammation, and mental health in type 2 diabetic patients	71
Women with low iron aged 18–40 years	To study the effectiveness of iron fortification in low-fat margarine	Incorporated 14 mg of iron as • micronized ground ferric pyrophosphate (MGFePP) or • sodium iron edetate (NaFeEDTA) or given a placebo daily	Both iron-fortified low-fat margarines improved the iron status in young women. NaFeEDTA fortification increased the body iron stores 2–3-fold compared with MGFePP fortification	72
Pregnant women from rural Vietnam	To compare the efficacy of iron supplementation in reducing anemia in pregnant women with tablets and iron-fortified cow's milk	The iron-fortified milk group was given 400 mL of milk with 15 mg of iron per day. The iron supplement tablet group received 60 mg of iron per day	Both iron supplementation through tablets and iron-fortified cow's milk enhanced the iron status of the target group. The distribution of fortified milk to pregnant women was far more expensive than the distribution of iron supplements	73
Postmenopausal women	To test the impact of bedtime consumption of fermented milk supplemented with calcium enhancer and	Control milk: fermented milk with 1.5% fat. Fermented milk supplemented with	The nocturnal bone resorption was reduced by bedtime consumption of fermented milk.	74

	calcium, and its role in inhibiting nocturnally increased bone resorption	1. calcium alone; 2. calcium, and calcium absorption enhancers (inulin-type fructans and casein phosphopeptides)	Supplemented calcium had no further effect without the addition of calcium absorption enhancers	
Zinc-deficient women	To assess the effect of the consumption of zinc-fortified bread on blood serum zinc and iron status	Non-fortified bread was prepared from non-fortified flour. Low- and high-zinc bread were made from flours fortified with 50 and 100 ppm of zinc sulfate, respectively	Fortification of flour with 50–100 ppm of zinc was an effective way to achieve adequate zinc intake and absorption in zinc-deficient subjects. Consumption of zinc-fortified bread improved iron absorption	75
Institutionalized elderly women	To investigate the outcome of dairy supplementation (soft plain cheese) moderately (17–25% of recommended allowance) fortified with vitamin D, calcium, and protein on bone metabolism	Soft plain cheese made of semi-skimmed milk provided 2.5 μg of vitamin D_3, 302 mg of Ca, 233 mg of P, and 14.2 g of proteins. Two servings per day for a 1 month period	The trial supplementation reduced bone resorption and improved bone formation, which can be concluded from the increase in the bone formation marker P1NP (amino-terminal propeptide of type 1 procollagen) and insulin-like growth factor-I	76
Women of reproductive age	Folic acid supplementation in pregnancy and its potential benefits	Folate, folic acid, and L-methylfolate	The demand for folate increased during pregnancy. Proper supplementation of folate is required to prevent anemia and peripheral neuropathy in mothers and congenital abnormalities such as neural tube defects in fetuses	77

Table 10.1 (*Continued*)

Targeted group	Key objective	Functional ingredient	Significant findings	Ref.
Nepalese men	To investigate the relationship between the average daily intake of dietary zinc, iron, and calcium on coronary artery disease	Less than *versus* equal to or more than the recommended daily allowance (RDA)	Dietary zinc intake equal to or above the RDA has potential for protection against the development of coronary artery disease. Further research is needed to draw any conclusions about the effect of dietary calcium and iron intake on the prevention of coronary artery disease	78
Overweight middle-aged men (40–65 years) with no history of bowel disease	To estimate the effect of high-fiber foods (whole-grain rye and wheat foods) and low-fiber foods on markers of bowel health, colon cancer risk, postprandial glucose, and insulin responses	The low-fiber diet provided 19 g of dietary fiber per day. The high-fiber wheat and rye diet provided 32 g of dietary fiber per day from the respective sources	Consumption of high-fiber rye and wheat food improved several markers of bowel and metabolic health compared with that of low-fiber food. Fiber from rye was more effective than that from wheat in the overall improvement of biomarkers of bowel health	79

Table 10.2 Commercially available gender-targeted personalized foods.

Brand name	Targeted gender tag line	Varieties available	Highlights
Kellogg's® Special K®	"Specifically designed with women's wellness in mind"	Kellogg's® Special K® Original	Source of fiber, protein (grain-based), vitamin D, calcium, iron, and B vitamins. Low in fat
		Kellogg's® Special K® High Fibre	High in fiber. Supports gut health. Source of protein
		Kellogg's® Special K® Fruit & Nut	Source of fiber, vitamin D, calcium, iron, and B vitamins
		Kellogg's® Special K® Forest Berries	
		Kellogg's® Special K® High Protein	High in protein (grain-, nut-, and seed-based protein). Good source of fiber, vitamins (B_2, niacin, B_6, folate, and D), and minerals (zinc, calcium, and iron)
		Kellogg's® Special K® Honey Almond	Good source of protein and fiber
		Kellogg's® Special K® Lower Sugar	Less than 4 g of sugar per serving
		Kellogg's® Special K® Gluten Free	Gluten free
		Kellogg's® Special K® Gluten Free Almond and Cranberry	

Table 10.2 (*Continued*)

Brand name	Targeted gender tag line	Varieties available	Highlights
Protein and vitamin supplements – several local brands	"To build muscle"	Powder and shakes	Support energy production. Enhance muscle strength. Increase lean body mass
Horlicks™	"Bone nutrition specialist with CALSEAL™ formula to help calcium reach your bones"	Horlicks Women's Plus™	100% RDA of calcium, vitamin D, and vitamin K_2 which helps in maintaining bone health. No added sugar
	"Designed keeping in mind the higher nutritional needs of pregnant and lactating mothers"	Horlicks Mother's Plus™	Contains 25 vital nutrients. Protein for healthy birth weight. docosahexaenoic acid for brain development. No added sugar

However, the requirements for other nutrients such as vitamin D, E, and K, biotin, calcium, phosphorus, and fluoride remain unaltered.[23] In the case of men, most of the clinical nutritional intervention studies were focused on cardiovascular health,[24,25] prostate cancer,[26,27] bowel health,[28,29] reproductive health,[30,31] muscle mass,[32] obesity, and weight and diabetes management.[33,34]

10.4 Functional Foods and Bioactive Components

In recent years, socioeconomic changes, and increasing public awareness regarding health and wellness, and the role of food in disease prevention have resulted in a lot of research effort to understand the relationship between food, health, and disease prevention. Major food companies have ventured to capitalize on this booming market trend and age-old philosophy Let food be thy medicine and medicine be thy food of Hippocrates. Functional foods are defined as whole foods along with fortified, enriched, or enhanced foods that have a potentially beneficial effect on health when consumed as part of a varied diet regularly at effective levels.[35] The term "nutraceuticals" has also been used in several instances interchangeably with "functional foods". However, the word nutraceuticals is derived from "nutrition" and "pharmaceuticals" and it was coined by Dr Stephen DeFelice in 1989 and is defined as any substance that is a food or a part of a food and provides medical or health benefits, including the prevention and treatment of disease.[36] Further, the consumption of various bioactive compounds present in foods or whole functional foods has been scientifically proven in many instances to treat or prevent certain diseases or improve overall health and well-being.[37] Table 10.3 provides a summary of various functional food ingredients that aid in the prevention of certain disease conditions and/or improve health.

10.5 Certain Foods and Their Functional Ingredients: Clinical Evidence in Disease Prevention, Treatment, and Reversal

Nowadays, various non-communicable metabolic disease conditions, mainly diabetes, obesity, cardiovascular diseases, cancers, and osteoporosis, are widely prevalent among the public and result in either morbidity and/or mortality. In this section, we summarize various clinically proven foods and their functional ingredients along with other positive lifestyle changes that can aid their prevention, treatment, and reversal.

10.5.1 Type 2 Diabetes Mellitus

Diabetes is a chronic metabolic disorder characterized by a persistent increase in blood glucose level (hyperglycemia) caused by defective insulin

Table 10.3 Summary of various functional food ingredients for preventing certain diseases conditions and /or improved health.

Food category	Functional food ingredients	Common food sources	Health claims and clinical significance	Ref.
Carbohydrates	Dietary fibers – RDI 38 and 25 g per day: Insoluble fibers: Cellulose Hemicellulose	Cellulose – bran, legumes, nuts, peas, roots, cabbage, and apple skins. Hemicellulose – bran, nuts, legumes, and whole grains	Reduce intestinal transit time of feces. Increase stool bulk and stimulate bowel movement. Aid in the prevention of gastrointestinal diseases, constipation, hemorrhoids, diverticulosis, and colorectal cancer	80
	Soluble dietary fiber: Inulin, pectin, and gums	Inulin – asparagus, bananas, burdock, chicory, dandelion root, garlic, artichokes, leeks, onions, and sugarcane. Pectin and gums – fruits, berries, and seeds	Decrease hunger, increase satiety and fullness. Slow the transit time of food waste in the bowel. Reduce the level of blood cholesterol and postprandial blood glucose. Considerably lessen the risks of developing diseases such as obesity, cardiovascular disease, hyperlipidemia, diabetes, and hypercholesterolemia	81 and 82
	β-Glucan	Barley, oats, and some varieties of mushrooms	Antioxidant, antiobesity, antidiabetic, antiviral, and antibacterial activities. Immunomodulatory, antitumor, anti-inflammatory, cardiovascular, and liver protective. Reduce appetite, enhance fullness, lower energy intake, and aid in improved body management	83–85

	Resistant starches – RDI 6–12 g per day	Primarily made of amylose and most abundantly present in cereal grains, seeds rich in starch and starchy foods subjected to cooling and heating	Gut microbes ferment resistant starches and produce short-chain fatty acids (such as butyric, propionic, and acetic acids), which prevent the proliferation of colon cancer. Improve insulin sensitivity, suppress appetite, and prevent blood sugar fluctuations. Lower glycemic response, lower digestion rate and rate of glucose absorption. Help with managing body weight, body fat storage, and cholesterol levels	86–88
Proteins	Food protein-derived hydrolyzates and bioactive peptides *via* enzyme treatment	Plant sources – walnuts, mulberry leaves, hazelnuts, rice bran, cowpea, hemp seed, flax seed, chia, quinoa, and millet. Other sources – include insects, algae, mushrooms, fish, sea animals, and animal proteins	Antioxidant proteins scavenge free radicals, reduce ferric iron to ferrous iron, bind metals (metal chelation), and inhibit lipid oxidation. Aid in the management and control of oxidative damage-related chronic diseases. They also provide various other health benefits such as antihypersensitive, antimicrobial, anti-inflammatory, anticancer, and immunomodulatory properties and prevent cardiovascular diseases	89–92

Table 10.3 (*Continued*)

Food category	Functional food ingredients	Common food sources	Health claims and clinical significance	Ref.
	Lectins	They are proteins or glycoproteins mainly derived from various mushrooms such as *Agaricus bisporus* and *Tricholoma mangolicum*	Regulate the immune system, promote the cell adhesion process, and support the activation of lymphocytes. Also possess strong antiproliferative properties and inhibit the proliferation of tumor cells, especially in breast cancer	93 and 94
Fats or lipids	Polyunsaturated fatty acids (PUFAs):		Aid in brain development and function of the nervous system. Alleviate inflammation caused by rheumatoid arthritis, and also help with lowering the risk of various diseases such as cancer, diabetes, heart attack, depression, and Alzheimer's disease. Also aid in lowering the concentration of low-density lipoprotein (LDL) cholesterol	95–97
	Omega-3 (n-3): α-linolenic acid (ALA, 18:3)	Seeds (chia, rapeseed, canola, flax, and walnuts), marine algae, vegetable oils, and natural herbs		
	Omega-6 (n-6): linoleic acid (LA, 18:2)	Seeds and oil of sunflower, wheat, corn, and soy		
	Conjugated linoleic acid (CLA)	Predominantly found in dairy products and meat		
	Long-chain PUFAs: eicosapentaenoic acid (EPA), docosahexaenoic acid (DHA), and docosapentaenoic acid (DPA)	Derived from omega-3 (n-3) of fish and animal products		

Antioxidants				
Carotenoids	α-Carotene and β-carotene	Fruits, vegetables, and carrots	Deactivate free radicals and thereby prevent cell damage. Aid in the prevention of various cancers. Play a vital role in the prevention of obesity and type 2 diabetes	98 and 99
	Lycopene	Red fruits and vegetables such as tomatoes, pink guavas, apricots, watermelons, and pink grapefruit. Tomato products (ketchup, sauces)	Exhibits antioxidant, anti-inflammatory antihypertensive, and antiaggregatory effects. Reduces the risk of prostate cancer, breast cancer, and cancer cell growth. Acts as a cardioprotective and neuroprotective agent	100 and 101
	Lutein	Green leafy vegetables: kale, broccoli, spinach, lettuce, and peas	Reduces the risk of muscular degeneration and has ophthalmological effects. Provides skin protective, neuroprotective, antimicrobial, cardioprotective, antiosteoporosis, and antiplasmodial effects. Aids in wound healing and cell proliferation	102
Phenolics	Anthocyanidins: cyanidin, delphinidin, pelargonidin, peonidin, malvidin, and petunidin	Fruits: berries, currants, grapes, and some tropical fruits. Vegetables: black carrots, purple corn, purple sweet potato, and red cabbage. Flowers and grain with red, purple, and blue colors	Neutralize free radicals. Reduce the risk of cancer. Provide antioxidative, antidiabetic, antiobesity, antimicrobial, and neuroprotective effects. Aid in the prevention of cardiovascular diseases and improve visual health	103

Table 10.3 (*Continued*)

Food category	Functional food ingredients	Common food sources	Health claims and clinical significance	Ref.
	Flavan-3-ols – catechins: epicatechin, epigallocatechin, epicatechin gallate, and epigallocatechin gallate	Teas, red wine, cocoa, chocolates, apples, blackberries, black grapes, apricots, pears, raspberries, and strawberries	Act as antioxidant, anticarcinogen, cardiopreventive, antimicrobial, antiviral, and neuroprotective agents	104
	Flavonones: naringenin, hesperetin, and eriodictyol	Citrus fruits and their peel: grapefruit, oranges, lemons, limes, *etc.* Tomato skin, kiwi fruit, and bananas	Neutralize free radicals. Reduce the risk of cancer, provide antioxidant and antidiabetic effects, and improve gastrointestinal health	105 and 106
	Flavones: apigenin, luteolin, chrysin, baicalin, vitexin, maysin, *etc.*	Fruits and vegetables: parsley, celery, maize silk, Chinese hawberry, *Scutellaria* root, blue passion fruit flower	Anti-inflammatory, antimicrobial and anticancer agents	107
	Lignans	Present in low concentrations in various seeds, grains, fruits, and vegetables and in high concentrations in flax, rye, and sesame	Aid in the prevention of cancer, cardiovascular diseases, and renal failure. Possess antimicrobial, anti-inflammatory, and antioxidant activities	108
	Tannins (proanthocyanidins)	Berries (lingonberry, cranberry, black elderberry, black chokeberry, blackcurrant, and blueberry), fruits (mulberry, plum, apricot, walnut, silverberry, and pomegranate), and cocoa products	Improve urinary tract health. Reduce the risk of cardiovascular diseases. Provide antioxidant, anticancer, antidiabetic, neuroprotective, and antimicrobial effects	109
	Flavonols: kaempferol, quercetin, myricetin, rhamnetin, morin,		They have impressive health benefits such as antioxidation,	110

	fisetin, galangin, azaleatin, and their respective glycosyl derivatives	Quercetin – raw capers, elderberry juice, chokeberry, dill, and coriander. Kaempferol – capers, saffron, arugula (rocket), and kale. Myricetin – carob fiber, fennel leaves, parsley, and goji berries	cardioprotection, antibacterial, antiviral, and anticancer activity	
	Isoflavones: daiazein and genistein	Soybeans and soy-based foods	Alleviate menopause symptoms such as hot flashes. Provide protection against heart disease and some cancers. Help with lowering LDL and total cholesterol	111
	Non-flavonoids: gallic acid, hydroxycinnamates (*p*-coumaric acid, caffeic acid, ferulic acid, chlorogenic acids), phenylvaleric acid, and stilbenes	Found in various fruits, berries, coffee, nuts, *etc.*	They are reported to inhibit or retard a wide variety of cardiovascular diseases and cancers. Increase stress resistance. Enhance longevity	112
Phytosterols	Stigmasterol	Corn, soy, wheat, wood. oils, various fruits (kiwi, papaya, and avocado), and vegetables (carrots, dill, parsley, and red beet)	Lowers blood cholesterol levels by inhibiting cholesterol absorption. Aids in the prevention of cancer, cardiovascular disease, osteoarthritis, and inflammation.	113
	β-Sitosterol	Lipid-rich plant-based foods such as nuts, legumes, rice bran, wheat germ, olive oil, peanuts, corn, seeds, and vegetable oils	Offers various beneficial effects such as anxiolytic and sedative effects, antioxidant, analgesic, immunomodulatory, and antidiabetic activities, antimicrobial and wound healing effects, and anticancer and anti-inflammatory activity.	114

Table 10.3 (*Continued*)

Food category	Functional food ingredients	Common food sources	Health claims and clinical significance	Ref.
			Aids with lipid lowering to protect the liver against non-alcoholic fatty liver disease. Aids in the prevention of respiratory diseases	
	Campesterol	Present in various fruits, nuts, vegetables, flowers, and oil seeds (canola and corn)	Provides anti-inflammatory effects, anticancer effects (especially ovarian cancer), reduces visceral fat deposition, and has cholesterol-lowering effects	115
Tocopherols/ tocotrienols	Vitamin E	Vegetable oils (especially corn, olive, palm, rice bran, and peanut), cereals, nuts, seeds, salad dressings, and margarine	Potential antioxidant. Aids in the prevention and treatment of cardiovascular, neurological, and aging-related diseases	116
Organosulfur compounds	Diallyl sulfide, allicin, methionine, *S*-allylcysteine, sulforaphane, lipoic acid, cysteine, and sulfonylureas	Cruciferous vegetables such as broccoli, cauliflower, and Brussels sprouts, and Allium vegetables such as cabbage, onions, garlic, and leeks	They have strong antioxidant activity. Beneficial in the treatment of various pathological conditions including neurodegenerative disorders, cardiovascular diseases, cancer, and diabetes. In addition, they have several other beneficial effects such as antiplatelet, antihyperlipidemic, antiaging, immunomodulatory, antimicrobial, antiparasitic, antihypertensive, anti-inflammatory, antiatherosclerotic and antiviral activities	117

secretion due to impaired β-cell function or its receptors leading to insulin resistance.[38] The International Diabetic Federation (IDF) has estimated that by 2045, about 783 million people might be affected by diabetes, and in 2021 about 537 million adults (aged 20–79 years) were living with diabetes.[39] Various factors, such as urbanization, sedentary work, and an unhealthy lifestyle (lack of exercise, poor eating habits, disturbed sleep patterns, smoking, and alcohol consumption), lead to obesity, diabetes, and cardiovascular diseases. If not treated well, diabetes can lead to various life-threatening complications, organ failure (heart, vascular, renal, neurons, eyes), and even death.[40] Apart from medications and insulin, various functional food ingredients such as herbs (*Ginkgo biloba*, ginseng, *Astragalus membranaceus*, *Gymnema sylvestre*, and *Polygonum multiflorum*), and functional foods (aloe vera, bitter melon/gourd, cinnamon, fenugreek, garlic, ginger, turmeric, green tea from *Camellia sinensis*, soybeans, polyunsaturated fatty acids, probiotics and prebiotics, and whole grains) have been used as an alternative or complementary therapy for managing diabetes and have been clinically proven to have antidiabetic/hypoglycemic effects.[41]

10.5.2 Obesity

Obesity, a global epidemic, has been increasing exponentially in recent years. Food, diet, lifestyle, insufficient physical activity, metabolic factors, and genetic predispositions play a key role in the manifestation of obesity.[42] Being overweight or obese puts subjects at major risk of developing various other disease conditions such as type 2 diabetes, cardiovascular diseases, obesity-induced inflammatory stress, several types of cancers, osteoarthritis, and early death.[43] In addition to positive lifestyle changes and physical activity, various bioactive compounds have been tested using *in vitro* cell culture models for potential anti-obesity activity. For example, flavonoids (quercitrin, afzelin, and isoquercitrin) from *Acer okamotoanum* were found to reduce triglyceride levels and significantly lowered the expression of lipogenesis-related proteins,[44] Indian gooseberry (*Phyllanthus emblica* L.) showed beneficial effects on adipogenesis and significantly decreased triglyceride accumulation,[45] the natural flavonoid kaempferol and the flavone hispidulin were found to reduce the accumulation of intracellular lipids,[46,47] curcumin metabolites had a beneficial effect on non-alcoholic fatty liver disease (NAFLD),[48,49] and resveratrol was shown to reverse free fatty acid (FFA)-induced insulin resistance.[50]

Various animal models and clinical trials in humans have been conducted using several foods and herbal products containing bioactive compounds such as epigallocatechin-3-gallate (EGCG) in green tea, diallyl disulfide in garlic, extracts of *Moringa oleifera*, *Curcuma longa*, and *Murraya koenigii*, sunflower seed extract, grape seed extract, cranberry extract, pigmented rice, bergamot extract, chia oil, anthocyanins, and hesperidin as potential candidates with anti-obesity effects.[51]

10.5.3 Heart Diseases

Cardiovascular diseases are the leading reason for sudden death globally. The roles of various dietary risk factors such as consumption of sodium and saturated fats have been investigated for their effect on the cardiovascular system through diet modification by reducing their content and through cohort studies with baseline dietary intake.[52] Further, in addition to dietary risk factors, there are various other cardiovascular risk factors, such as (i) non-modifiable risk factors (advancing age, male gender, family history/genotype), (ii) lifestyle risk factors (diet, smoking, physical activity), and (iii) metabolic risk factors (hypertension, hyperlipidemia, diabetes mellitus, metabolic syndrome, obesity/overweight).[53] Various studies have revealed that the consumption of several functional foods such as nuts (tocopherols, omega-3 fatty acids), whole cereals and legumes (phytochemicals, fiber, and polyphenols), soy proteins (genistein and daidzein), dark chocolate (flavonoids), fish (omega-3 fatty acids), fruits and vegetables (folate, phytochemicals, polyphenols, vitamin A and C), and green and black teas (tea polyphenols) provides potential cardiovascular protective effects.[54] All of the above-mentioned functional foods confer their beneficial effects through several potential mechanisms, namely by lowering blood triglycerides and blood cholesterol, lowering blood homocysteine, inhibiting the oxidation of LDL cholesterol, antioxidant action, anti-inflammatory action, reducing blood pressure, and preventing platelet aggregation.[55]

10.5.4 Cancer

Cancer remains a major health problem worldwide. Diet is the major and significantly modifiable risk factor in the case of many cancers. Common causative factors of cancer are oxidative stress, metabolic syndrome, inflammation, and excess weight. Further, several foods and food ingredients, such as saturated fats, artificial sweeteners, processed red meat, refined grains, acrylamide, charred meat, sugary drinks, and additives, are food/diet-related risk factors for cancer.[56] A Mediterranean diet is a highly recommended diet type as it prevents or reduces the risk of developing chronic illnesses, including cancer (especially breast and colon cancers); the diet consists predominantly of fish, fruits, nuts, vegetables, whole grains, legumes, and extra virgin olive oil.[57] In recent years, functional food interventional trials have been carried out along with cancer treatments and they have been demonstrated to be beneficial in reducing cancer occurrence and reoccurrence rates. The polyphenol-rich foods/ingredients soy isoflavones, EGCG, a flavonoid mixture (20 mg of apigenin plus 20 mg of EGCG), and curcumin were found to be beneficial in the treatment of prostate, breast, colorectal, and colon cancer, respectively.[56] Further, various functional ingredients such as fiber, proteins, probiotics, vitamins, and minerals have been extensively investigated in cancer prevention and treatment.[58] In general, increased consumption of fruits, vegetables, mushrooms, beverages such as tea and coffee, and whole grains is beneficial for cancer prevention.[59]

10.6 Tailor-made Foods for Specific Health Benefits and Conditions: Strategies and Techniques

Tailor-made foods are being designed for people with various health conditions, especially for elderly subjects, people with malnutrition and micronutrient deficiencies, persons with food allergies, and those with special preferences. The aging population suffers from dysphagia, which is a swallowing disorder that causes difficulty in swallowing or the inability to form or move a food bolus safely from the mouth to the stomach. It is estimated that by 2050 the global elderly population (aged 65 years and older) will reach 1.6 billion.[60] Dysphagia is prevalent in the elderly, namely about 13–38% of those who live independently, 68% of those who live in nursing homes, and 30% of those who are hospitalized.[61] Hence in recent years a lot of research has been conducted into designing personalized foods for the elderly and other people with swallowing difficulties by using different texture modification techniques, such as using thickening agents (modified starch and hydrocolloids/gum based), mostly in combination with maltodextrin, high-pressure processing (HPP), high-hydrodynamic pressure processing, pulsed electric field treatment, plasma processing, ultrasound-assisted processing, irradiation, and 3D food printing.[60,62] Especially 3D food printing can provide individually tailored foods for the elderly with improved sensorial characteristics and covers special nutritional needs such as specific micronutrients, allergies or intolerances, health status, and comorbidities.[63]

In general, around 33–50% of the global population suffers from various micronutrient deficiencies (vitamins and essential trace minerals) and "hidden hunger". Several factors such as food supply, including cropping systems, soil fertility, method of cooking, diet deficiency in specific nutrients, and inherent malabsorption or low bioavailability are responsible for micronutrient deficiencies. The micronutrient deficiencies can be prevented by using various strategies such as biofortification (through conventional plant breeding or genetic engineering), food fortification (mass fortification of common food supplies such as salt fortified with iodine and rice and flour fortified with zinc and iron), and commercially fortified foods such as fortified biscuits, breakfast cereals, pasta, and milk.[64–66] Nanotechnology also plays a prominent role in designing personalized nanosized foods by encapsulating one or more bioactive agents to produce a potential delivery system that protects the nutrients and delivers them in a bioavailable form at targeted sites after ingestion.[67] Various functional foods and beverages containing bioactive agents such as curcumin, resveratrol, coenzyme Q10, and probiotics encapsulated in colloidal particles such as microemulsions, liposomes, solid lipid nanoparticles, oil bodies, and microgels have been explored recently.[68]

Nowadays, there is a segment of the population that has allergies and intolerances to various foods such as cow's milk, eggs, peanuts, tree nuts, soy, shellfish, and finned fish, and food ingredients such as gluten and lactose.[69] To accommodate the needs of this segment of the population, specific care and personalized products are being produced and commercialized, such as gluten-free product ranges and lactose-free milk-based products or plant-based alternatives. The allergens are clearly labeled on the packages to avoid any adverse events.[70]

10.7 Conclusion

In this chapter, we have summarized nutritional needs based on age and gender, personalized foods available in the market, current research evidence regarding functional foods and ingredients that aid in the inhibition and treatment of various disease conditions, and also different techniques to develop tailor-made foods for a targeted population group with special needs. The concept of personalized food and nutrition emerged to reduce the burden of diseases and improve health and well-being. Further, the personalization of foods can be performed throughout the food value chain from the selection of raw materials to consumption – it is not only about the nutritional content, it encompasses complex variables including the selection of foods, other combination ingredients, processing method, final product, sensory profile, and bioavailability of nutrients. However, there are many challenges in implementing and following a personalized recommended diet due to various factors such as food security, social status, availability, culture, demography, awareness, and behavior that affect food choice and diet. Creating awareness and educating the public with concrete evidence and allowing them to make changes in their food choices and guiding them to choose better food options have a much more positive impact than blindly enforcing people to consume or cut down on certain foods. Further, in recent years several online business platforms have appeared that provide valuable and sustainable nutrition tips and meal plans designed specifically for an individual's customized needs.

References

1. P.-J. Chen and M. Antonelli, Conceptual models of food choice: influential factors related to foods, individual differences, and society, *Foods*, 2020, **9**(12), 1898.
2. M. Vabø and H. Hansen, The relationship between food preferences and food choice: a theoretical discussion, *Int. J. Bus. Soc. Sci.*, 2014, **5**(7), 145–157.
3. V. De Cosmi, A. Mazzocchi, G. P. Milani and C. Agostoni, Dietary Patterns *vs.* Dietary Recommendations, *Front. Nutr.*, 2022, **9**, 883806.
4. Ø. Ueland, T. Altintzoglou and B. Kirkhus, *et al.*, Perspectives on personalised food, *Trends Food Sci. Technol.*, 2020, **102**, 169–177.
5. S. H. Adams, J. C. Anthony and R. Carvajal, *et al.*, Perspective: guiding principles for the implementation of personalized nutrition

approaches that benefit health and function, *Adv. Nutr.*, 2020, **11**(1), 25–34.

6. I. M. de Hoogh, B. L. Winters and K. M. Nieman, *et al.*, A Novel Personalized Systems Nutrition Program Improves Dietary Patterns, Lifestyle Behaviors and Health-Related Outcomes: Results from the Habit Study, *Nutrients*, 2021, **13**(6), 1763.
7. J. B. Marcus, *Culinary Nutrition: The Science and Practice of Healthy Cooking*, Academic Press, 2013.
8. C. R. Martin, P.-R. Ling and G. L. Blackburn, Review of infant feeding: key features of breast milk and infant formula, *Nutrients*, 2016, **8**(5), 279.
9. F. Mosca and M. L. Giannì, Human milk: composition and health benefits, *Pediatr. Med. Chir.*, 2017, **39**(2), 47–52.
10. H. J. H. Blewett, M. C. Cicalo, C. D. Holland and C. J. Field, The immunological components of human milk, in *Advances in Food and Nutrition Research*, 2008, vol. 54, pp. 45–80.
11. P. Christian, E. R. Smith, S. E. Lee, A. J. Vargas, A. A. Bremer and D. J. Raiten, The need to study human milk as a biological system, *Am. J. Clin. Nutr.*, 2021, **113**(5), 1063–1072.
12. G. Kent, Global infant formula: monitoring and regulating the impacts to protect human health, *Int. Breastfeed. J.*, 2015, **10**(1), 1–12.
13. C. Prell and B. Koletzko, Breastfeeding and complementary feeding: recommendations on infant nutrition, *Dtsch. Ärzteblatt Int.*, 2016, **113**(25), 435.
14. J. L. Hollis, C. E. Collins, F. DeClerck, L. K. Chai, K. McColl and A. R. Demaio, Defining healthy and sustainable diets for infants, children and adolescents, *Global Food Secur.*, 2020, **27**, 100401.
15. D. Hargreaves, E. Mates and P. Menon, *et al.*, Strategies and interventions for healthy adolescent growth, nutrition, and development, *Lancet*, 2022, **399**(10320), 198–210.
16. K. Norman, U. Haß and M. Pirlich, Malnutrition in older adults—recent advances and remaining challenges, *Nutrients*, 2021, **13**(8), 2764.
17. J. M. Ordovas and S. Berciano, Personalized nutrition and healthy aging, *Nutr. Rev.*, 2020, **78**(Supplement_3), 58–65.
18. L. Donini, Control of food intake in aging, in *Food for the Aging Population*, Elsevier, 2017, pp. 25–55.
19. M. Gettings, Food safety and older people, in *Food for the Aging Population*, Elsevier, 2017, pp. 299–321.
20. N. Plastow, Gender and Food in Later Life: Shifting Roles and Relationships, in *Food for the Aging Population*, Elsevier, 2017, pp. 99–115.
21. National Academies of Sciences, Engineering, and Medicine, *Current Status and Response to the Global Obesity Pandemic: Proceedings of a Workshop*, 2019.
22. E. W. Kimani-Murage, C. K. Nyamasege and S. Mutoni, *et al.*, Personalized Nutrition for Women, Infants, and Children, in *Trends in Personalized Nutrition*, 2019, pp. 169–194.

23. O. A. Ladipo, Nutrition in pregnancy: mineral and vitamin supplements, *Am. J. Clin. Nutr.*, 2000, **72**(1), 280S–290S.
24. J. Delgado-Lista, J. F. Alcala-Diaz and J. D. Torres-Peña, *et al.*, Long-term secondary prevention of cardiovascular disease with a Mediterranean diet and a low-fat diet (CORDIOPREV): a randomised controlled trial, *Lancet*, 2022, **399**(10338), 1876–1885.
25. M. J. Banez, M. I. Geluz and A. Chandra, *et al.*, A systemic review on the antioxidant and anti-inflammatory effects of resveratrol, curcumin, and dietary nitric oxide supplementation on human cardiovascular health, *Nutr. Res.*, 2020, **78**, 11–26.
26. P.-H. Lin, W. Aronson and S. J. Freedland, Nutrition, dietary interventions and prostate cancer: the latest evidence, *BMC Med.*, 2015, **13**(1), 1–15.
27. M. Matsushita, K. Fujita and N. Nonomura, Influence of diet and nutrition on prostate cancer, *Int. J. Mol. Sci.*, 2020, **21**(4), 1447.
28. S. K. Johnson, V. Chua, R. S. Hall and A. L. Baxter, Lupin kernel fibre foods improve bowel function and beneficially modify some putative faecal risk factors for colon cancer in men, *Br. J. Nutr.*, 2006, **95**(2), 372–378.
29. J. Keamey, E. Giovannucci and E. B. Rimm, *et al.*, Calcium, vitamin D, and dairy foods and the occurrence of colon cancer in men *Am. J. Epidemiol.*, 1996, **143**(9), 907–917.
30. J. La, N. H. Roberts and F. A. Yafi, Diet and men's sexual health, *Sex. Med. Rev.*, 2018, **6**(1), 54–68.
31. X. Zhou, F. Liu and S. Zhai, Effect of L-carnitine and/or L-acetyl-carnitine in nutrition treatment for male infertility: a systematic review, *Asia Pac. J. Clin. Nutr.*, 2007, **16**, 383.
32. A. Chanet, S. Verlaan and J. Salles, *et al.*, Supplementing breakfast with a vitamin D and leucine–enriched whey protein medical nutrition drink enhances postprandial muscle protein synthesis and muscle mass in healthy older men, *J. Nutr.*, 2017, **147**(12), 2262–2271.
33. T. Han, F. Wu and M. Lean, Obesity and weight management in the elderly: a focus on men, *Best Pract. Res., Clin. Endocrinol. Metab.*, 2013, **27**(4), 509–525.
34. S. Ash, M. Reeves, S. Yeo, G. Morrison, D. Carey and S. Capra, Effect of intensive dietetic interventions on weight and glycaemic control in overweight men with Type II diabetes: a randomised trial, *Int. J. Obes.*, 2003, **27**(7), 797–802.
35. K. M. Crowe and C. Francis, Position of the academy of nutrition and dietetics: functional foods, *J. Acad. Nutr. Diet.*, 2013, **113**(8), 1096–1103.
36. A. Cencic and W. Chingwaru, The role of functional foods, nutraceuticals, and food supplements in intestinal health, *Nutrients*, 2010, **2**(6), 611–625.
37. G. Dable-Tupas, M. C. B. Otero and L. Bernolo, Functional foods and health benefits, in *Functional Foods and Nutraceuticals: Bioactive Components, Formulations and Innovations*, 2020, pp. 1–11.

38. P. Li, Y. Tang, L. Liu, D. Wang, L. Zhang and C. Piao, Therapeutic potential of buckwheat hull flavonoids in db/db mice, a model of type 2 diabetes, *J. Funct. Foods*, 2019, **52**, 284–290.
39. Atlas ID, 10th. Available at www.diabe tesat las.org. Accessed; 2022.
40. P. Mirmiran, Z. Bahadoran and F. Azizi, Functional foods-based diet as a novel dietary approach for management of type 2 diabetes and its complications: A review, *World J. Diabetes*, 2014, **5**(3), 267.
41. K. Venkatakrishnan, H.-F. Chiu and C.-K. Wang, Popular functional foods and herbs for the management of type-2-diabetes mellitus: a comprehensive review with special reference to clinical trials and its proposed mechanism, *J. Funct. Foods*, 2019, **57**, 425–438.
42. M. Green, K. Arora and S. Prakash, Microbial medicine: prebiotic and probiotic functional foods to target obesity and metabolic syndrome, *Int. J. Mol. Sci.*, 2020, **21**(8), 2890.
43. K. Venkatakrishnan, H.-F. Chiu and C.-K. Wang, Extensive review of popular functional foods and nutraceuticals against obesity and its related complications with a special focus on randomized clinical trials, *Food Funct.*, 2019, **10**(5), 2313–2329.
44. J. H. Kim, S. Lee and E. J. Cho, Flavonoids from Acer okamotoanum inhibit adipocyte differentiation and promote lipolysis in the 3T3-L1 cells, *Molecules*, 2020, **25**(8), 1920.
45. S. R. Balusamy, K. Veerappan and A. Ranjan, *et al.*, Phyllanthus emblica fruit extract attenuates lipid metabolism in 3T3-L1 adipocytes *via* activating apoptosis mediated cell death, *Phytomedicine*, 2020, **66**, 153129.
46. D. Torres-Villarreal, A. Camacho, H. Castro, R. Ortiz-Lopez and A. De la Garza, Anti-obesity effects of kaempferol by inhibiting adipogenesis and increasing lipolysis in 3T3-L1 cells, *J. Physiol. Biochem.*, 2019, **75**, 83–88.
47. S. G. Lee, J. S. Kim, K. Min, T. K. Kwon and J.-O. Nam, Hispidulin inhibits adipogenesis in 3T3-L1 adipocytes through PPARγ pathway, *Chem.-Biol. Interact.*, 2018, **293**, 89–93.
48. L.-Y. Wu, C.-W. Chen, L.-K. Chen, H.-Y. Chou, C.-L. Chang and C.-C. Juan, Curcumin attenuates adipogenesis by inducing preadipocyte apoptosis and inhibiting adipocyte differentiation, *Nutrients*, 2019, **11**(10), 2307.
49. J.-W. Chen, Z.-L. Kong, M.-L. Tsai, C.-Y. Lo, C.-T. Ho and C.-S. Lai, Tetrahydrocurcumin ameliorates free fatty acid-induced hepatic steatosis and improves insulin resistance in HepG2 cells, *J. Food Drug Anal.*, 2018, **26**(3), 1075–1085.
50. D. J. Den Hartogh, F. Vlavcheski, A. Giacca and E. Tsiani, Attenuation of free fatty acid (FFA)-induced skeletal muscle cell insulin resistance by resveratrol is linked to activation of AMPK and inhibition of mTOR and p70 S6K, *Int. J. Mol. Sci.*, 2020, **21**(14), 4900.
51. G. Sandner, A. König, M. Wallner and J. Weghuber, Functional foods-dietary or herbal products on obesity: Application of selected bioactive

compounds to target lipid metabolism, *Curr. Opin. Food Sci.*, 2020, **34**, 9–20.

52. M. Jessri, D. Hennessey and A. B. Eddeen, *et al.*, Sodium, added sugar and saturated fat intake in relation to mortality and cardiovascular disease events in adults: Canadian National Nutrition Survey linked with vital statistics and health administrative databases, *Br. J. Nutr.*, 2022, 1–33.
53. E. M. Alissa and G. A. Ferns, Functional foods and nutraceuticals in the primary prevention of cardiovascular diseases, *J. Nutr. Metab.*, 2012, **2012**, 569486.
54. S. Asgary, A. Rastqar and M. Keshvari, Functional food and cardiovascular disease prevention and treatment: a review, *J. Am. Coll. Nutr.*, 2018, **37**(5), 429–455.
55. I. Rudkowska and P. J. Jones, Functional foods for the prevention and treatment of cardiovascular diseases: cholesterol and beyond, *Expert Rev. Cardiovasc. Ther.*, 2007, **5**(3), 477–490.
56. S. R. Islam and T. J. Siddiqua, Functional foods in cancer prevention and therapy: Recent epidemiological findings, in *Functional Foods in Cancer Prevention and Therapy*, 2020, pp. 405–433.
57. M. C. Mentella, F. Scaldaferri, C. Ricci, A. Gasbarrini and G. A. D. Miggiano, Cancer and Mediterranean diet: a review, *Nutrients*, 2019, **11**(9), 2059.
58. M. Aghajanpour, M. R. Nazer, Z. Obeidavi, M. Akbari, P. Ezati and N. M. Kor, Functional foods and their role in cancer prevention and health promotion: a comprehensive review, *Am. J. Cancer Res.*, 2017, **7**(4), 740.
59. Y.-W. Zeng, J.-Z. Yang and X.-Y. Pu, *et al.*, Strategies of functional food for cancer prevention in human beings, *Asian Pac. J. Cancer Prev.*, 2013, **14**(3), 1585–1592.
60. P. Methacanon, C. Gamonpilas, A. Kongjaroen and C. Buathongjan, Food polysaccharides and roles of rheology and tribology in rational design of thickened liquids for oropharyngeal dysphagia: A review, *Compr. Rev. Food Sci. Food Saf.*, 2021, **20**(4), 4101–4119.
61. L. Sura, A. Madhavan, G. Carnaby and M. A. Crary, Dysphagia in the elderly: management and nutritional considerations, *Clin. Interventions Aging*, 2012, 287–298.
62. S. Sungsinchai, C. Niamnuy, P. Wattanapan, M. Charoenchaitrakool and S. Devahastin, Texture modification technologies and their opportunities for the production of dysphagia foods: A review, *Compr. Rev. Food Sci. Food Saf.*, 2019, **18**(6), 1898–1912.
63. A. Escalante-Aburto, G. Trujillo-de Santiago, M. M. Álvarez and C. Chuck-Hernández, Advances and prospective applications of 3D food printing for health improvement and personalized nutrition, *Compr. Rev. Food Sci. Food Saf.*, 2021, **20**(6), 5722–5741.
64. S. S. Dhaliwal, V. Sharma and A. K. Shukla, *et al.*, Biofortification—A frontier novel approach to enrich micronutrients in field crops to encounter the nutritional security, *Molecules*, 2022, **27**(4), 1340.

65. S. M. T. Gharibzahedi and S. M. Jafari, The importance of minerals in human nutrition: Bioavailability, food fortification, processing effects and nanoencapsulation, *Trends Food Sci. Technol.*, 2017, **62**, 119–132.
66. M. Di Cairano, R. Tolve, N. Cela, L. Sportiello, T. Scarpa, F. Galgano, Functional Cereal-Based Bakery Products, Breakfast Cereals, and Pasta Products, in *Functional Cereals and Cereal Foods: Properties, Functionality and Applications*, Springer, 2022, pp. 215–249.
67. D. Rawat, V. Tomer, S. Pandey and S. K. Flora, Role of Nanotechnology in Nutrient Delivery, in *Nanotechnology Horizons in Food Process Engineering*, Apple Academic Press, 2023, pp. 133–160.
68. D. J. McClements, Nano-enabled personalized nutrition: Developing multicomponent-bioactive colloidal delivery systems, *Adv. Colloid Interface Sci.*, 2020, **282**, 102211.
69. J. Muthukumar, P. Selvasekaran, M. Lokanadham and R. Chidambaram, Food and food products associated with food allergy and food intolerance–An overview, *Food Res. Int.*, 2020, **138**, 109780.
70. R. Gupta, M. Kanaley, O. Negris, A. Roach and L. Bilaver, Understanding precautionary allergen labeling (PAL) preferences among food allergy stakeholders, *J. Allergy Clin. Immunol.: Pract.*, 2021, **9**(1), 254–264e1.
71. M. A. Farhangi, A. Z. Javid, B. Sarmadi, P. Karimi and P. Dehghan, A randomized controlled trial on the efficacy of resistant dextrin, as functional food, in women with type 2 diabetes: Targeting the hypothalamic–pituitary–adrenal axis and immune system, *Clin. Nutr.*, 2018, **37**(4), 1216–1223.
72. M. Andersson, W. Theis and M. B. Zimmermann, *et al.*, Random serial sampling to evaluate efficacy of iron fortification: a randomized controlled trial of margarine fortification with ferric pyrophosphate or sodium iron edetate, *Am. J. Clin. Nutr.*, 2010, **92**(5), 1094–1104.
73. P. T. Hoa, N. C. Khan, C. van Beusekom, R. Gross, W. L. Conde and H. D. Khoi, Milk Fortified with Iron or Iron Supplementation to Improve Nutritional Status of Pregnant Women: An Intervention Trial from Rural Vietnam, *Food Nutr. Bull.*, 2005, **26**(1), 32–38.
74. B. Adolphi, K. E. Scholz-Ahrens, M. de Vrese, Y. Açil, C. Laue and J. Schrezenmeir, Short-term effect of bedtime consumption of fermented milk supplemented with calcium, inulin-type fructans and caseinphosphopeptides on bone metabolism in healthy, postmenopausal women, *Eur. J. Nutr.*, 2009, **48**, 45–53.
75. A. Badii, N. Nekouei, M. Fazilati, M. Shahedi and S. Badiei, Effect of consuming zinc-fortified bread on serum zinc and iron status of zinc-deficient women: a double blind, randomized clinical trial, *Int. J. Prev. Med.*, 2012, **3**(Suppl 1), S124–S130.
76. J.-P. Bonjour, V. Benoit, O. Pourchaire, M. Ferry, B. Rousseau and J.-C. Souberbielle, Inhibition of markers of bone resorption by consumption of vitamin D and calcium-fortified soft plain cheese by institutionalised elderly women, *Br. J. Nutr.*, 2009, **102**(7), 962–966.

77. J. A. Greenberg, S. J. Bell, Y. Guan and Y.-H. Yu, Folic acid supplementation and pregnancy: more than just neural tube defect prevention, *Rev. Obstet. Gynecol.*, 2011, **4**(2), 52.
78. T. B. Basnet, G. Srijana, R. Basnet, B. Neupane and G. Thapa, Causal effects of dietary calcium, zinc and iron intakes on coronary artery disease in men: G-estimation and inverse probability of treatment weighting (IPTW) analyses, *Clin. Nutr. ESPEN*, 2021, **42**, 73–81.
79. G. H. McIntosh, M. Noakes, P. J. Royle and P. R. Foster, Whole-grain rye and wheat foods and markers of bowel health in overweight middle-aged men, *Am. J. Clin. Nutr.*, 2003, **77**(4), 967–974.
80. J. Liu, S. Willför and C. Xu, A review of bioactive plant polysaccharides: Biological activities, functionalization, and biomedical applications, *Bioact. Carbohydr. Diet. Fibre*, 2015, **5**(1), 31–61.
81. A. T. Kunzmann, H. G. Coleman, W.-Y. Huang, C. M. Kitahara, M. M. Cantwell and S. I. Berndt, Dietary fiber intake and risk of colorectal cancer and incident and recurrent adenoma in the Prostate, Lung, Colorectal, and Ovarian Cancer Screening Trial, *Am. J. Clin. Nutr.*, 2015, **102**(4), 881–890.
82. I. H. Ho, L. Matia-Merino and L. M. Huffman, Use of viscous fibres in beverages for appetite control: a review of studies, *Int. J. Food Sci. Nutr.*, 2015, **66**(5), 479–490.
83. F. Zhu, B. Du and B. Xu, A critical review on production and industrial applications of beta-glucans, *Food Hydrocolloids*, 2016, **52**, 275–288.
84. G. A. Soliman, Dietary Fiber, Atherosclerosis, and Cardiovascular Disease, *Nutrients*, 2019, 1155.
85. S. Pentikäinen, L. Karhunen and L. Flander, *et al.*, Enrichment of biscuits and juice with oat β-glucan enhances postprandial satiety, *Appetite*, 2014, **75**, 150–156.
86. J. Wang, H. H. Huang and P. S. Chen, Structural and physicochemical properties of banana resistant starch from four cultivars, *Int. J. Food Prop.*, 2017, **20**(6), 1338–1347.
87. K. L. Marlatt, U. A. White and R. A. Beyl, *et al.*, Role of resistant starch on diabetes risk factors in people with prediabetes: Design, conduct, and baseline results of the STARCH trial, *Contemp. Clin. Trials*, 2018, **65**, 99–108.
88. M. I. Zafar, K. E. Mills and J. Zheng, *et al.*, Low-glycemic index diets as an intervention for diabetes: a systematic review and meta-analysis, *Am. J. Clin. Nutr.*, 2019, **110**(4), 891–902.
89. I. D. Nwachukwu and R. E. Aluko, Structural and functional properties of food protein-derived antioxidant peptides, *J. Food Biochem.*, 2019, **43**(1), e12761.
90. C. C. Udenigwe and R. E. Aluko, Food protein-derived bioactive peptides: production, processing, and potential health benefits, *J. Food Sci.*, 2012, **77**(1), R11–R24.
91. C. E. Ejike, S. A. Collins, N. Balasuriya, A. K. Swanson, B. Mason and C. C. Udenigwe, Prospects of microalgae proteins in producing

peptide-based functional foods for promoting cardiovascular health, *Trends Food Sci. Technol.*, 2017, **59**, 30–36.
92. A. Jakubczyk, M. Karaś, K. Rybczyńska-Tkaczyk, E. Zielińska and D. Zieliński, Current trends of bioactive peptides—New sources and therapeutic effect, *Foods*, 2020, **9**(7), 846.
93. M. A. A. Hassan, R. Rouf, E. Tiralongo, T. W. May and J. Tiralongo, Mushroom lectins: specificity, structure and bioactivity relevant to human disease, *Int. J. Mol. Sci.*, 2015, **16**(4), 7802–7838.
94. R. Sarup Singh, H. Preet Kaur and J. Rakesh Kanwar, Mushroom lectins as promising anticancer substances, *Curr. Protein Pept. Sci.*, 2016, **17**(8), 797–807.
95. C. Ruxton, S. Reed, M. Simpson and K. Millington, The health benefits of omega-3 polyunsaturated fatty acids: a review of the evidence, *J. Hum. Nutr. Diet.*, 2007, **20**(3), 275–285.
96. N. Siriwardhana, N. S. Kalupahana and N. Moustaid-Moussa, Health benefits of n-3 polyunsaturated fatty acids: eicosapentaenoic acid and docosahexaenoic acid, in *Advances in Food and Nutrition Research*, 2012, vol. 65, pp. 211–222.
97. J. H. Kim, Y. Kim, Y. J. Kim and Y. Park, Conjugated linoleic acid: potential health benefits as a functional food ingredient, *Annu. Rev. Food Sci. Technol.*, 2016, **7**, 221–244.
98. G. Marcelino, D. J. Machate and K. D. C. Freitas, *et al.*, β-Carotene: preventive role for type 2 diabetes mellitus and obesity: a review *Molecules*, 2020, **25**(24), 5803.
99. R. K. Saini, P. Prasad and V. Lokesh, *et al.*, Carotenoids: Dietary sources, extraction, encapsulation, bioavailability, and health benefits—A review of recent advancements, *Antioxidants*, 2022, **11**(4), 795.
100. U. M. Khan, M. Sevindik and A. Zarrabi, *et al.*, Lycopene: Food sources, biological activities, and human health benefits, *Oxid. Med. Cell. Longevity*, 2021, **2021**, 2713511.
101. S. Przybylska, Lycopene–a bioactive carotenoid offering multiple health benefits: a review, *Int. J. Food Sci. Technol.*, 2020, **55**(1), 11–32.
102. S. Mitra, A. Rauf and A. M. Tareq, *et al.*, Potential health benefits of carotenoid lutein: An updated review, *Food Chem. Toxicol.*, 2021, **154**, 112328.
103. H. E. Khoo, A. Azlan, S. T. Tang and S. M. Lim, Anthocyanidins and anthocyanins: Colored pigments as food, pharmaceutical ingredients, and the potential health benefits, *Food Nutr. Res.*, 2017, **61**(1), 1361779.
104. P. M. Aron and J. A. Kennedy, Flavan-3-ols: Nature, occurrence and biological activity, *Mol. Nutr. Food Res.*, 2008, **52**(1), 79–104.
105. B. Tanwar and R. Modgil, Flavonoids: Dietary occurrence and health benefits, *Spatula DD*, 2012, **2**(1), 59–68.
106. Y. Stevens, E. V. Rymenant and C. Grootaert, *et al.*, The intestinal fate of citrus flavanones and their effects on gastrointestinal health, *Nutrients*, 2019, **11**(7), 1464.

107. N. Jiang, A. I. Doseff and E. Grotewold, Flavones: from biosynthesis to health benefits, *Plants*, 2016, **5**(2), 27.
108. C. Rodríguez-García, C. Sánchez-Quesada, E. Toledo, M. Delgado-Rodríguez and J. J. Gaforio, Naturally lignan-rich foods: A dietary tool for health promotion?, *Molecules*, 2019, **24**(5), 917.
109. A. Rauf, M. Imran and T. Abu-Izneid, *et al.*, Proanthocyanidins: A comprehensive review, *Biomed. Pharmacother.*, 2019, **116**, 108999.
110. D. Barreca, D. Trombetta and A. Smeriglio, *et al.*, Food flavonols: Nutraceuticals with complex health benefits and functionalities, *Trends Food Sci. Technol.*, 2021, **117**, 194–204.
111. B. P. Singh, D. Yadav and S. Vij, Soybean bioactive molecules: current trend and future prospective, in *Bioactive Molecules in Food*, Springer, 2019, pp. 267–294.
112. A. Crozier, I. B. Jaganath and M. N. Clifford, Dietary phenolics: chemistry, bioavailability and effects on health, *Nat. Prod. Rep.*, 2009, **26**(8), 1001–1043.
113. R. Ashraf and H. N. Bhatti, Stigmasterol, in *A Centum of Valuable Plant Bioactives*, Elsevier, 2021, pp. 213–232.
114. S. Babu and S. Jayaraman, An update on β-sitosterol: A potential herbal nutraceutical for diabetic management, *Biomed. Pharmacother.*, 2020, **131**, 110702.
115. H. Bae, S. Park, C. Yang, G. Song and W. Lim, Disruption of endoplasmic reticulum and ROS production in human ovarian cancer by campesterol, *Antioxidants*, 2021, **10**(3), 379.
116. F. Shahidi, A. C. C. Pinaffi-Langley, J. Fuentes, H. Speisky and A. C. de Camargo, Vitamin E as an essential micronutrient for human health: Common, novel, and unexplored dietary sources, *Free Radicals Biol. Med.*, 2021, **176**, 312–321.
117. A. M. P. Walag, O. Ahmed and J. Jeevanandam, *et al.*, Health benefits of organosulfur compounds, in *Functional Foods and Nutraceuticals: Bioactive Components, Formulations and Innovations*, 2020, pp. 445–472.

Subject Index